Lecture Notes in Energy

Volume 43

Lecture Notes in Energy (LNE) is a series that reports on new developments in the study of energy: from science and engineering to the analysis of energy policy. The series' scope includes but is not limited to, renewable and green energy, nuclear, fossil fuels and carbon capture, energy systems, energy storage and harvesting, batteries and fuel cells, power systems, energy efficiency, energy in buildings, energy policy, as well as energy-related topics in economics, management and transportation. Books published in LNE are original and timely and bridge between advanced textbooks and the forefront of research. Readers of LNE include postgraduate students and non-specialist researchers wishing to gain an accessible introduction to a field of research as well as professionals and researchers with a need for an up-to-date reference book on a well-defined topic. The series publishes single- and multi-authored volumes as well as advanced textbooks.

Indexed in Scopus and EI Compendex The Springer Energy board welcomes your book proposal. Please get in touch with the series via Anthony Doyle, Executive Editor, Springer (anthony.doyle@springer.com)

More information about this series at https://link.springer.com/bookseries/8874

J. W. Day · Rachael G. Hunter · H. C. Clark
Editors

Energy Production in the Mississippi River Delta

Impacts on Coastal Ecosystems and Pathways to Restoration

Editors
J. W. Day
Department of Oceanography and Coastal Sciences
Louisiana State University
Baton Rouge, LA, USA

Rachael G. Hunter
Comite Resources, Inc.
Baton Rouge, LA, USA

H. C. Clark
Rice University
Houston, TX, USA

ISSN 2195-1284 ISSN 2195-1292 (electronic)
Lecture Notes in Energy
ISBN 978-3-030-94528-2 ISBN 978-3-030-94526-8 (eBook)
https://doi.org/10.1007/978-3-030-94526-8

© The Editor(s) (if applicable) and The Author(s), under exclusive license to Springer Nature Switzerland AG 2022, corrected publication 2022
This work is subject to copyright. All rights are solely and exclusively licensed by the Publisher, whether the whole or part of the material is concerned, specifically the rights of translation, reprinting, reuse of illustrations, recitation, broadcasting, reproduction on microfilms or in any other physical way, and transmission or information storage and retrieval, electronic adaptation, computer software, or by similar or dissimilar methodology now known or hereafter developed.
The use of general descriptive names, registered names, trademarks, service marks, etc. in this publication does not imply, even in the absence of a specific statement, that such names are exempt from the relevant protective laws and regulations and therefore free for general use.
The publisher, the authors and the editors are safe to assume that the advice and information in this book are believed to be true and accurate at the date of publication. Neither the publisher nor the authors or the editors give a warranty, expressed or implied, with respect to the material contained herein or for any errors or omissions that may have been made. The publisher remains neutral with regard to jurisdictional claims in published maps and institutional affiliations.

This Springer imprint is published by the registered company Springer Nature Switzerland AG
The registered company address is: Gewerbestrasse 11, 6330 Cham, Switzerland

The original version of the book was revised: In chapters 1,2,3,4,5,6,7 and 8, the author's affiliation has been amended and Index has been included. The correction to the book is available at https://doi.org/10.1007/978-3-030-94526-8_10

Contents

1 **Introduction** 1
   John W. Day, Rachael G. Hunter, and H. C. Clark

2 **Environmental Setting of the Mississippi River Delta** 7
   John W. Day and Rachael G. Hunter

3 **The Geology of the Mississippi River Delta and Interactions
   with Oil and Gas Activities** 39
   H. C. Clark and Charles Norman

4 **The Regulatory and Legal Framework—Oil and Gas Influence
   Over Environmental Management in Louisiana** 83
   Paul H. Templet

5 **Impacts of Oil and Gas Activity in the Mississippi River Delta** 93
   John W. Day, Rachael G. Hunter, and H. C. Clark

6 **Chemical and Toxin Impacts of Oil and Gas Activities
   on Coastal Systems** 133
   John H. Pardue and Vijaikrishnah Elango

7 **The Impact of Oil and Gas Activities on the Value of Ecosystem
   Goods and Services of the Mississippi River Delta** 155
   David Batker and Tania Briceno

8 **Restoring Coastal Ecosystems Impacted by Oil and Gas Activity** 193
   Charles Norman, John W. Day, and Rachael G. Hunter

9 **Summary and Conclusions** 223
   John W. Day, Rachael G. Hunter, and H. C. Clark

Correction to: Energy Production in the Mississippi River Delta C1
J. W. Day, Rachael G. Hunter, and H. C. Clark

Appendix 225

Index 237

Contributors

David Batker Batker Consulting, LLC, Tacoma, WA, USA

Tania Briceno Conservation Strategy Fund (CSF), Washington, DC, USA

H. C. Clark Department of Earth, Environmental and Planetary Sciences, Rice University, Houston, TX, USA

John W. Day Department of Oceanography and Coastal Science, Louisiana State University, Baton Rouge, LA, USA

Vijaikrishnah Elango Department of Civil and Environmental Engineering, Louisiana State University, Baton Rouge, LA, USA

Rachael G. Hunter Comite Resources, Inc, Baton Rouge, LA, USA

Charles Norman Charles Norman & Associates, Lake Charles, LA, USA

John H. Pardue Department of Civil and Environmental Engineering, Louisiana State University, Baton Rouge, LA, USA

Paul H. Templet Rancho de Taos, NM, USA

Chapter 1
Introduction

John W. Day, Rachael G. Hunter, and H. C. Clark

The Mississippi River Delta (MRD) has produced more oil and gas than any other region of the United States, resulting in over 600 fields with more than 56,000 wells and 15,000 km of canals dredged through wetlands for exploration, extraction and transportation (Fig. 1.1; Day et al. 2014). As a result of canal dredging, and other anthropogenic activities such as levee construction and creation of waterways for navigation, over 4,500 km^2 of coastal wetlands in the MRD have been lost during the twentieth century (Fig. 1.2; Couvillion et al. 2017).

Wetlands in the MRD are particularly susceptible to oil- and gas-related activities for a number of reasons, including their dependency on natural hydrological flows and the disruption of this hydrology by levees, canals, and impoundments; induced subsidence due to oil and gas production; toxic impacts of oil spills and produced water discharge; and abandoned infrastructure (Ko and Day 2004; Morton et al. 2002, 2005a, 2005b; Chang et al. 2014; Arnold 2020; Day et al. 2020). Analysis of oil and gas production and wetland loss demonstrate a strong causal relationship between the two (Fig. 1.3; Morton et al. 2005a).

The original version of this chapter was revised: The author "Rachael G. Hunter's" affiliation has been updated. The correction to this chapter is available at
https://doi.org/10.1007/978-3-030-94526-8_10

J. W. Day (✉)
Department of Oceanography and Coastal Sci, Louisiana State University, 2005 Olive St., Baton Rouge, LA 70806-6660, USA
e-mail: johnday@lsu.edu

R. G. Hunter
Comite Resources, Inc, Baton Rouge, LA, USA

H. C. Clark
Department of Earth, Enviromental and Planetary Sciences, Rice University, Houston, TX 77005, USA
e-mail: hcclark@rice.edu

© The Author(s), under exclusive license to Springer Nature Switzerland AG 2022, corrected publication 2022
J. W. Day et al. (eds.), *Energy Production in the Mississippi River Delta*,
Lecture Notes in Energy 43, https://doi.org/10.1007/978-3-030-94526-8_1

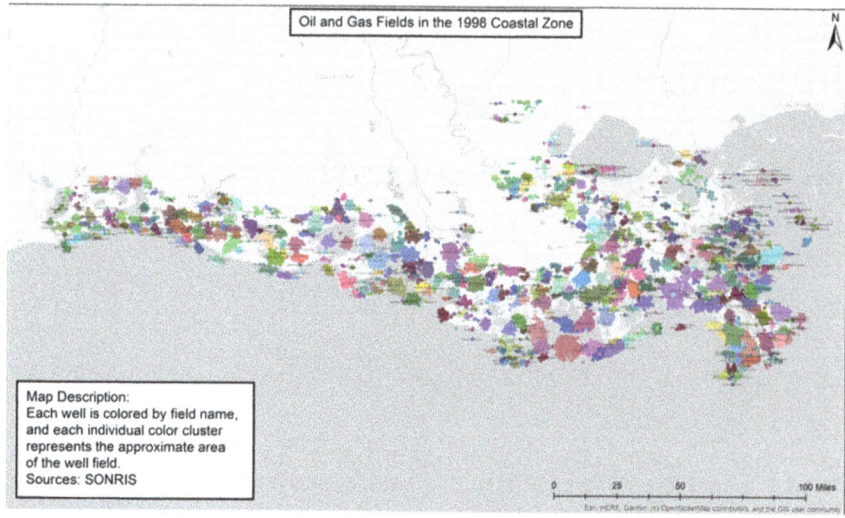

Fig. 1.1 Location of oil and gas fields in coastal Louisiana (LDNR SONRIS)

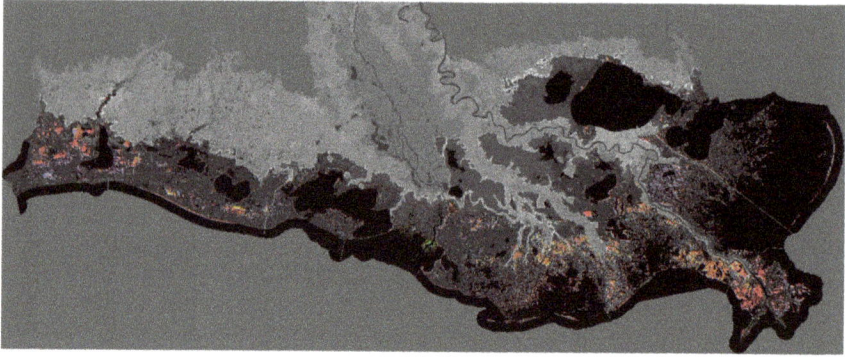

Fig. 1.2 Wetland loss in coastal Louisiana from 1932 to 2016. Red and yellow areas have high land loss rates. Note that land loss is low in the central coast and in the northeastern flank of the delta. (*Source* Couvillion et al. 2011, 2017, The map can be downloaded at https://pubs.usgs.gov/sim/3381/sim3381.pdf for detailed examination of specific areas of change. See also https://pubs.er.usgs.gov/publication/sim3381)

To understand the significance of oil and gas (O&G) exploration, extraction, and transport on Louisiana coastal wetlands, it is important to recognize that wetland degradation in oil and gas fields is occurring on a much more rapid time scale than the natural delta building and abandonment cycle (Day et al. 2000, 2020; Blum and Roberts 2012). Delta lobe building and abandonment, along with sea-level rise, and natural subsidence and erosion of deltaic and Chenier plain wetlands, occurred over centuries to millennia (Gould and McFarlan 1959; Penland and Suter 1989; Roberts 1997) and these natural processes did not change significantly in the mid to

1 Introduction

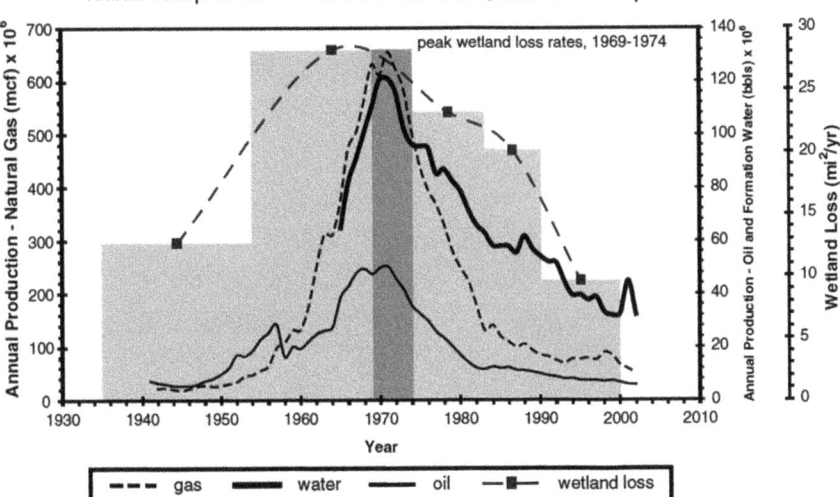

Fig. 1.3 Composite histories of fluid production from oil and gas fields and wetland loss in south Louisiana. Production data from the Louisiana Department of Natural Resources and the PI/Dwights PLUS database [ref]. Wetland loss is from Britch and Dunbar (1993) and John Barras (unpublished data). These historical data, integrated across the delta plain, show close temporal and spatial correlations between rates of wetland loss and rates of fluid production (Morton et al. 2005a). Note that "water" in the Legend refers to produced water

late twentieth century. Naturally-occurring geologic faulting, sediment compaction, changes in river discharge, global sea-level change, wave erosion and storms such as hurricanes have also been shaping the MRD for hundreds to thousands of years and, in the case of geologic faulting, millions of years (Williams et al. 1994; Kulp 2000; Reed 2002; Gagliano et al. 2003; Reed and Wilson 2004). By the late twentieth century, however, and in a very short period of time compared to the natural processes that shaped the Louisiana coast, oil and gas wells were drilled and canals dredged throughout the MRD (Fig. 1.4; Turner and Cahoon 1987; USACE 2004; Turner and McClenachan 2018).

Both anthropogenic and natural processes negatively impact MRD wetlands but it is difficult to quantify the amount of land loss attributed to a specific process (Turner and Cahoon 1987; Johnston et al. 2009; Cahoon et al. 2020). One direct impact, such as dredging a canal, can cause several different types of indirect impacts such as saltwater intrusion and increased flooding These impacts are affected by geologic setting, habitat type, subsidence rate, and/or previously existing impacts such as prior hydrologic modifications. The interaction of these factors influences the severity of the indirect impact (Johnston et al. 2009; Day et al. 2020).

This book discusses the impacts of oil and gas activities in the Louisiana coastal zone which encompasses much of the MRD. In Chap. 2, the environmental setting of the MRD is described, along with factors that enhance wetland sustainability.

Fig. 1.4 Canals dredged through coastal wetlands for oil and gas production and transportation and associated wetland loss. (Photos courtesy of J. Day)

Chapter 3 deals with the ancient and modern geology of the delta after the opening of the Gulf of Mexico 165 million years ago. Geology is important in terms of the formation of both oil and natural gas and the reservoirs where O&G is found. This chapter introduces the process of induced subsidence due to the production of O&G. Chapter 4 addresses the regulatory and legal framework and how the O&G industry affected environmental management. Chapter 5 discusses surface and subsurface impacts associated with exploration, production and transport of oil and natural gas. Chapter 6 addresses chemical and toxin pollution related with O&G activities. Chapter 7 describes the concept of the value of ecosystem goods and services of the Mississippi Delta and how they are impacted by O&G activities. Chapter 8 addresses restoration of coastal wetland ecosystems impacted by O&G activities in terms of restoration of hydrology, sediment management and decontamination. Chapter 9 summarizes the findings and conclusions of the book.

Acknowledgements The authors acknowledge that they have served as experts in litigation related to oil and gas activities in the Mississippi Delta. No direct funding for the preparation of this book came from this service.

References

Arnold JT (2020) A thousand ways denied: The environmental legacy of oil in Louisiana. LSU Press, Baton Rouge, Louisiana. 282 pp

Blum MD, Roberts HH (2012) The Mississippi delta region: past, present, and future. Annu Rev Earth Planet Sci 40:655–683

Britsch LD, Dunbar JB (1993) Land loss rates: Louisiana Coastal Plain. J Coastal Res 9:16

Cahoon DR, McKee KL, Morris JT (2020) How plants influence resilience of salt marsh and mangrove wetlands to sea-level rise. Estuar Coasts 44:883–898

Chang C, Mallman E, Zoback M (2014) Time-dependent subsidence associated with drainage-induced compaction in Gulf of Mexico shales bounding a severely depleted gas reservoir. AAPG Bull 98(6):1145–1159. https://doi.org/10.1306/11111313009

Couvillion BR, Beck H, Schoolmaster D, Fischer M (2017) Land area change in coastal Louisiana (1932–2016). U.S. Geolog Surv Sci Invest Map 3381:16. pamphlet, https://doi.org/10.3133/sim 3381.

Day JW Jr, Britsch LD, Hawes SR, Shaffer GP, Reed DJ, Cahoon D (2000) Pattern and process of land loss in the Mississippi Delta: a spatial and temporal analysis of wetland habitat change. Estuaries 23:425–438

Day JW, Kemp GP, Freeman A, Muth DP (eds) (2014) Perspectives on the restoration of the Mississippi Delta: the once and future delta, estuaries of the world. Springer Netherlands. https://doi.org/10.1007/978-94-017-8733-8

Day JW, Clark HC, Chang C, Hunter R, Norman CR (2020) Life cycle of oil and gas fields in the Mississippi River Delta: a review. Water 12:30

Gagliano SM, Kemp EB, Wicker KM, Wiltenmuth KS (2003) Active geological faults and land change in Southeastern Louisiana, final report. New Orleans, LA: Coastal Environments, Inc., United States Army Corps of Engineers. Biolog Rep 89(22):247–277

Gould HR, McFarlan E Jr (1959) Geologic history of the Chenier Plain, southwestern Louisiana. Trans-Gulf Coast Assoc Geolog Soc 9:261–270

Johnston JB, Cahoon DR, La Peyre MK (2009) Outer continental shelf (OCS)-related pipelines and navigation canals in the Western and Central Gulf of Mexico: relative impacts on wetland habitats and effectiveness of mitigation. U.S. Dept. of the Interior, Minerals Management Service, Gulf of Mexico OCS Region, New Orleans, LA. OCS Study MMS 2009-048, p 200

Ko JY, Day JW (2004) A review of ecological impacts of oil and gas development on coastal ecosystems in the Mississippi Delta. Ocean Coast Manag 47:597–623

Kulp MA (2000) Holocene stratigraphy, history, and subsidence: Mississippi River delta region, north-central Gulf of Mexico. PhD thesis, University of Kentucky, Lexington, p 336

Morton RA, Buster NA, Krohn MD (2002) Subsurface controls on historical subsidence rates and associated wetland loss in southcentral Louisiana. Gulf Coast Assoc Geolog Soc Trans 52:767–778

Morton RA, Bernier JC, Barras JA, Ferina NF (2005a) Rapid subsidence and historical wetland loss in the south-central Mississippi delta plain: likely causes and future implications. U.S. Geological Survey Open-file Report 2005-1216. http://www.pubs.usgs.gov/of/2005/1216

Morton RA, Bernier JC, Barras JA, Ferina NF (2005) Historical subsidence and wetland loss in the Mississippi Delta plain. Gulf Coast Assoc Geolog Soc Trans 55:555–571

Penland S, Sutor JR (1989) The geomorphology of the Mississippi River Chenier Plain. Mar Geol 90:231–258

Reed DJ (2002) Sea-level rise and coastal marsh sustainability: geological and ecological factors in the Mississippi delta plain. Geomorphology 48:233–243. https://doi.org/10.1016/S0169-555 X(02)00183-6

Reed DJ, Wilson L (2004) Coast 2050: a new approach to restoration of Louisiana coastal wetlands. Phys Geogr 25:4–21

Roberts HH (1997) Dynamic changes of the holocene Mississippi river delta plain: the delta cycle. J Coastal Res 13:605–627

Turner RE, Cahoon DR (eds) (1987) Causes of wetland loss in the coastal Central Gulf of Mexico, volume II: technical narrative. Prepared for Minerals Management Service, New Orleans, LA. Contract No. 14-12-0001-30252. OCS Study/MMS 87-0120, p 400

Turner RE, McClenachan G (2018) Reversing wetland death from 35,000 cuts: opportunities to restore Louisiana's dredged canals. PLoS One 13(12):e0207717. https://doi.org/10.1371/journal.pone.0207717

Williams SJ, Penland S, Roberts HH (1994) Processes affecting coastal wetland loss in the Louisiana deltaic plain. In: Williams SJ, Cichon HA (eds) Processes of coastal wetlands loss in Louisiana. Presented at Coastal Zone '93, New Orleans, Louisiana: USGS Open-File Report 94-0275, pp 21–29

U.S. Dept of the Army Corps of Engineers (USACE) (2004) Louisiana Coastal Area (LCA), Louisiana Ecosystem Restoration Study, Final Volume 1—LCA Study, Main Report. http://www.lca.gov/final_report.aspx

Chapter 2
Environmental Setting of the Mississippi River Delta

John W. Day and Rachael G. Hunter

2.1 Development of the Mississippi River Delta (MRD)

After the end of the last glaciation, sea level rose about 150 m and stabilized at approximately its present level about 6000 years ago. In the northern Gulf of Mexico, the rising sea level flooded a broad area of low relief continental margin. After sea level stabilized, river discharge formed the MRD that grew to approximately 25,000 km^2 of wetlands, shallow water bodies and low relief uplands (Fig. 2.1, Roberts 1997; Roberts et al. 2015; Day et al. 1995, 2000, 2007, 2019; Hijma et al. 2017). The Holocene delta is about 15,000 km^2 while flat prairie terraces occupy about 10,000 km^2. In the twentieth century, however, pervasive changes in the MRD caused the loss of about a quarter of the coastal wetlands (Britsch and Dunbar 1993; Couvillon et al. 2017).

There has been considerable discussion of the causes of the wetland loss (Gagliano and van Beek 1970; Turner 1997; Turner and McClenachan 2018; Day et al. 2000, 2019, 2020; Morton et al. 2005a, b; Penland et al. 2005; Wilson and Allison 2008; Kolker et al. 2011; Twilley et al. 2016). This discussion has included the relative role of "natural" versus anthropogenic impacts and consideration of specific human impacts such as isolation of most coastal wetlands from riverine input and pervasive alteration of the hydrology of the deltaic plain. However, the fact

---

The original version of this chapter was revised: The author "Rachael G. Hunter's" affiliation has been updated. The correction to this chapter is available at
https://doi.org/10.1007/978-3-030-94526-8_10

J. W. Day (✉)
Department of Oceanography and Coastal Science, Louisiana State University, 2005 Olive St., Baton Rouge, LA 70806-6660, USA
e-mail: johnday@lsu.edu

R. G. Hunter
Comite Resources, Inc, Baton Rouge, LA, USA

© The Author(s), under exclusive license to Springer Nature Switzerland AG 2022,
corrected publication 2022
J. W. Day et al. (eds.), *Energy Production in the Mississippi River Delta*,
Lecture Notes in Energy 43, https://doi.org/10.1007/978-3-030-94526-8_2

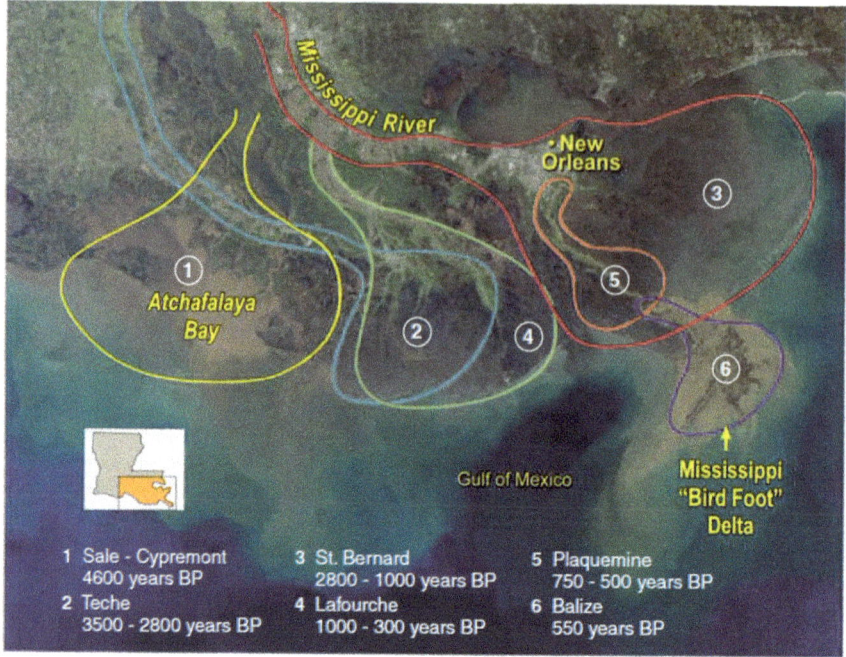

Fig. 2.1 Delta lobes of the Mississippi River Delta and times of active growth for each delta (Day et al. 2007)

that the change in MRD wetland area went from positive to negative during the twentieth century indicates that human activities shifted the balance from forces that led to delta growth and sustainability to a situation of decline. Because O&G activities extend over the entire coastal area and impact coastal ecosystems in so many ways, the role of O&G activities in MRD wetland loss has attracted considerable study (Gagliano and van Beek 1970; Davis 1973; Turner and Cahoon 1988; Turner 1990, 1997; Day et al. 2000, 2007; Ko and Day 2004; Morton et al 2006; Olea and Coleman 2014; Condrey et al. 2014; Turner and McClenachan 2018; Day et al. 2019, 2020).

The Mississippi River is one of the largest rivers in the world and drains a watershed of 3.2 million km^2 of the continental United States and Canada (Blum and Roberts 2012). The MRD was formed over the past 7,000 years by the building of large deltaic lobes through Mississippi River sediment deposition and wetland formation (Roberts 1997, Day et al. 2007, 2019; Mikhailov and Mikhailov 2010; Hijma et al. 2017). Historically, seasonal overbank flooding, crevasse formation, and minor distributaries introduced large amounts of river water with associated sediments and nutrients into the interdistributary basins of the MRD plain, forming and sustaining wetlands by directly promoting vertical accretion and providing nutrients that enhance vegetation productivity and result in organic soil formation through increased root growth and organic matter deposition (Day et al. 2011, 2019; Day and Erdman 2018; Lane et al. 2018). Vertical growth of the wetland surface offsets natural subsidence of the MRD that can be as high as 1.5 cm y^{-1} (Cahoon et al. 1995a, b, 1999). The construction of flood control levees and closure of distributary channels began soon after colonization of New Orleans by the French in 1718 (Welder 1959; Boesch 1996; Colten 2000; Colten and Day 2018). After the great flood of 1927, levees were upgraded and made continuous along most of the lower Mississippi River, isolating the majority of deltaic wetlands from riverine input containing valuable nutrients and sediments (Kesel 1988, 1989; Mossa 1996).

It is important to understand that the processes that developed and help sustain coastal wetlands occur on different spatial and temporal scales and have been occurring since the development of the MRD began 7,000 years ago. *It is, therefore, extremely important to put the recent impacts of oil and gas activities into the context of these long-term forcings*. Since the major impacts of oil and gas activities occurred from the 1940s through the 1990s and continue to the present; it is important to ascertain how these activities impacted or were impacted by the different forcings that regulated delta development and sustainability over time scales ranging from weekly to millennia (Fig. 2.2). It is also important to understand the timing of other anthropogenic impacts on the MRD (e.g., reduction of suspended sediments in the river, levee construction) relative to oil and gas activities.

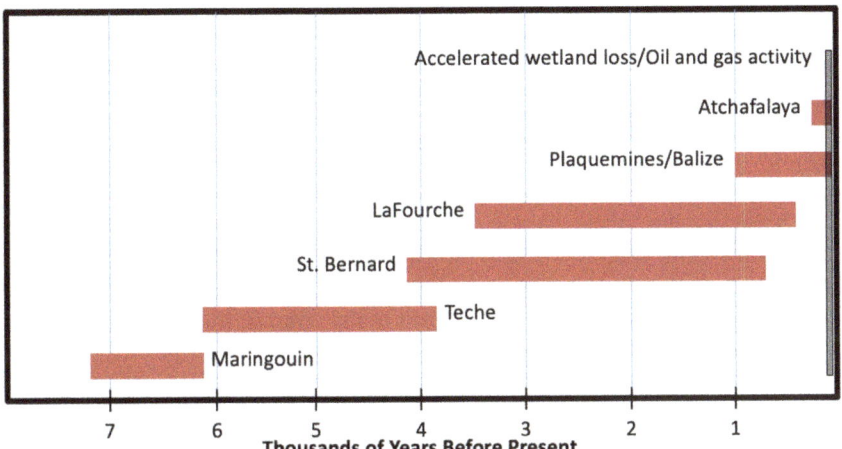

Fig. 2.2 Mississippi River Delta lobe formation over time and the forcing or pulsing events that affect the formation and sustainability of the delta (Table adapted from Day et al. 1997, 2007). Note that all the factors that impacted the development of the Mississippi River Delta have been active for thousands of years while oil and gas activity and high rates of land loss have occurred in less than a century (shown in gray bar)

The development, functioning, and sustainability of all deltas, including the MRD, results from episodic external and internal inputs of energy and materials (Fig. 2.2; Day et al. 1997, 2007, 2019). These inputs occur as pulses in a hierarchical manner that produce benefits over different spatial and temporal scales (Odum et al. 1995; Day et al. 1997, 2007, 2019). Inputs range from waves and daily tides to switching of major delta lobes, which occur on the order of hundreds to over a thousand years, and include crevasse formation, minor distributaries, frontal passages, river floods of varying magnitude, strong tropical storms and associated storm surges, and winter frontal passage. Infrequent events, such as channel switching, crevasse formation, great river floods and strong storms, control the location and rate of sediment delivery to the MRD and impact geomorphology. Pulsed events are especially important considering projections of sea-level rise (e.g., Day et al. 2016a, b; Day and Rybczyk 2019). Annual river floods, seasonal storms, and daily tidal flows help maintain salinity gradients, deliver sediments and nutrients and regulate biogeochemical and biological processes (Day et al. 1989, 2009).

Currently, only two distributaries of the Mississippi are functioning—the main channel of the Mississippi River and the Atchafalaya River, which carries about one third of the river flow. However, under natural conditions, hundreds of major and minor distributaries distributed river water over the entire delta.

Condrey et al. (2014) summarized the functioning of the MRD prior to any significant human impact. Their observations document just how profoundly the delta has changed since colonial times. Condrey et al. used maps and journals of European explorers to describe what they called the last natural delta of the Mississippi that existed just before European settlement. The delta was a seaward-advancing arc that occupied, through four distributaries, all of the five most recent delta complexes of the Mississippi River (Teche, St. Bernard, Lafourche, Modern, and Atchafalaya) and extended across the deltaic plain. It was characterized by plumes of fresh water that extended for more than 10 km into the Gulf of Mexico during the spring flood of the river, along with a vast offshore oyster reef. They suggested that much of the Louisiana coast was advancing into the sea at the onset of European colonization, and that colonial and post-colonial modification of the Mississippi River resulted in the loss of much of this potential. The natural delta described by Condrey et al. was formed and sustained by the hierarchical series of energetic forcings described above.

2.2 Freshwater and Sediment Dynamics in the Mississippi River Delta

Freshwater input, water level variability and sediment dynamics play important roles in regulating the health and sustainability of coastal wetlands in the MRD. Coastal water level variability depends on river discharge, tidal fluctuations, wind waves, climate, atmospheric forcings such as storms, topography, and sea-level; all occurring on different time scales (Baumann 1987; Allison and Meselhe 2010; Hiatt et al. 2019). Atmospheric processes are important drivers of water level variability and flux in the MRD and weather and climate play central roles in determining the impact of these processes. Climate variation ranging from daily temperature change, strong precipitation events, and storms ranging from frontal passages to tropical cyclones all impact coastal water level variability (Hiatt et al. 2019). Water levels are generally lower in the winter when winds are predominantly from the north depressing nearshore Gulf of Mexico water levels. Onshore winds during the warmer part of the year, in combination with thermal expansion, result in higher water levels. Mean monthly water levels in September and October can be up to 20 cm higher than those in January, causing water levels that are significantly higher in the growing season that may stress wetland vegetation. Impacts, such as those associated with O&G activities, can increase relative water levels and further stress vegetation.

There are multi-decadal water level changes along the northern Gulf of Mexico coast that are related to a series of global forcings affecting the Atlantic Basin including the Atlantic Multidecadal Oscillation (AMO) and the North Atlantic Oscillation (NAO) (Dima and Lohmann 2007; Kennedy et al. 2011). An analysis of tide records from Grand Isle shows a long-term relative sea level rise (RSLR), the combined effect of natural and induced subsidence and eustatic sea-level rise (ESLR; twentieth century mean of 1–2 mm yr^{-1}, Gornitz et al. 1982; currently about 3–4 mm yr^{-1}, Hansen et al. 2015), of about 1 cm yr^{-1} or more. If these data are expressed as annual means of detrended water level (i.e., the long-term rise is removed), interannual variations in water levels up to 10 cm above and below the long-term detrended mean become apparent (Fig. 2.3). These cycles last from 2 to 3 decades. These water level changes are related to decadal patterns of inundation and wetland loss in the Mississippi delta as well as the AMO and the NAO (Hiatt et al. 2019). In addition, over the past century and a half ESLR has accelerated due to climate change (Church et al. 2013). This rise is due to melting land-based ice masses (glaciers and the ice packs of Greenland and Antarctica) and thermal expansion of ocean waters (IPCC 2013). Sea-level rise currently is between 3 and 4 mm/yr and by 2100 is projected to be as much as a meter or more (Horton et al. 2014; IPCC 2017).

The Mississippi River is directly or indirectly the source of almost all mineral sediments in the MRD. Over the past seven millennia, sediments from the river have built major delta lobes that include natural levee distributary ridges and crevasses as well as barrier islands and Chenier ridges. These elevated features formed a skeletal network that protected interior wetlands in the delta from physical forcing origination from the Gulf (e.g., hurricanes) as well as salt-water intrusion. The Mississippi River originally provided over 500 M metric tons of sediment annually, but this has decreased by more than half due to human activities (e.g., dams, diversions, pervasive hydrological alteration of the delta plain; Blum and Roberts 2009; Kemp et al. 2016).

Winter storm fronts pass every seven to ten days from November through March resulting in frequent flooding and draining of marshes. The strong frontal winds resuspend shallow bottom sediments resulting in TSS concentrations of 400 to as high as 2,000 mg L^{-1} and high deposition of mineral sediments in wetlands (Perez

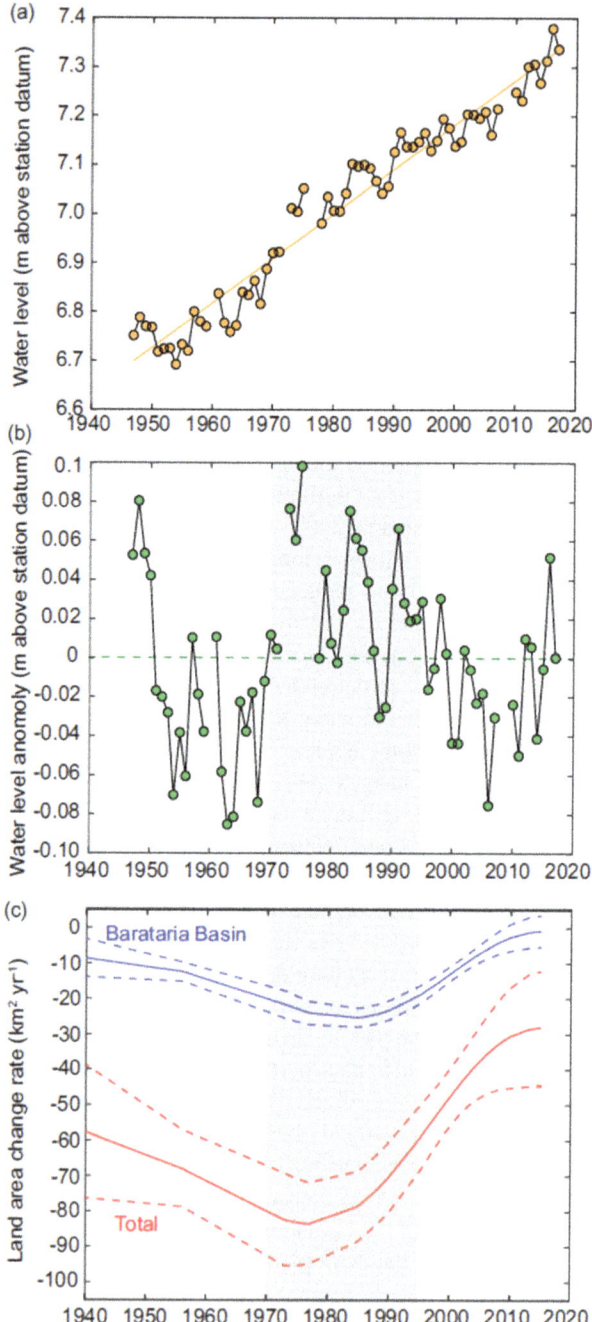

◀**Fig. 2.3 a** Annual water level data at NOAA Station No. 8761724 at Grand Isle (Barataria Basin, LA) from 1947 to 2017. There was a long-term water level increase of about 0.9 **cm**/yr. **b** Annual average detrended water levels at Grand Isle. **c** Land area changes in coastal LA (red) and Barataria Basin (blue) as calculated by Couvillion et al. (2017). The dashed lines represent the 95% confidence bounds for the land area change calculations. Land loss rates increased until reaching their maximum in the 1960s through 1980s (shaded region in **b** and **c**) and have steadily decreased since 1990 (from Hiatt et al. 2019)

et al. 2000; Day et al. 2011). Tropical cyclones, especially hurricanes, are infrequent but regular events that cause extreme water level changes along the Gulf coast. Hurricane surge is commonly greater than 1–2 m but can be as high as 10 m as occurred near the Louisiana-Mississippi border during Hurricane Katrina (Shaffer et al. 2009). Hurricanes can lead to high levels of sediment deposition in coastal wetlands but can also cause significant erosion (Conner et al. 1989; Turner et al. 2007; Shaffer et al. 2009; Howes et al. 2010; Morton and Barras 2011; Baustian and Mendelssohn 2015; Smith et al. 2015). Even areas that normally do not have high riverine sediment inputs can have extremely high concentrations in association with meteorologically driven events. For example, Bayou Chitigue in northwestern Terrebonne Bay does not directly receive riverine sediments, yet Murray et al. (1993) recorded sediment concentrations over 2000 mg/L during a severe winter storm and attributed the high levels to channel scour and resuspension of bay sediment. Baustian and Mendelssohn (2015) reported that up to 12 cm of sediment was deposited on a coastal Louisiana marsh surface during the 2008 hurricane season.

Marshes and beach ridges (cheniers) in western coastal Louisiana developed over time in response to episodic periods of riverine sediment input from the Mississippi and Red Rivers. When the Mississippi River changed course to a more westerly route, large quantities of riverine-derived suspended sediment were moved westerly by longshore currents and deposited along the Chenier Plain shore, resulting in seaward movement of the shoreline (Fig. 2.4; McBride et al. 2007). As the Mississippi River changed to a more easterly course, the shoreline retreated due to reduced sediment supply via littoral transport. These geologic processes of beach building and retreating formed a wide zone of coastal marshes and beach ridges (Louisiana Coastal Wetlands Conservation and Restoration Task Force 1993; Owen 2008).

2.3 Hurricanes

Hurricanes have affected the MRD for centuries and millennia and were likely a positive factor in the growth of the delta. When a hurricane makes landfall, it can impact landforms, ecosystems, and sedimentary and hydrological processes (Fig. 2.5). Only in the twentieth century have hurricanes caused significant negative impacts due to interaction with human impacts such as oil and gas activity. Conner et al. (1989) reviewed impacts of hurricanes on coastal wetlands in the Gulf of Mexico and found

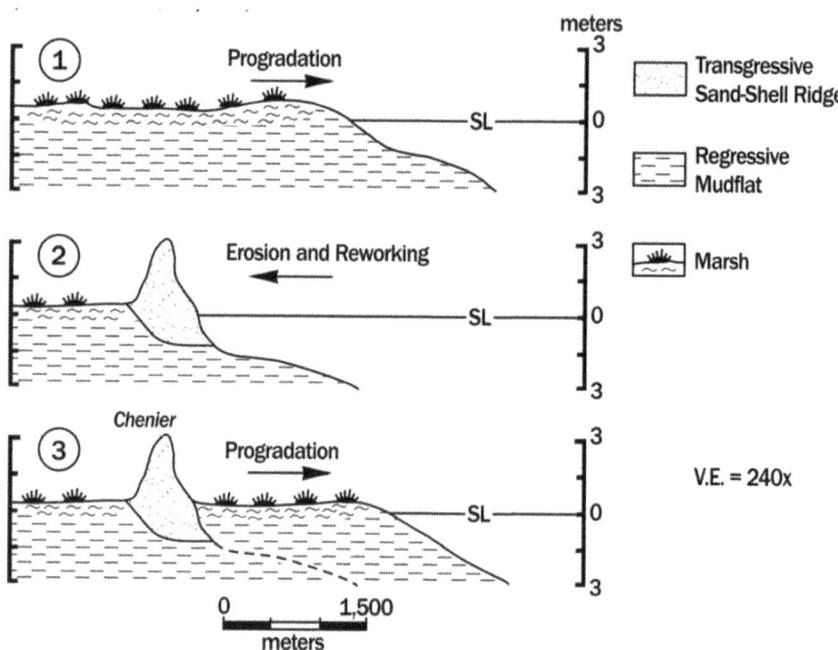

Fig. 2.4 Depositional model illustrating chenier plain development through mudflat progradation (1), wave erosion and reworking to create a transgressive ridge (2), followed by mudflat progradation, which completes chenier genesis (3). SL = sea level (McBride et al. 2007)

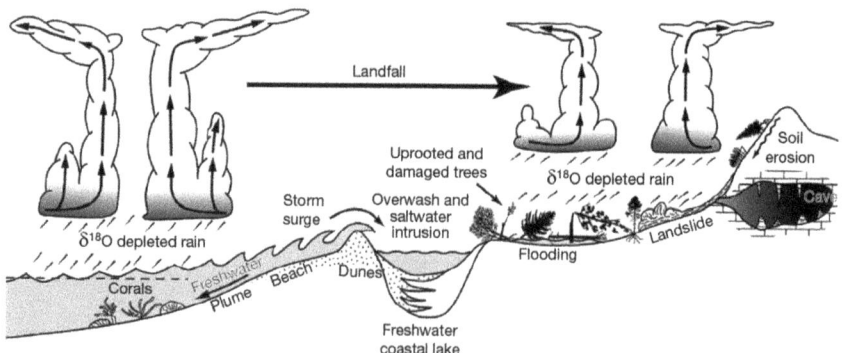

Fig. 2.5 Environmental impacts of catastrophic hurricanes. A landfalling hurricane may cause a storm surge that overtops beach barriers, resulting in the formation of an overwash fan and the deposition of a sand layer in the sediments of a back-barrier lake or marsh. The strong wind may cause massive damage or mortality to trees, leaving behind a paleoecological record of disturbance and succession, including the occurrence of posthurricane fires. Heavy precipitation may cause flooding in the lowlands, and soil erosion and landslide in the uplands. The δ18O-depleted signal in the hurricane rains may be recorded in the cellulose of tree rings and the calcium carbonate of speleotherms and coral skeletons if it is not attenuated by hydrological processes (Liu 2004)

that they generally do not produce long-term detrimental impacts to unmodified coastal systems and they often provide net benefits (e.g., redistribution of sediments into interior areas, freshwater from heavy rains to offset salinity intrusion) that contribute to long-term wetland sustainability, habitat diversity, and vegetation productivity.

Morton and Barras (2011) reviewed information on a half century of hurricane impacts on wetlands in coastal Louisiana. They reported that there were repeated patterns, both temporally and spatially, of wetland loss due to hurricanes. These patterns were a result of the impact of high winds and storm surge on different types of wetland vegetation. Erosional features included formation of elongated and amorphous ponds, expansion of ponds, folded and denuded marsh, and shoreline erosion. Rapid drainage of flood waters formed dendritic incisions around pond margins and braided channels. There were extensive zones of deposited organic wrack and sandy washover terraces and interior-marsh mud deposition. There were also compressed marsh and displaced marsh mats and marsh balls. Prolonged inundation and salinization were also common. These wetland patterns of wetland change often became legacies of prior storm impacts and influenced subsequent storm damages. There was often wetland recovery. The degree of persistent wetland damage due to hurricanes depends on impact duration and the distance the hurricane passes over wetlands. If repeated hurricane damage in an area is not too frequent or severe, wetlands can recover.

In coastal wetlands with anthropogenic modifications such as canals and spoil banks, hurricane impacts can be detrimental because canals provide preferential pathways for high-velocity flows and spoil banks reduce sediment deposition and impact vegetation growth by impeding post-storm floodwater drainage and prolonging saltwater intrusion (Harris and Chabreck 1958; Morgan et al. 1958; Meeder 1987; Austin 2006; Suhayda and Jacobsen 2008; Howes et al. 2010; Morton and Barras 2011). For example, impounded marshes in the Chenier plain region of coastal Louisiana remained flooded for over nine months following Hurricane Rita (Barras 2007). Similarly, Morton and Barras (2011) noted that Hurricane Audrey in 1957 caused saltwater to flood an impounded marsh at Grand Chenier that killed the vegetation and allowed salt-tolerant vegetation to colonize the area, although with lower productivity than the original vegetation. After four years, however, soil salinity had dropped and the plant composition returned to pre-storm conditions.

Hurricane impacts are particularly detrimental to vegetation already stressed from anthropogenic activities. Hurricane Katrina traversed the Breton Sound estuary and caused massive wetland loss (>100 km^2) in the area (Barras 2006; Day et al. 2007; Shaffer et al. 2009). Inside the Gentilly oil and gas field, the impact of Hurricane Katrina was much greater in the northern areas where production and extraction were occurring compared to adjacent marsh areas in the southeastern portion of the field where little production occurred (Fig. 2.6).

In addition to directly affecting vegetation through removal or re-distribution, hurricanes also impact wetland soil elevation. Cahoon (2003) identified eight

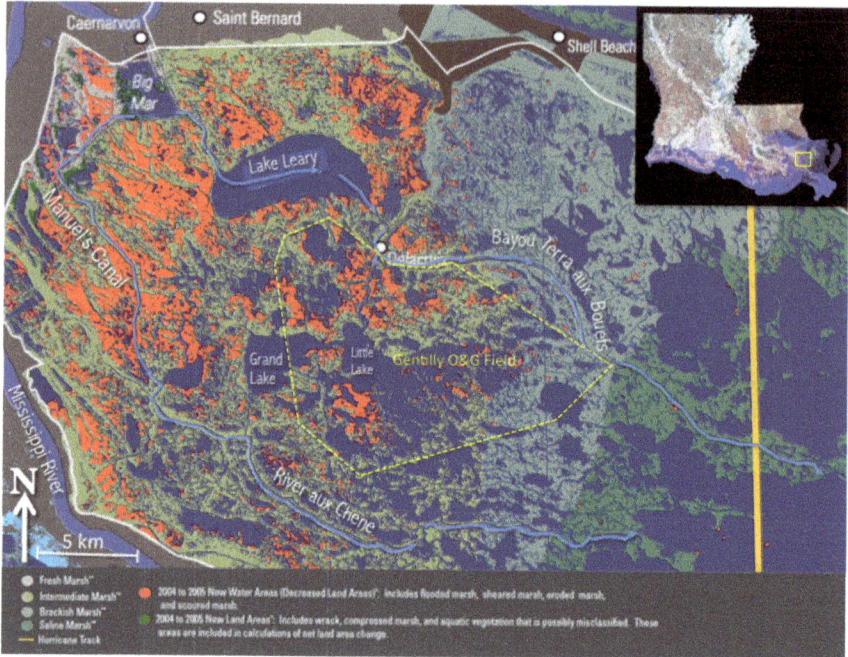

Fig. 2.6 Land loss (in red) associated with Hurricane Katrina (modified from Barras 2006). Note high wetland loss in the vicinity of the Gentilly oil and gas field (yellow dashed circle)

processes through which major storms can impact wetland elevation, including sediment deposition, sediment erosion, sediment compaction, soil shrinkage, root decomposition (following tree mortality from high winds), root growth, soil swelling, and lateral folding of the marsh root mat. In a review of 26 elevation responses to 15 major storms at 17 study sites, Cahoon (2006) found that storm surge affected soil elevation through both surface and subsurface processes, as described above, and was the primary mechanism driving elevation change in 23 of the 26 elevation responses studied. The passage of a hurricane may result in a high rate of sediment deposition on the surface of coastal wetlands. Hurricane Andrew was a major storm that enhanced accretion at wetlands in Terrebonne Basin (Table 2.1; Cahoon et al. 1995a, b; Nyman et al. 1995a). Baustian and Mendelssohn (2015) reported that hurricane sedimentation was strongly related to marsh primary production (Fig. 2.7). *S. alterniflora* showed a strong relationship with the depth of hurricane sediments deposition with production increasing by a factor of three with deposition up to nine cm. Other species showed about a doubling of production. Thus, hurricane sediment deposition is an important source of sediments in natural wetlands helping to offset subsidence.

Table 2.1 Accretion following Hurricane Andrew (August 26, 1992) at coastal Louisiana marshes, Terrebonne Basin

Location	Accretion (cm)	References
Bayou Blue	4.1	Cahoon et al. (1995a, b)
Carenco Bayou	6.0	Cahoon et al. (1995a, b)
Old Oyster Bayou	1.8–3.5	Cahoon et al. (1995a, b), Nyman et al. (1995a)
Bayou Chitigue	1.7–3.0	Cahoon et al. (1995a, b), Nyman et al. (1995a)
Upstream Hard Bayou	3.2	Nyman et al. (1995a)
Downstream Hard Bayou	3.5	Nyman et al. (1995a)
Blue Hammock Bayou	9.0	Nyman et al. (1995a)
Bayou DuLarge	3.3	Nyman et al. (1995a)
King Lake	3.6	Nyman et al. (1995a)
Grand Pass	6.5	Nyman et al. (1995a)
Madison Bay	4.0	Nyman et al. (1995a)

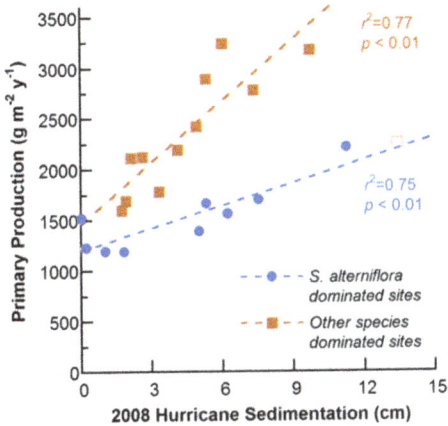

Fig. 2.7 Aboveground primary production in response to 2008 hurricane sedimentation. The open square was identified as an outlier and omitted from the regression (from Baustian and Mendelssohn 2015)

2.3.1 Paleo Hurricanes

Hurricanes have impacted coastal Louisiana in the recent past but also for hundreds and thousands of years and there is considerable literature about paleo tropical cyclones for both tropical (McCloskey and Keller 2009; McCloskey and Liu 2012) and temperate coastal zones (Liu and Fearn 1993, 2000). Liu (2013) reviewed the literature on paleotempestology, which is the study of paleo-tropical cyclone impacts. In different parts of the world tropical cyclones are called hurricanes, typhoons, and cyclones. Paleotempestology studies these storms before the period of instrumentation measurements (going back about a century and a half) and spans a time of several centuries to six millennia in the past. Paleo-hurricanes focus mainly on catastrophic hurricanes of category 4 and 5. Two main sources of data are available for reconstructing past hurricane activity to beyond the instrumental period—geological proxy records and historical documentary records.

A number of geological and ecological proxies have been used to document past hurricanes including overwash sand deposits and storm-formed beach ridges. Oxygen isotope values in stalagmites and tree rings have also been used to build a record of past tropical activity. Historical documentary records of hurricanes contain information about hurricane activity (e.g., Mock et al. 2004; Garcia Herrera et al. 2005). These records range from a few centuries to a thousand years. Thus, many studies document that hurricanes have continuously but episodically impacted tropical to temperate coastal areas long before the modern era. Therefore, any suggestion that hurricanes are an important factor leading to twentieth century wetland loss has to explain why hurricanes caused wetland loss in recent decades but the MRD developed and grew over a several thousand year period when hurricanes were a quasi-continuous atmospheric forcing on the developing delta. In other words, what made twentieth century deltaic wetlands more susceptible to hurricanes. It is clear that oil and gas impacts played a role in this.

2.4 Coastal Wetland Sustainability

Coastal wetlands in the MRD exist in dynamic equilibrium between forces that lead to wetland establishment and maintenance and those that lead to wetland deterioration (Mendelssohn and Morris 2000; Lane et al. 2006; Day et al. 2011; Elsey-Quirk et al. 2019). If these forces are not in balance, a coastal wetland can become stressed and degrade to open water (Fig. 2.8). In the vertical dimension, one of the most important processes affecting coastal wetlands is RSLR of 2–10 mm yr^{-1} or more (Penland and Ramsey 1990). If wetlands are to survive RSLR, they must grow vertically at a rate such that surface elevation gain is sufficient to offset the rate of water level rise (Reed and Cahoon 1992; Cahoon et al. 1995a, b; Day et al. 1997). In the horizontal plane, high wave energy on exposed marsh shores can lead to shoreline retreat, scour of the surface of the marsh and, to a lesser extent, enlargement of interior marsh ponds and lakes (Day et al. 1997, 1999, 2011; Morton and Barras 2011). Saltwater intrusion into lower salinity areas can stress or kill fresh and low salinity wetlands (Mendelssohn and Morris 2000; Day et al. 2000; Shaffer et al. 2009, 2016). Even in saline marshes, high salinity waters can combine with other stressors (e.g., excessive inundation, high hydrogen sulfide concentrations, low oxygen concentrations) to reduce above- and belowground growth and/or kill vegetation (Mendelssohn and McKee 1988; Mendelssohn and Morris 2000).

The two primary processes that maintain coastal wetlands by affecting elevation are input of mineral sediments and in situ production of above- and belowground biomass that leads to organic soil formation (DeLaune et al. 1990; Day et al. 2000, 2011). Suspended mineral sediments in surface water can be advected over wetlands and settle on the wetland soil surface to promote elevation gain and an increase in

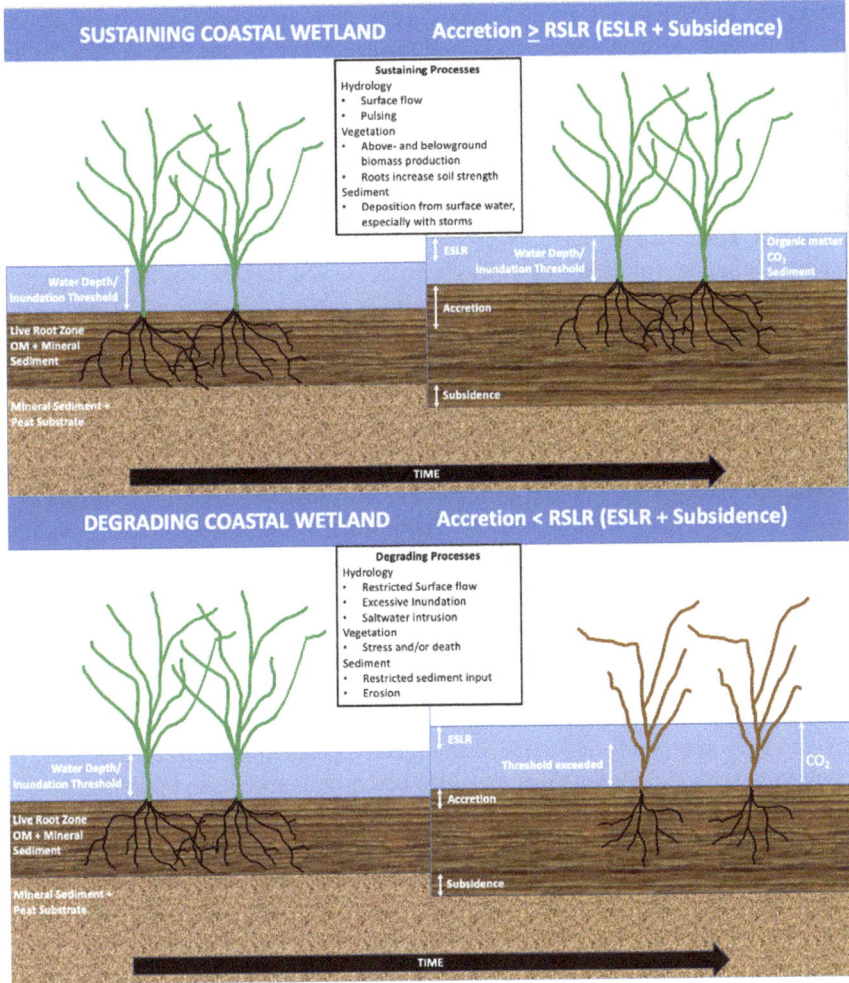

Fig. 2.8 For a coastal marsh to sustain itself over time, accretion must be greater than relative sea level rise (RSLR) which is the combination of eustatic sea level rise (ESLR and subsidence). In degrading coastal wetlands, accretion is less than RSLR

wetland soil strength (Perez et al. 2000; Day et al. 2011, 2016b; Jafari et al. 2019). Sources of these sediments include river water, eroded sediments from surrounding areas, sediments that are resuspended during frontal passages and hurricanes, and wave resuspension along the coastline. Above- and belowground wetland biomass production contributes to elevation gain through the formation of organic soils (Callaway et al. 1997; Elsey-Quirk et al. 2019). When aboveground biomass senesces, some of it is deposited onto the marsh surface where it accumulates and increases soil elevation. As a growing plant produces roots, belowground biomass accumulates (both dead and live roots) and increases soil elevation. In a sustainable marsh, accretion of mineral sediments and organic soil formation keeps pace with or exceeds the factors causing elevation loss (e.g., ESLR and subsidence; Nyman et al. 1990).

In a degrading marsh, where accretion of mineral sediments and organic matter is less than RSLR, the accretionary deficit ultimately leads to marsh loss and conversion to open water (Fig. 2.8; Cahoon 1994; Mendelssohn and Morris 2000; Day et al. 2007, 2011). The change in equilibrium can result from either a decrease in mineral sediments, a decrease in vegetation productivity and resulting organic soil formation, or both (DeLaune et al. 1990), as well as increases in RSLR (Day et al. 2020). Changes in sediment coming into a wetland typically occur from alterations in hydrology (e.g., levees along a river, canal spoil banks which prevent sheet flow, water control structures) that prevent movement of surface water, containing sediments, across a marsh (Reed et al. 1997; Cahoon 1994; Boumans and Day 1994).

Louisiana coastal marsh vegetation species have evolved to adapt to specific salinity regimes, tidal ranges, inundation frequency and duration, and elevation and even small changes to any of these factors can impact species composition and productivity (Mendelssohn and Morris 2000; Mitsch and Gosselink 2015; Nyman 2014). Couvillion and Beck (2013) used remote sensing, field data, elevation data and mean water level data for saline, brackish and intermediate marshes to identify a range of elevations at which marsh communities typically occur and are most productive and those where marsh collapse is likely to occur (Couvillion and Beck 2013). These data demonstrate the importance of elevation to coastal marsh productivity and sustainability. When a disturbance lowers marsh elevation, vegetation becomes stressed and open water may replace emergent vegetation that has died because of the narrow vertical flooding range in which herbaceous species in Louisiana coastal marshes have evolved (Sasser 1977; Mendelssohn et al. 1981; Day et al. 2011; Nyman 2014).

In a marsh impounded or semi-impounded by spoil banks, surface water exchange with adjacent open water is reduced but water that does enter impounded areas drains more slowly and flooding duration is greater than in areas without spoil banks. This reduction in water exchange is especially important in reducing sediment input during frontal passages when high sediment concentrations occur (Swenson and Turner 1987; Boumans and Day 1994; Cahoon 1994; Day et al. 2000; Perez et al. 2000). Prolonged inundation causes water logging, anoxic soils, root oxygen deficiency, sulfide toxicity, and/or salt stress in areas of fresh, intermediate, brackish, or even saline vegetation (Mendelssohn and McKee 1988; Mendelssohn and Morris 2000; Mendelssohn and Batzer 2006). As anoxic stress becomes more severe, plants become less efficient in nutrient uptake (Mendelssohn and McKee 1988). These stressors impact vegetation health and lead to a reduction in above- and belowground biomass that, in turn, decreases the accretion rate due to organic soil formation (e.g., Snedden et al. 2015). Over time, the decrease in accretion, in combination with RSLR, reduces elevation gain, leading to even longer periods of inundation that further stresses vegetation and further decreases elevation gain. This negative feedback loop continues until vegetation dies and, once this occurs, there is additional loss of soil volume due to loss of root turgor and decomposition of root organic matter which leads to elevation collapse (Nyman et al. 2006; Day et al. 2011). In brackish and saline marshes, revegetation often cannot then occur because of the low elevation and weak soil strength and the wetland converts to open water (Ko et al. 2004; Day et al. 2011; Nyman 2014). In coastal wetlands impacted by canal dredging and spoil banks, the pattern of marsh degradation is typically seen as patches of open water in impounded or semi-impounded marshes that expand and combine to form larger open water bodies until the only remaining vegetated landscape features are dominated by the elevated spoil banks (Turner 1987).

Soil shear strength "defines the ability of soils to resist displacement or deformation when subjected to shear stress" and is indicative of soil resistance to erosion (Teal et al. 2012). Shear strength is influenced by soil composition (including vegetation root biomass), void ratio, water content, pore water chemistry, soil structure and loading conditions (Day et al. 2011; Teal et al. 2012; Jafari et al. 2019). In Louisiana coastal marshes, shear strength generally increases with an increase in live belowground biomass because roots add strength due to the tensile force required to break roots and rhizomes (Fig. 2.9; Teal et al. 2012; Sasser et al. 2013, 2018; Nyman 2014). Thus, in these marshes, factors that affect vegetation belowground productivity will directly impact soil strength and thus susceptibility to erosion. The high degree of scatter in Fig. 2.9 indicates that other processes are also important.

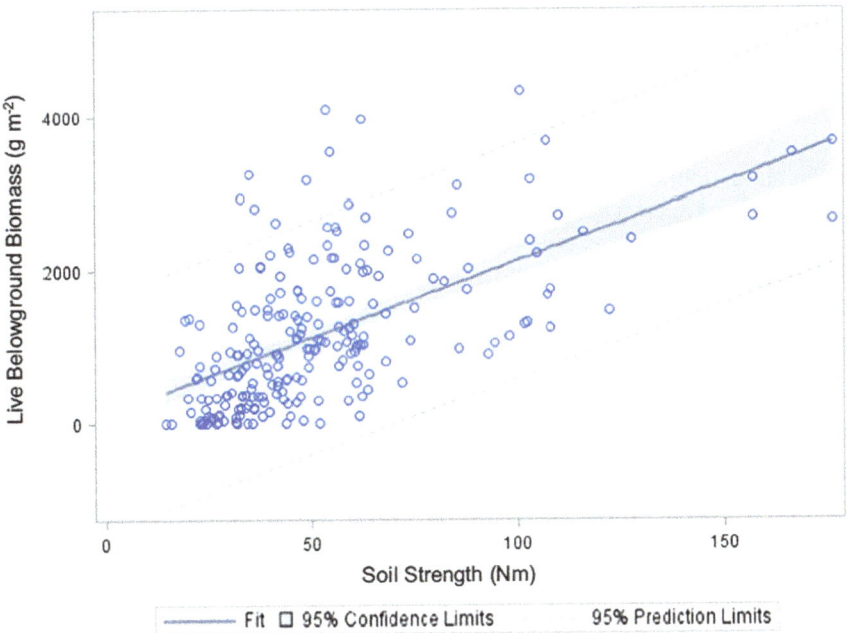

Fig. 2.9 Relationship of soil strength versus live belowground biomass for Louisiana coastal marshes (from Sasser et al. 2018)

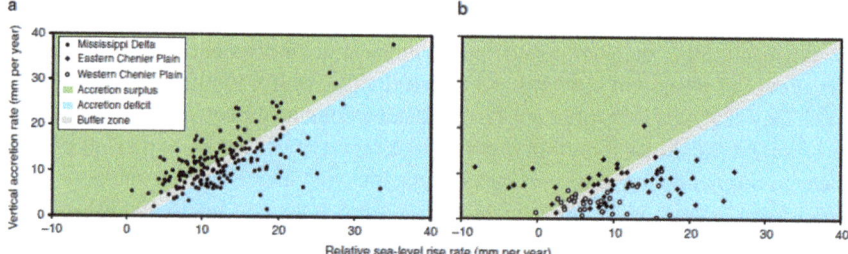

Fig. 2.10 Vulnerability of coastal Louisiana wetlands to present-day rates of RSLR. **a** Mississippi Delta; **b** Chenier Plain. Sites that fall within the accretion surplus field have vertical accretion rates that exceed the rate of RSLR. Sites within the grey buffer zone have an accretion deficit that is <2 mm per year and are assumed herein not to be vulnerable (although this is uncertain). Sites that fall below this buffer zone have an accretion deficit >2 mm per year and are considered vulnerable given current rates of RSLR (from Jankowski et al. 2017)

The Coastwide Reference Monitoring System (CRMS) was implemented in Louisiana in 2006 to monitor a range of ecological conditions across a variety of habitats (swamp forest and fresh, intermediate, brackish, and salt marsh). There are approximately 390 CRMS sites across coastal Louisiana that are currently monitored for hydrology (percent time flooded, water level range), water salinity, vegetation species, percent cover, basal area, soil accretion and elevation change, and other parameters. These data are used to map conditions across the coast and to monitor the effectiveness of restoration efforts. Jankowski et al. (2017) analyzed surface elevation change, vertical accretion, shallow subsidence, and relative sea level rise rate in a subset of 274 CRMS sites. They found that 35% of wetland sites in the Mississippi Delta and 58% of the sites in the Chenier Plain have an accretion deficit that makes them particularly vulnerable to modern rates of RSLR (Fig. 2.10). Because most CRMS sites are near a water body, these results likely underestimate the vulnerability of coastal marshes to rising water levels because interior marshes are at a lower elevation than streamside marshes. Factors that further stress wetland vegetation such as spoil banks and induced subsidence such as occurs due to O&G activities can negatively impact wetlands.

Peat collapse is the process of rapid wetland mortality and collapse of soil structure as a result of vegetation death and soil organic matter decomposition. Peat collapse results from factors such as excessive inundation, plant stress, and salinity intrusion. Hydrologic disruption due to canals, spoil banks and induced subsidence exacerbates collapse. DeLaune et al. (1994) first used the term "peat collapse" to describe a rapid localized loss of marsh elevation and conversion to open water triggered by an event, such as saltwater intrusion or herbivory, that stresses or kills vegetation. They hypothesized that the decrease in marsh elevation occurred due to structural collapse and death of the living root network and/or an increase in the decomposition rate of belowground organic matter following plant mortality. Wetland vegetation roots have a gas-filled arenchyma system that occupies a significant amount of root volume (e.g., 29–35% in *Spartina alterniflora* and *S. patens*) and loses turgor with plant death.

More recently, peat collapse has been defined as "a specific type of shallow subsidence unique to highly organic soils in which a loss of soil strength and structural integrity contributes to a decline in elevation below the lower limit for emerging plant growth and natural recovery" (Chambers et al. 2019). Similarly, Sklar et al (2019) described peat collapse as a dramatic shift in the soil carbon balance that leads to a net loss of organic matter, a loss of soil elevation, and conversion of marsh to open water. When peat collapse occurs, it is unlikely that the marsh will recover naturally because the area that subsided will be permanently inundated. The collapse typically manifested as patches of open water in a marsh, rather than the loss of a large area all at once (Day et al. 2011; Fig. 2.11). But these patches can coalesce over time to form larger areas of open water.

Marsh soils are composed of mineral matter, organic matter, and pore spaces that may be filled with gas and/or water (Reddy and DeLaune 2008). Organic matter content in Louisiana marsh soils varies greatly dependent upon proximity to a mineral sediment sources such as rivers, tidal creeks, and open bays. Interior marshes without defined hydrologic links to an outside sediment source will typically have higher organic matter content than marshes with hydrologic exchange. In mineral soils about 50% of the soil volume is solids and the remaining 50% is gas or water filled poor spaces whereas in organic soils much of the volume (up to 95%) is composed of gas or water filled pore spaces with organic matter and mineral solids typically less than 20% (Reddy and DeLaune 2008; Chambers et al. 2019). In wetlands that experience wet and dry periods, soil bulk density is typically higher than wetlands

Fig. 2.11 Peat collapse will initially manifest as a "patchy" deterioration of a wetland, rather than the loss of a large area all at once. Right image from the Everglades (Chambers et al. 2019), left image from marshes in the Mississippi delta that are isolated from any Mississippi River water input (Day et al. 2011)

that are permanently inundated because drying allows mineral and organic matter to become incorporated into the soil structure.

Based on a review of nine studies that quantified peat collapse, Chambers et al. (2019) found that the process occurred over a few months to a few years in the upper 15–50 cm of soils in coastal marshes and was driven by the severe stress or death of vegetation due to some type of disturbance. Chambers et al. (2019) proposed four processes that contribute to peat collapse following vegetation stress or death, including (1) compression of gas-filled pore spaces within the soil during dry-down conditions; (2) deconsolidation of excessively waterlogged peat, followed by transport; (3) compaction of aerenchyma tissue in wetland plant roots (i.e., loss of root turgor), and possibly collapse of root channels; and (4) acceleration of soil mineralization due to the addition of labile carbon (dying roots), oxygen (decreased flooding), nutrients (eutrophication), or sulfate (saltwater intrusion).

Peat collapse can have more significant repercussions in freshwater marshes than in intermediate, brackish, or saline marshes because the former typically have more soil organic matter and are much less tolerant to increased salinity than the latter. Lane et al. (2016) conducted a study to determine greenhouse gas emissions and elevation change with vegetation death (herbicide treatment) in freshwater, brackish, and saltwater marshes in Louisiana. Total carbon stocks in the upper 50 cm of the soil horizon in this study were 15.0 ± 0.5 kg C/m^2 at the freshwater site, 11.1 ± 0.9 kg C/m^2 at the brackish site, and 8.5 ± 1.4 kg C/m^2 at the saltwater site. At the end of the 1.5 year study, elevation change was -4.24 ± 0.57, -1.57 ± 0.36, and -1.48 ± 0.34 cm in the fresh, brackish, and salt marshes, respectively, which equated to a mass loss of soil organic carbon of 1273 g C/m^2, 389 g C/m^2 and 207 g C/m^2, respectively, in fresh, brackish, and saline marshes. Baustian et al. (2017) analyzed total carbon in Louisiana coastal marshes and found mean percent total carbon in the upper 10 cm of soils was 32, 17, 16, and 9% in fresh, intermediate, brackish, and saline marsh types, respectively.

As mentioned previously, once peat collapse occurs it is usually permanent and cannot be reversed without human intervention such as an increase in elevation through sediment addition. Nyman et al. (1993) and DeLaune et al. (1994) reported about 10 cm elevation loss in an interior marsh that died as a result of excessive flooding. Day et al. (2011) also reported that collapse led to an elevation loss of 10–15 cm over a few months. Nyman et al. (1995b) killed marsh clumps with an herbicide and measured a loss in elevation over 2 years. They also reported that physical removal or erosion in the months following collapse of sediments was not significant as the entire ^{137}Cs inventory in the upper 50 cm of the initial soil column was retained after the collapse. Once open water forms these areas may increase in size through edge erosion, bank slumping, or soil creep, a viscous-like slow deformation resulting in a net downslope transport (Day et al. 2011; Chambers et al. 2019; Mariotti et al. 2019).

The above information indicates that once marsh collapse occurs, it is highly unlikely that marsh recovery will take place because the elevation of the marsh is below the elevation range where vegetation can grow. Oil and gas activities lead to pervasive surface alteration of hydrology that can lead to reduction of sediment input, excessive waterlogging, salinity intrusion, and lower above- and belowground productivity that cause stress and mortality of coastal marshes and subsequent marsh collapse and loss of elevation. Thus, the wetland loss caused by oil and gas activities is nearly always permanent and wetlands will not recover without restoration.

2.5 Implications for Restoration

Healthy coastal marshes and forested wetlands in MRD are dependent on specific elevation and salinity regimes. A general goal for restoration is to achieve optimum conditions for promoting sustainable wetlands. To increase elevation all sources of potential sediments should be utilized. If elevation is very low, then marsh creation using dredged sediments will be necessary. The hydrologic network should be optimized to maximize input of available sediment resources including river input, sediments resuspended by frontal passages, hurricanes, and wave activity on shorelines. Thin layer nourishment can be used to increase elevation as marshes subside. Hydrologic management should also be used to minimize salinity intrusion. Details of these and other approaches are addressed in Chap. 8.

References

Adams RD, Banas PJ, Baumann RH, Blackmon JH, McIntire WG (1978) Shoreline erosion in coastal Louisiana: inventory and assessment. Final Report to Louisiana Department of Transportation and Development, p 149

Allison MA, Meselhe EA (2010) The use of large water and sediment diversions in the lower Mississippi river (Louisiana) for coastal restoration. J Hydrol 387:346–360

API (2000) Produced water impacts on Louisiana wetlands. Health and Environmental Sciences API Publication Number 4517. Washington, D.C

Asano T (1995) Sediment transport under sheet-flow conditions. J Waterway Port Coastal Ocean Eng ASCE 121:239–246

Austin D (2006) Coastal exploitation, land loss, and hurricanes: a recipe for disaster. Am Anthropol 108:671–691

Barras JA (2006) Land area changes in coastal Louisiana after the 2005 hurricanes: a series of three maps. U.S. Geological Survey Open-File Report 06–1274

Barras JA (2007) Satellite images and aerial photographs of the effects of Hurricanes Katrina and Rita on coastal Louisiana: U.S. Geological Survey Data Series 281, https://pubs.usgs.gov/ds/2007/281

Barras J, Beville S, Britsch D, Hartley S, Hawes S, Johnston J, Kemp P, Kinler Q, Martucci A, Porthouse J, Reed D, Roy K, Sapkota S, Suhayda J (2003) Historical and projected coastal Louisiana land changes: 1978–2050: USGS Open File Report 03–334, p 39

Barras JA, Bernier JC, Morton RA (2008) Land area change in coastal Louisiana—a multidecadal perspective (from 1956 to 2006). U.S. Geological Survey Scientific Investigations Map 3019, scale 1:250,000, p 14 Pamphlet

Bass AS (1998) Accidental oilfield brine spill impact study: A survey of the magnitude and distribution of accidental brine spills in the Louisiana coastal wetlands conservation area. Final Report. Louisiana Geological Survey Report submitted to the Louisiana Coastal Management Division, Louisiana Department of Natural Resources, Baton Rouge, LA, p 55

Bass AS, Turner RE (1997) Relationships between salt marsh loss and dredged canals in three Louisiana estuaries. J Coastal Res 13:895–903

Baumann RH (1987) Physical variables. In: Conner WH, Jr Day JW (eds) The ecology of Barataria Basin, Louisiana: an estuarine profile. US Fish Wildl Serv Biol Rep 85(7.13):8–17

Baustian JJ, Mendelssohn IA (2015) Hurricane-induced sedimentation improves marsh resilience and vegetation vigor under high rates of relative sea level rise. Wetlands 35:795–802

Baustian JJ, Mendelssohn IA, Hester MW (2012) Vegetation's importance in regulating surface elevation in a coastal salt marsh facing elevated rates of sea level rise. Glob Change Biol 18:3377–3382. https://doi.org/10.1111/j.1365-2486.2012.02792.x

Baustian MM, Stagg CL, Perry CL, Moss LC, Carruthers TJB, Allison M (2017) Relationships between salinity and short-term soil carbon accumulation rates from marsh types across a landscape in the Mississippi River Delta. Wetlands 37:313–324

Blum MD, Roberts HH (2009) Drowning of the Mississippi Delta due to insufficient sediment supply and global sea-level rise. Nat Geosci 2:488–491

Blum MD, Roberts HH (2012) The Mississippi delta region: past, present, and future. Annu Rev Earth Planet Sci 40:655–683

Boesch DF (1996) Science and management in four U.S. coastal ecosystems dominated by land–ocean interactions. J Coastal Conserv 2:103–114

Boesch DF, Rabalais NN (eds) (1987) Long-term environmental effects of offshore oil and gas development. NewYork, Elsevier Science, p 710

Boesch DF, Rabalais NN (eds) (1989) Produced waters in sensitive coastal habitats: An analysis of impacts, central coastal Gulf of Mexico. OCS Report/MMS89-0031. U.S. Dept. of the Interior, Minerals Management Service, Gulf of Mexico OCS Regional Office, New Orleans, Louisiana, 157 pp

Boesch DF, Josselyn MN, Mehta AJ, Morris JT, Nuttle WK, Simenstad CA, Swift DJP (1994) Scientific assessment of coastal wetland loss, restoration and management. J Coastal Res Special 20:1–103

Boumans RM, Day JW (1994) Effects of two Louisiana marsh management plans on water and materials flux and short-term sedimentation. Wetlands 14:247–261

Britsch LD, Dunbar JB (1993) Land loss rates: Louisiana coastal plain. J Coastal Res 9:324–338

Brown & Root, Inc. (1992) Conceptual engineering report for Freshwater Bayou Canal bank stabilization, Vermilion Parish, Louisiana. Prepared for Department of Natural Resources/Coastal Restoration Division. Brown and Root, Inc., Belle Chase, La 26 pp

Brunn P (1962) Sea level rise as a cause of shore erosion. Am Soc Civil Eng Proc J Waterways Harbor Div 88:117–130

Bryant JC, Chabreck RH (1998) Effects of impoundment on vertical accretion of coastal marsh. Coastal Estuarine Res Federation 21:416–422

Cahoon DR (1990) Field monitoring of structural marsh management. In: Cahoon DR, Groat CG (eds) A study of marsh management practice in coastal Louisiana, vol III, ecological evaluation. Final report submitted to Minerals Management Service (MMS), New Orleans, Louisiana, pp 357–368. Contract Number 14–12–0001–30410. Outer Continental Shelf Study/MMS 90–0077

Cahoon DR (1994) Recent accretion in two managed marsh impoundments in coastal Louisiana. Ecol Appl 4:166–176

Cahoon DR (2003) Storms as agents of wetland elevation change: Their impact on surface and subsurface sediment processes. In: Proceedings of the international conference on coastal sediments 2003. CD-ROM, World Scientific Publishing Corporation and East Meets West Productions, Corpus Christi, Texas

Cahoon DR (2006) A review of major storm impacts on coastal wetland elevations. Estuaries Coasts 29:889–898

Cahoon DR, Groat CG (1990) A study of marsh management practice in Coastal Louisiana, vol 1. executive summary. U.S. Dept. of the Interior, Minerals Management Service, Gulf of Mexico OCS Region, New Orleans, Louisiana, 36 pp

Cahoon DR, Turner Jr RE (1989) Accretion and canal impacts in a rapidly subsiding wetland: II. Feldspar marker horizon technique. Estuaries 12(4):260–268

Cahoon D, Reed D, Day J, Steyer G, Boumans R, Lynch J, McNally D, Latif N (1995a) The influence of Hurricane Andrew on sediment distribution in Louisiana coastal marshes. J Coastal Res SI 18:280–294

Cahoon D, Reed D, Day J (1995b) Estimating shallow subsidence in microtidal saltmarshes of the southeastern United States: Kaye and Barghoorn revisited. Mar Geol 30:1–9

Cahoon DR, Day JW, Reed DJ (1999) The influence of surface and shallow subsurface soil processes on wetland elevation: a synthesis. In: Current topics in wetland biogeochemistry 3:72–88

Callaway JC, DeLaune RD, Patrick WH Jr (1997) Sediment accretion rates from four coastal wetlands along the Gulf of Mexico. J Coastal Res 13:181–191

Chambers LG, Reddy KR, Osborne TZ (2011) Short-term response of carbon cycling to salinity pulses in a freshwater wetland. Soil Sci Soc Am J 75:2000–2007

Chambers LG, Osborne TZ, Reddy KR (2013) Effect of salinity-altering pulsing events on soil organic carbon loss along an intertidal wetland gradient: a laboratory experiment. Biogoechemisty 115:363–383

Chambers LG, Steinmuller HE, Breithaupt JL (2019) Toward a mechanistic understanding of "peat collapse" and its potential contribution to coastal wetland loss. Ecology 100(7):e02720. https://doi.org/10.1002/ecy.2720

Chan AW, Zoback MD (2007) The role of hydrocarbon production on land subsidence and fault reactivation in the Louisiana coastal zone. J Coastal Res 23:771–786

Charles SP, Kominoski JS, Troxler TG, Gaiser EE, Servais S, Wilson BJ, Davis SE, Sklar FH, Coronado-Molina C, Madden CJ, Kelly S, Rudnick DT (2019) Experimental saltwater intrusion drives rapid soil elevation and carbon loss in freshwater and brackish Everglades marshes. Estuaries Coasts 42:1868–1881

Childers DL, Day JW Jr (1990) Marsh-water column interactions in two Louisiana estuaries. I. sediment dynamics. Coastal Estuarine Res Federation 13:393–403

Childers DL, Iwaniec D, Rondeau D, Rubio G, Verdon E, Madden CJ (2006) Responses of sawgrass and spikerush to variation in hydrologic drivers and salinity in Southern Everglades marshes. Hydrobiologia 569:273–292. https://doi.org/10.1007/s10750-006-0137-9

Church JA, Clark PU, Cazenave A, Gregory JM, Jevrejeva S, Levermann A, Merrifield MA, Milne GA, Nerem RS, Nunn PD, Payne AJ, Pfeffer WT, Stammer D, Unnikrishnan AS (2013) Sea level change. In: Stocker TF, Qin D, Plattner G-K, Tignor M, Allen SK, Boschung J, Nauels A, Xia Y, Bex V, Midgley PM (eds) Climate change 2013: the physical science basis. Contribution of working group I to the fifth assessment report of the intergovernmental panel on climate change. Cambridge University Press, Cambridge, United Kingdom and New York, NY, USA

Colten CE (2000) Levees and the making of a dysfunctional floodplain. In: Day JW, Erdman JA (eds) Mississippi delta restoration. Springer, Cham, Switzerland, pp 29–37

Colten C, Day J (2018) Resilience of natural systems and human communities in the Mississippi Delta: moving beyond adaptability due to shifting baselines. In: Mossop E (ed) Sustainable coastal design and planning. CRC Press, United States, pp 195–207

Conner WH, Day JW Jr, Baumann RH, Randall JM (1989) Influence of hurricanes on coastal ecosystems along the northern Gulf of Mexico. Wetland Ecosyst Manage 1:45–56

Condrey RE, Hoffman PE, Evers DE (2014) The last naturally active delta complexes of the Mississippi River (LNDM): discovery and implications. In: Day JW, Kemp GP, Freemen AM, Muth DP (eds) Perspectives on the restoration of the Mississippi delta. Springer Netherlands, pp 33–50

Couvillion BR, Beck H (2013) Marsh collapse thresholds for Coastal Louisiana estimated using elevation and vegetation index data. J Coastal Res 63:58–67

Couvillion BR, Beck H, Schoolmaster D, Fischer M (2017) Land area change in coastal Louisiana (1932 to 2016): U.S. Geological Survey Scientific Investigations Map 3381, 16 p. https://doi.org/10.3133/sim3381

Craig NJ, Turner RE, Day Jr JW (1979) Land loss in coastal Louisiana (U.S.A.). Environ Manage 3:133–144

Davis DW (1973) Louisiana canals and their influence on wetland development. Ph.D. Dissertation. Louisiana State University, 234 pp

Davis DW (2000) Historical perspective on crevasses, levees, and the Mississippi River. In: Colten CE (ed) Transforming New Orleans and its Environs. University of Pittsburgh Press, Pittsburgh, pp 84–106

Day JW Jr, Shaffer GP (2016) Effects of the Mississippi River Gulf outlet on coastal wetlands and other ecosystems in Southeastern Louisiana. Expert report prepared for: Plaintiffs in: Biloxi Marsh Lands Corporation and Lake Eugenie Land & Development, Inc. versus The United States of America, Case Number 12-382, consolidated with The Borgnemouth Realty Co., Limited and The Livaudais Company, L.L.C. versus The United States of America, case Number 14-0003, in the United States Court of Federal Claims

Day J, Erdman J (eds) (2018) Mississippi Delta restoration pathways to a sustainable future. Springer, Cham, Switzerland

Day J, Rybczyk J (2019) Global change Impacts on the future of coastal systems: Perverse interactions among climate change, ecosystem degradation, energy scarcity and population. In: Wolanski E, Day J, Elliott M, Ramachandran R (eds) Coasts and Estuaries - The Future. Elsevier, Amsterdam, Netherlands. pp. 635–654.

Day J, Clark H, Chang C et al (2020) Life cycle of oil and gas fields in the Mississippi River Delta: A review. Water 12:1492. https://doi.org/10.3390/w12051492

Day JW, Hall C, Kemp M, Yáñez-Arancibia A (1989) Estuarine ecology. Wiley and Sons, New York, p 558

Day R, Holz R, Day J (1990) An inventory of wetland impoundments in the coastal zone of Louisiana, USA: historical trends. Environ Manage 14(2):229–240

Day J, Pont D, Hensel P, Ibañez C (1995) Impacts of sea-level rise on deltas in the Gulf of Mexico and the Mediterranean: the importance of pulsing events to sustainability. Estuaries 18(4):636–647

Day JW Jr, Martin JF, Cardoch L, Templet PH (1997) System functioning as a basis for sustainable management of deltaic ecosystems. Coastal Manage 25:115–153

Day JW, Rybczyk J, Scarton F, Rismondo A, Are D, Cecconi G (1999) Soil accretionary dynamics, sea-level rise and the survival of wetlands in Venice Lagoon: a field and modelling approach. Estuar Coast Shelf Sci 49:607–628

Day JW Jr, Britsch LD, Hawes SR, Shaffer GP, Reed DJ, Cahoon D (2000) Pattern and process of land loss in the Mississippi Delta: a spatial and temporal analysis of wetland habitat change. Estuaries 23:425–438

Day JW, Boesch DF, Clairain EJ, Kemp GP, Laska SB, Mitsch WJ, Orth K, Mashriqui H, Reed DJ, Shabman L, Simenstad CA, Streever BJ, Twilley RR, Watson CC, Wells JT, Whigham DF (2007) Restoration of the Mississippi delta: lessons from hurricanes Katrina and Rita. Science 315:1679–1684

Day J, Christian R, Boesch D, Yanez A, Morris J, Twilley R, Naylor L, Schaffner L, Stevenson C (2008) Consequences of climate change on the ecogeomorphology of coastal wetlands. Estuaries Coasts 37:477–491

Day JW, Cable JE, Cowan JH, Delaune R, de Mutsert K, Fry B, Mashriqui H, Justic D, Kemp P, Lane RR, Rick J, Rozas LP, Snedden G, Swenson E, Twilley RR, Wissel B (2009) The impacts of pulsed reintroduction of river water on a Mississippi Delta Coastal Basin. J Coastal Res 54:225–243

Day JW, Kemp GP, Reed DJ, Cahoon DR, Boumans RM, Suhayda JM, Gambrell R (2011) Vegetation death and rapid loss of surface elevation in two contrasting Mississippi delta salt marshes: the role of sedimentation, autocompaction and sea-level rise. Ecol Eng 37:229–240

Day JW, Cable JE, Lane RR, Kemp GP (2016) Sediment deposition at the Caernarvon crevasse during the great Mississippi Flood of 1927: implications for coastal restoration. Water 3(38):1–12

Day J, Lane R, D'Elia C, Wiegman A, Rutherford J, Shaffer G, Brantley C, Kemp G (2016) Large infrequently operated river diversions for Mississippi delta restoration. Estuar Coast Shelf Sci. https://doi.org/10.1016/j.ecss.2016.05.001

Day J, Dominguez AL, Herrera-Silveira J, Kemp P (2019) Climate change in areas of the Gulf of Mexico with high freshwater input – A review of impacts and potential mitigation. Jaina Costas y Mares ante el Cambio Cimático. 1(1):87–108

DeLaune RD, Pezeshki SR, Pardue JH, Whitcomb JH, Patrick WH Jr (1990) Some influences of sediment addition to a deteriorating salt marsh in the Mississippi River deltaic plain: a pilot study. J Coastal Res 6:181–188

DeLaune RD, Nyman JA, Patrick WH Jr (1994) Peat collapse, ponding and wetland loss in a rapidly submerging coastal marsh. J Coastal Res 10:1021–1030

DeLaune RD, Lindau CW, Gambrell RP (eds) (1999) Effect of produced-water discharge on bottom sediment chemistry. U.S. Dept. of the Interior, Minerals Management Service, Gulf of Mexico OCS Region, New Orleans, LA. OCS Study MMS 99–0060, 47 pp

Dima M, Lohmann G (2007) A hemispheric mechanism for the Atlantic multidecadal oscillation. J Climate 20:2706–2719

Doiron LN, Whitehurst CA (1974) Geomorphic processes active in the Southeast Louisiana Canal, LaFourche Parish, Louisiana. Research Monographs, Division of Engineering Research – RM 5, 48 pp

Dozier MD (1983) Assessment of change in the marshes of southwestern Barataria basin, Louisiana, using historical aerial photographs and a spatial information system. Master's Thesis, Louisiana State University, Baton Rouge

Elsey-Quirk T, Graham S, Mendelssohn I, Snedden G, Day J, Sharp L, Twilley R, Pahl J, Lane R (2019) Synthesis of wetland responses to freshwater, sediment, and nutrient inputs: Will Mississippi River sediment diversions enhance wetland sustainability? Est Coast Shelf. https://doi.org/10.1016/j.ecss.2019.03.002

EPA (Environmental Protection Agency) (1996) Water quality benefits analysis of final effluent limitations guidelines and standards for the coastal subcategory of the oil and gas extraction point source category. EPA-821-R-96-024

Finkelstein K, Hardaway CS (1988) Late holocene sedimentation and erosion of estuarine fringing marshes, York river, Virginia. J Coastal Res 4:447–456

FitzGerald DM, Fenster MS, Argow BA, Buynevich IV (2008) Coastal impacts due to sea-level rise. Ann Revue Earth Planetary Sci 36:601–647

Gagliano SM (1973) Canals, dredging and land reclamation in the Louisiana coastal zone. Hydrologic and Geologic Studies of Coastal Louisiana. Report no. 14. Center for Wetland Resources, Louisiana State University, Baton Rouge, 104 pp

Gagliano SM, van Beek JL (1970) Geologic and geomorphic aspects of deltaic processes, Mississippi River system. Hydrological and geological studies of coastal Louisiana volume 1. Center for wetland resources, Louisiana State University, Baton Rouge, LA.

Gagliano SM, Wicker KM (1988) Processes of wetland erosion in the Mississippi River deltaic plain. In: Duffy WG, Clark D (eds) Marsh management in coastal Louisiana: effects and issues—proceedings of a symposium. U.S. Fish and Wildlife Service and Louisiana Department of Natural Resources. U.S. Fish and Wildlife Service Biological Report 89(22), pp 28–48

Gagliano SM, Wicker KM (1989) Processes of wetland erosion in the Mississippi River deltaic plain. In: Duffy WG, Clark D (eds) Marsh Management in Coastal Louisiana: Effects and Issues—Proceedings of a Symposium. U.S. Fish and Wildlife Service and Louisiana Department of Natural Resources. U.S. Fish and Wildlife Service Biological Report 89(22), pp 28–48

Gagliano SM, Kemp EB, Wicker KM, Wiltenmuth KS (2003) Active geological faults and land change in Southeastern Louisiana, Final Report. New Orleans, LA: Coastal Environments, Inc., United States Army Corps of Engineers. Biol Report 89(22):247–277

Garcia-Herrera R, Gimeno L, Ribera P, Hernandez E (2005) New records of Atlantic hurricanes from Spanish documentary sources. J Geophys Res 110. https://doi.org/10.1029/2004jD005272

Gascuel-Odux C, Cros-Cayot S, Durand P (1996) Spatial variations of sheet flow and sediment transport on an agricultural field. Earth Surface Processes Landforms 21:843–851

Georgiou IY, Fitzgerald D, Stone GW (2005) The impact of physical processes along the Louisiana coast. J Coastal Res 21:72–89

Gornitz V, Lebedeff S, Hansen J (1982) Global sea level trend in the past century. Science 215:1611–1614

Gosselink JG, Cordes CL, Parsons JW (1979) An ecological characterization study of the Chenier Plain coastal ecosystem of Louisiana and Texas. 3 vols. U.S. Fish and Wildlife Service, Office of Biological Services. FWS/OBS-78/9 through 78/11

Gould HR, McFarlan E Jr (1959) Geologic history of the Chenier Plain, southwestern Louisiana. Trans Gulf Coast Assoc Geol Soc 9:261–270

Graham SA, Mendelssohn IA (2013) Functional assessment of differential sediment slurry applications in a deteriorating brackish marsh. Ecol Eng 51:264–274

Graham SA, Mendelssohn IA (2014) Coastal wetland stability maintained through counterbalancing accretionary responses to chronic nutrient enrichment. Ecology 95:3271–3283

Hanor JS, Bailey JE, Rogers MC, Milner LR (1986) Regional variations in physical and chemical properties of South Louisiana oil field brines. Trans Gulf Coast Assoc Geol Soc 36:143–149

Hansen J, Sato M, Hearty P, Ruedy R, Kelley M, Masson-Delmotte V, Russell G, Tselioudis G, Cao J, Rignot E, Velicogna I, Kandiano E, von Schuckmann K, Kharecha P, Legrande AN, Bauer M, Lo K-W (2015) Ice melt, sea level rise and superstorms: evidence from paleoclimate data, climate modeling, and modern observations that 2 °C global warming is highly dangerous. Atmosp Chem Phys Discussions 15:20059–20179

Haque SM (1993) Effects of surface brine disposal on the marshes of coastal Louisiana. Proceedings of Coastal Zone

Harris VT, Chabreck RH (1958) Some effects of Hurricane Audrey to the marsh at Marsh Island, Louisiana. Proc Louisiana Acad Sci 21:47–50

Hiatt M, Snedden G, Day JW, Rohli RV, Nyman JA, Lane RR, Sharp LA (2019) Drivers and impacts of water level fluctuations in the Mississippi River delta: Implications for delta restoration. Estuar Coast Shelf Sci 224:117–137

Hijma MP, Shen Z, Törnqvist TE, Mauz B (2017) Late Holocene evolution of a coupled, mud-dominated delta plain–chenier plain system, coastal Louisiana, USA. Earth Surf Dyn 5:689–710. https://doi.org/10.5194/esurf-5-689-2017

Horton BP, Rahmstorf S, Engelhart S, Kemp A (2014) Expert assessment of sea level rise by AD 2100 and AD 2300. Quat Sci Rev 84:1–6

Howes NC, FitzGerald DM, Hughes ZJ, Georgiou IY, Kulp MA, Miner MD, Smith JM, Barras JA (2010) Hurricane-induced failure of low salinity wetlands. Proc Natl Acad Sci 107:14014–14019

IPCC (Intergovernmental Panel on Climate Change) (2013) Climate change 2013: the physical science basis. In: Stocker TF, Qin D, Plattner GK, Tignor M, Allen SK, Boschung J, Nauels A, Xia Y, Bex V, Midgley PM (eds) Contribution of working group 1 to the fifth assessment report of the intergovernmental panel on climate change. Cambridge, United Kingdom and New York, NY, USA, 1535 pp

IPCC (Intergovernmental Panel on Climate Change) (2017) Special report on global warming of 1.5°C. Assessment report of the Intergovernmental panel on climate change, Cambridge, UK and New York, NY, USA.

Jafari N, Harris B, Cadigan J et al (2019) Wetland soil strength with emphasis on the impact of nutrients and sediments. Est Coast Shelf Sci 229. https://doi.org/10.1016/j.ecss.2019.106394

Jankowski K, Tornqvist T, Fernandes A (2017) Vulnerability of Louisiana's coastal wetlands to present-day relative sea-level rise. Nat Commun 8: 14792. https://doi.org/10.1038/nscomms14792

Jarvis JC (2010) Vertical accretion rates in coastal Louisiana: a review of the scientific literature. ERDC/EL TN-10-5. U.S. Army Engineer Research and Development Center, Vicksburg, MS. 15 pp

Johnson WB, Gosselink JG (1982) Wetland loss directly associated with canal dredging in the Louisiana coastal zone. In: Boesch D (ed) Proceedings of the land loss conference. Baton Rouge, Louisiana, pp 60–72

Johnston JB, Cahoon DR, La Peyre MK (2009) Outer continental shelf (OCS)-related pipelines and navigation canals in the Western and Central Gulf of Mexico: relative impacts on wetland habitats and effectiveness of mitigation. U.S. Dept. of the Interior, Minerals Management Service, Gulf of Mexico OCS Region, New Orleans, LA. OCS Study MMS 2009–048, 200 pp

Jones SF, Stagg CL, Krauss KW, Hester MW (2016) Tidal saline wetland regeneration of sentinel vegetation types in the North Gulf of Mexico: an overview. Estuar Coast Shelf Sci 174:A1–A10

Kemp P, Day J, Rogers D et al (2016) Enhancing mud supply to the Mississippi River delta: dam bypassing and coastal restoration. Est Coast Shelf Sci 183:304–313

Kennedy J, Morice C, Parker D (2011) Global and regional climate in 2010. Weather 66:188–194

Kesel RH (1988) The decline in the suspended load of the lower Mississippi River and its influence on adjacent wetlands. Environ Geol Water Sci 11:271–281

Kesel RH (1989) The role of the lower Mississippi River in wetland loss in southeastern Louisiana, USA. Environ Geol Water Sci 13:183–193

Ko JY, Day JW (2004) A review of ecological impacts of oil and gas development on coastal ecosystems in the Mississippi Delta. Ocean Coast Manag 47:597–623

Ko JY, Day J, Barras J, Morton R, Johnston J, Kemp G, Clairain E, Theriot R (2004) Impacts of oil and gas activities on coastal wetland loss in the Mississippi Delta

Kolker AS, Allison MA, Hameed S (2011) An evaluation of subsidence rates and sea-level variability in the northern Gulf of Mexico. Geophys Res Letters 38:L21404. https://doi.org/10.1029/2011GL049458

Kuhn NL, Mendelssohn IA, Reed DJ (1999) Altered hydrology effects on Louisiana salt marsh function. Wetlands 19:617–626

Kulp MA (2000) Holocene stratigraphy, history, and subsidence: Mississippi River delta region, north-central Gulf of Mexico, Ph.D. Thesis, Univ. of Ky., Lexington, 336 pp

Lane RR, Day JW Jr, Day JN (2006) Wetland surface elevation, vertical accretion, and subsidence at three Louisiana estuaries receiving diverted Mississippi River water. Wetlands 26:1130–1142

Lane RR, Day JW Jr, Marx B, Hyfield E, Day JN, Reyes E (2007) The effects of riverine discharge on temperature, salinity, suspended sediment and chlorophyll a in a Mississippi delta estuary measured using a flow-through system. Estuar Coast Shelf Sci 74:145–154

Lane RR, Mack S, Day JW, DeLaune RD, Madison MJ, Precht PR (2016) Fate of soil organic carbon during wetland loss. Wetlands 36(6):1167–1181

Lane RR, Mack SK, Day JW, Kempka R, Brady LJ (2017) Carbon sequestration at a forested wetland receiving treated municipal effluent. Wetlands 37(5):861–873

Lane RR, Kemp GP, Day JW (2018) A brief history of delta formation and deterioration. In: Day JW, Erdman JA (eds) Mississippi delta restoration. Springer, Cham, Switzerland, pp 11–27

Liu KB (2004) Paleotempestology: principles, methods, and examples from Gulf Coast lake sediments. In: Murnane RJ, Kiu K-B(eds) Hurricanes and typhoons: past, present, and future. Columbia University Press, New York, pp 13–57

Liu KB (2013) Paleoclimate reconstruction: paleotempestology. In: Elias SA, Mock CJ (eds) Encyclopedia of quaternary science, 2nd ed. Elsevier, Amsterdam, Netherlands. pp 209–221

Liu KB, Fearn ML (1993) Lake-sediment record of late Holocene hurricane activities from coastal Alabama. Geology 21: 793–796

Liu KB, Fearn ML (2000) Reconstruction of prehistoric landfall frequencies of catastrophic hurricanes in northwestern Florida from lake sediment records. Quat Res 54:238–245

Louisiana Coastal Wetlands Conservation and Restoration Task Force (2002) Hydrologic investigation of the Louisiana Chenier plain. Louisiana Department of Natural Resources, Coastal Restoration Division, Baton Rouge, LA, 135 pp

Macek P, Rejmankova E (2007) Response of emergent macrophytes to experimental nutrient and salinity additions. Funct Ecol 21:478–488. https://doi.org/10.1111/j.1365-2435.2007.01266.x

Madden C, Day J, Randall J (1988) Coupling of freshwater and marine systems in the Mississippi deltaic plain. Limnol Oceanogr 4:982–1004

Mariotti G, Kearney WS, Fagherazzi S (2019) Soil creep in a mesotidal salt marsh channel bank: fast, seasonal, and water table mediated. Geomorphology 334:126–137

Mendelssohn IA, McKee KL, Patrick Jr WH (1981) Oxygen deficiency in Spartina alterniflora roots: Metabolic adaptation to anoxia. Science 214: 439–441

McBride RA, Taylor MJ, Byrnes MR (2007) Coastal morphodynamics and Chenier-Plain evolution in southwestern Louisiana, USA: A geomorphic model. Geomorphology 88:367–422

McCloskey TA, Keller G (2009) 5000 year sedimentary record of hurricane strikes on the central coast of Belize. Quat Intl 195: 53–68

McCloskey, TA, Liu KB (2012) A 7000 year record of paleohurricane activity from a coastal wetland in Belize. The Holocene 23: 278–291

McGenity TJ, Folwell BD, McKew BA, Sanni GO (2012) Marine crude-oil biodegradation: a central role for interspecies interactions. Aquatic Biosyst 8:10

McGinis JT, Ewing RA, Willingham CA, Rogers SE, Douglass DH, Morrison DL (1972) Environmental aspects of gas pipeline operations in the Louisiana coastal marshes. Battelle Columbus Laboratories

McKee KL, Mendelssohn IA (1989) Response of a freshwater marsh plant community to increased salinity and increased water level. Aquat Bot 34:301–316

Meade RH, Moody JA (2010) Causes for the decline of suspended-sediment discharge in the Mississippi River system, 1940–2007. Hydrol Process 24:35–49

Meeder J (1987) Variable effects of hurricanes on the coast and adjacent marshes: a problem for marsh managers. In: Brodtman NV (ed) Proceedings of the fourth water quality and wetlands management conference, pp 337–374

Mendelssohn IA, Batzer D (2006) Abiotic constraints for wetland plants and animals. In: Batzer DP, Sharitz RR (eds) Ecology of freshwater and estuarine wetlands. University of California Press, Berkeley, CA, pp 82–114

Mendelssohn IA, Kuhn NL (2003) Sediment subsidy: effects on soil plant responses in a rapidly submerging coastal salt marsh. Ecol Eng 21:115–128

Mendelssohn IA, McKee KL (1988) *Spartina alterniflora* die-back in Louisiana: time-course investigation of soil waterlogging effects. J Ecol 76:509–521

Mendelssohn IA, Morris JT (2000) Eco-physiological controls on the productivity of Spartina alterniflora loisel. In: Weinstein MP, Kreeger DA (eds) Concepts and controversies in tidal marsh ecology. Kuwer Acedemic Publishers, Boston, MA, USA, pp 59–80

Mendelssohn IA, Flynn KM, Wilsey BJ (1990) The relationship between produced water discharges, and plant biomass and species composition, in three Louisiana marshes. Oil Chem Pollut 7:317–335

Miao SL, Zou CB (2012) Effects of inundation on growth and nutrient allocation of six major macrophytes in the Florida Everglades. Ecol Eng 42:10–18

Mikhailov VN, Mikhailov MV (2010) Delta formation processes at the Mississippi River mouth. Water Resour 37:595–610

Mitsch WJ, Gosselink JG (2015) Wetlands. Wiley, Hoboken, N.J

Mock CJ (2004) Tropical cyclone reconstructions from documentary records: Examples from South Carolina. In: Murnane RJ, Liu KB (eds) Hurricanes and typhoons: past, present and future. Columbia University Press, New York, NY. pp 121–148

Monte JA (1978) The impact of petroleum dredging on Louisiana's coastal landscape—a plant biogeographical analysis and resource assessment of spoil bank habitats in the Bayou Lafourche delta. Ph. D. dissertation, Louisiana State University, Baton Rouge, LA, 321 pp

Morgan JP, Nichols LG, Wright M (1958) Morphological effects of Hurricane Audrey on the Louisiana coast. Louisiana State University Coastal Studies Institute Technical Report Number 10, Baton Rouge, Louisiana

Morris JT, Sundareshwar PV, Nietch CT, Kjerfve B, Cahoon DR (2002) Responses of coastal wetlands to rising sea level. Ecology 83:2869–2877. https://doi.org/10.2307/3072022

Morton RA, Barras JA (2011) Hurricane impacts on coastal wetlands: a half-century record of storm-generated features from Southern Louisiana. J Coastal Res 27(6A):27–43

Morton RA, Buster NA, Krohn MD (2002) Subsurface controls on historical subsidence rates and associated wetland loss in southcentral Louisiana. Gulf Coast Assoc Geol Soc Trans 52:767–778

Morton RA, Bernier JC, Barras JA, Ferina NF (2005a) Historical subsidence and wetland loss in the Mississippi Delta plain. Gulf Coast Assoc Geol Soc Trans 55:555–571

Morton RA, Bernier JC, Barras JA, Ferina NF (2005b) Rapid subsidence and historical wetland loss in the south-central Mississippi delta plain: likely causes and future implications. U.S. Geological Survey Open-file Report 2005–1216. http://www.pubs.usgs.gov/of/2005/1216

Morton RA, Bernier JC, Barras JA (2006) Evidence of regional subsidence and associated interior wetland loss induced by hydrocarbon production, Gulf Coast region, USA. Environ Geol 50:261–274

Mossa J (1996) Sediment dynamics in the lowermost Mississippi River. Eng Geol 45:457–479

Murray SP, Walker ND, Adams CE Jr (1993) Impacts of winter storms on sediment transport within the Terrebonne Bay marsh complex. In: Laska S, Puffer A (eds) Coastlines of the Gulf of Mexico. American Society of Civil Engineers, New York, pp 56–70

Neff J, Lee K, DeBlois EM (2011) Produced water: overview of composition, fates, and effects. Produced water. Springer, New York, NY, pp 3–54

Neill C, Turner RE (1987) Backfilling canals to mitigate wetland dredging in Louisiana coastal marshes. Environ Manage 11:823–836

Neto AG, Costa CSB (2009) Survival and growth of the dominant salt marsh grass Spartina alterniflora in an oil industry saline wastewater. Int J Phytorem 11(7):640–650

Neubauer SC, Franklin RB, Berrier DJ (2013) Saltwater intrusion into tidal freshwater marshes alters the biogeochemical processing of organic carbon. Biogeosciences 10:8171–8183

Nichols LG (1958) Erosion of canal banks on the Rockefeller Wildlife Refuge, LA. Wildlife and Fisheries Commission, Refuge Division. New Orleans, LA

Nichols LG (1961) Erosion of canal banks on the Rockefeller wildlife refuge. La. Wildlife and Fisheries Comm, New Orleans, LA, p 9

Nyman JA (2014) Integrating successional ecology and the delta lobe cycle in wetland research and restoration. Estuaries Coasts 37:1490–1505

Nyman JA, DeLaune RD, Patrick WH Jr (1990) Wetland soil formation in the rapidly subsiding Mississippi River Deltaic Plain: mineral and organic matter relationships. Estuar Coast Shelf Sci 31:57–69

Nyman JA, DeLaune RD, Roberts HH, Patrick WH Jr (1993) Relationship between vegetation and soil formation in a rapidly submerging coastal marsh. Mar Ecol Prog Ser 96:269–279

Nyman J, DeLaune R, Pezeshki R, Patrick W (1995a) Organic matter fluxes and marsh stability in a rapidly submerging estuarine marsh. Estuaries 18:207–217

Nyman JA, Crozier CR, DeLaune RD (1995b) Roles and patterns of hurricane sedimentation in an estuarine marsh landscape. Est Coastal Shelf Sci 40:665–679

Nyman JA, Walters RJ, Delaune RD, Patrick WH Jr (2006) Marsh vertical accretion via vegetative growth. Estuar Coast Shelf Sci 69:370–380. https://doi.org/10.1016/j.ecss.2006.05.041

Odum WE, Odum EP, Odum HT (1995) Nature's pulsing paradigm. Est Coasts 18: 547–555

Odum WE, Odum EP, Odum HT (2014) Nature's pulsing paradigm. Estuaries Coasts 18:547–555

Olea RA, Coleman JL Jr (2014) A synoptic examination of causes of land loss in Southern Louisiana as related to the exploitation of subsurface geologic resources. J Coastal Res 30:1025–1044

O'Neil T (1949) The muskrat in the Louisiana marshes. Louisiana Wildlife and Fisheries Commission, New Orleans, LA

Owen DE (2008) Geology of the chenier plain of cameron parish, Southwestern Louisiana. In: Moore G (ed) Geological society of America field guide 14. pp 27–38

Penland S, Sutor JR (1989) The geomorphology of the Mississippi River Chenier Plain. Mar Geol 90:231–258

Penland S, Ramsey K (1990) Relative sea-level rise in Louisiana and the Gulf of Mexico, 1908–1988. J Coastal Res 6:323–342

Penland S, Boyd R, Nummendal D, Roberts HH (1981) Deltaic barrier development on the Louisiana coast. Gulf Coast Assoc Geol Soc Trans 31:471–465

Penland S, Ramsey KE, McBride RA, Moslow TF, Westphal KA (1989) Relative sea level rise and subsidence in Louisiana and the Gulf of Mexico. Louisiana Geological Survey, Coastal Geology Technical Report 3, Baton Rouge, Louisiana, 65 pp

Penland S, Conner PF, Beall A et al (2005) Changes in Louisiana's shoreline: 1855–2002. J Coast Res 44:7–39

Perez B, Day J, Rouse L, Shaw R, Wang M (2000) Influence of Atchafalaya River discharge and winter frontal passage on suspended sediment concentration and flux in Fourleague Bay, Louisiana. Estuar Coast Shelf Sci 50:271–290

Perez BC, Day JW Jr, Justic D, Twilley RR (2003) Nitrogen and phosphorus transport between Fourleague Bay, Louisiana and the Gulf of Mexico: the role of winter cold fronts and Atchafalaya river discharge. Estuar Coast Shelf Sci 57:1065–1078

Pethick J (1993) Shoreline adjustments and coastal management: physical and biological processes under accelerated sea-level rise. Geogr J 159:162–168

Pezeshki SR, DeLaune RD, Patrick WH Jr (1987a) Response of the freshwater marsh species, *Panicum hemitomon* Schult., to increased salinity. Freshw Biol 1:195–200

Pezeshki SR, Delaune RD, Patrick WH (1987b) Effects of flooding and salinity on photosynthesis of *Sagittaria lancifolia*. Marine Ecol Progress Series 41:87–91. https://doi.org/10.3354/meps04 1087

Pezeshki SR, Delaune RD, Patrick WH (1989) Assessment of saltwater intrusion impact on gas exchange behavior of Louisiana Gulf Coast wetland species. Wetland Ecol Manage 1:21–30

Phillips JD (1986) Coastal submergence and marsh fringe erosion. J Coastal Res 2:427–436

Rabalais NN (2005) Relative contribution of produced water discharge in the development of hypoxia. OCS Study MMS 2005-044. U.S. Dept. of the Interior, Minerals Management Service, Gulf of Mexico Region, New Orleans, LA, 56 pp

Rabalais NN, Harper DE (1992) Studies of benthic biota in areas affected by moderate and severe hypoxia. In: Proceedings of national oceanic and atmospheric administration workshop on nutrient enhanced coastal ocean productivity. Texas A&M University Sea Grant Program, College Station, Texas

Rabalais NN, McKee BA, Reed DJ, Means JC (1991). Fate and effects of nearshore discharges of OCS produced waters, vol 1. executive summary. OCS Study/MMS 91-0004. U.S. Dept. of the Interior, Minerals Management Service, Gulf of Mexico OCS Regional Office, New Orleans, LA, 48 pp

Rabalais NN, McKee BA, Reed DJ, Means JC (1992) Fate and effects of produced water discharges in coastal Louisiana, Gulf of Mexico, USA. Produced Water. Springer, Boston, MA, pp 355–369

Rabalais NN, Smith LE, Overton EB, Zoeller AL (1993) Influence of hypoxia on the interpretation of effects of petroleum production activities. U.S. Dept. of the Interior, Minerals Management Service, Gulf of Mexico OCS Region, New Orleans, LA. OCS Study MMS 93-0022, 158 pp

Rayle MF, Mulino MM (1992) Produced water impacts in Louisiana coastal waters. In: Produced water. Springer, Boston, MA, pp 343–354

Reddy KR, DeLaune RD (2008) Biogeochemistry of wetlands: science and applications. CRC Press, Boca Raton, Florida, p 774

Reed DJ (1992) Effect of weirs on sediment deposition in Louisiana coastal marshes. Environ Manage 16:55–65

Reed DJ (1995) Status and Historical Trends of Hydrologic Modification, Reduction in Sediment Availability, and Habitat Loss/Modification in the Barataria and Terrebonne Estuarine System. BTNEP Publ. No. 20, Barataria-Terrebonne National Estuary Program, Thibodaux, Louisiana, 338 pp

Reed DJ (2002) Sea-level rise and coastal marsh sustainability: Geological and ecological factors in the Mississippi delta plain. Geomorphology 48:233–243. https://doi.org/10.1016/S0169-555 X(02)00183-6

Reed DJ, Cahoon DR (1992) The relationship between marsh surface topography and vegetation parameters in a deteriorating Louisiana *Spartina alterniflora* salt marsh. J Coastal Res 8:77–87

Reed DJ, Wilson L (2004) Coast 2050: a new approach to restoration of Louisiana coastal wetlands. Phys Geogr 25:4–21

Reed DJ, De Luca N, Foote AL (1997) Effect of hydrologic management on marsh surface sediment deposition in coastal Louisiana. Estuaries 20:301–311

Reimold RJ, Hardisky MA, Adams PC (1978) The effects of smothering a *Spartina alterniflora* salt marsh with dredged material. US Army Corps of Engineers, Washington, DC, Technical Report D-78-38

Rejmankova E, Macek P (2008) Response of root and sediment phosphatase activity to increased nutrients and salinity. Biogeochemistry 90:159–169. https://doi.org/10.1007/s10533-008-9242-3

Roberts HH (1997) Dynamic changes of the holocene Mississippi river delta plain: the delta cycle. J Coastal Res 13:605–627

Roberts HH, DeLaune RD, White JR, Li C, Sasser C, Braud D, Weeks E, Khalil S (2015) Floods and cold front passages: impacts on coastal marshes in a river diversion setting (Wax Lake Delta Area, Louisiana). J Coast Res 312:1057–1068

Rogers DR, Rogers BD, Herke WH (1992) Effects of a marsh management plan on fishery communities in coastal Louisiana. Wetlands 12:53–62

Rosen PS (1977) Increasing shoreline erosion rates with decreasing tidal range in the Virginia Chesapeake Bay. Chesapeake Sci 18:383–386

Sasser CE, Dozier MD, Gosselink JG, Hill JM (1986) Spatial and temporal changes in Louisiana's Barataria basin marshes, 1945–1980. Environ Manage 10(5):671–680

Sasser CE, Evers-Hebert E, Holm GO Jr, Milan B, Sasser JB, Peterson EF, DeLaune RD (2018) Relationships of marsh soil strength to belowground vegetation biomass in Louisiana coastal marshes. Wetlands 38:401–409. https://doi.org/10.1007/s13157-017-0977-2

Sasser CE, Evers-Hebert E, Milan B, Holm GO, Jr (2013) Relationships of marsh soils strength to vegetation biomass. Final Report to the Louisiana Coastal Protection and Restoration Authority through State of Louisiana Interagency Agreement No. 2503–11–45, 73 pp

Sasser CA (1977) Distribution of vegetation in Louisiana coastal marshes as response to tidal flooding. Master's thesis, Louisiana State University, Baton Rouge

Saucier RT (1963) Recent geomorphic history of the Pontchartrain basin. Louisiana State University Press, Baton Rouge, LA, USA

Scaife WW, Turner RE, Costanza R (1983) Coastal Louisiana recent land loss and canal impacts. Environ Manage 7:433–442

Schwimmer RE (2001) Rates and processes of marsh shoreline erosion in Rehoboth Bay, Delaware, USA. J Coastal Res 17:672–683

Servais S, Kominoski JS, Charles SP, Gaiser EE, Mazzei V, Troxler TG, Wilson BJ (2019) Saltwater intrusion and soil carbon loss: Testing effects of salinity and phosphorus loading on microbial functions in experimental freshwater wetlands. Geoderma 337:1291–1300

Shaffer G, Day J, Kandalepas D et al (2016) Decline of the Maurepas swamp, Pontchartrain Basin, Louisiana and approaches to restoration, Water 7. https://doi.org/10.3390/w70x000x

Shaffer G, Day J, Mack S, Kemp P, van Heerden I, Poirrier M, Westphal K, FitzGerald D, Milanes A, Morris C, Bea R, Penland S (2009) The MRGO navigation project: a massive human-induced environmental, economic, and storm disaster. J Coastal Res 54:206–224

Sklar FH, Meeder JF, Troxler TG, Dreschel T, Davis SE, Ruiz PL (2019) The everglades: at the forefront of transition. In: Wolanski E, Day J (eds) Coasts and estuaries: the future. Elsevier, Oxford, United Kingdom, pp 277–292

Slocum MG, Mendelssohn IA, Kuhn NL (2005) Effects of sediment slurry enrichment on salt marsh rehabilitation—plant and soil responses over seven years. Estuaries 28:519–528

Smith JE, Bentley SJ, Snedden GA, White C (2015) What role do hurricanes play in sediment delivery to subsiding river deltas. Sci Rep 5:17582. https://doi.org/10.1038/srep17582

Snedden GA, Cretini K, Patton B (2015) Inundation and salinity impacts to above- and belowground productivity in *Spartina patens* and *Spartina alterniflora* in the Mississippi River deltaic plain: implications for using river diversions as restoration tools. Ecol Eng 81:133–139

St. Pe KM (1990) An assessment of produced water impacts to low-energy brackish water systems in southeast Louisiana. Louisiana Department of Environmentally Quality, Water Pollution Control Division, Baton Rouge, Louisiana, 204 pp

Stagg CL, Mendelssohn IA (2010) Restoring ecological function to a submerged salt marsh. Restor Ecol 18:10–17

Stagg CL, Mendelssohn IA (2011) Controls on resilience and stability in a sediment-subsidized saltmarsh. Ecol Appl 21:1731–1744

Stagg CL, Schoolmaster DR, Krauss KW, Cormier N, Conner WH (2017) Causal mechanisms of soil organic matter decomposition: deconstructing salinity and flooding impacts in coastal wetlands. Ecology 98(8):2003–2018

Stern MK, Day JW, Teague KG (1986) Seasonality of materials transport through a coastal freshwater marsh: through riverine versus tidal forcing. Estuaries 9:301–308

Stern M, Day J, Teague K (1991) Nutrient transport in a riverine-influenced tidal freshwater bayou in Louisiana. Estuaries 14:382–394

Steward KK, Ornes WH (1975) The autoecology of sawgrass in the Florida Everglades. Ecology 56:162–171

Stone JH, Bahr LM, Day JW Jr (1978) Effects of canals on freshwater marshes in coastal Louisiana and implications for management. In: Good RE, Whigham DF, Simpson RL (eds) Freshwater wetlands—Ecological processes and management potential. Academic Press, New York, pp 291–321

Suhayda J, Jacobsen B (2008) Interaction of hurricanes and coastal landscape features: a literature review. Prepared for U. S. Army Corps of Engineers, Engineer Research Development Center, Coastal and Hydraulics Laboratory, Vicksburg, Mississippi

Swenson EM, Turner RE (1987) Spoil banks: effects on a coastal marsh water-level regime. Estuar Coast Shelf Sci 24:599–609

Swenson EM (1983) Marsh hydrological studies, 1982–1983 Data report. Coastal Ecology Institute, Center for Wetland Resources, Louisiana State University, Baton Rouge. Prepared for the National Marine Fisheries Service, Southeast Region, St. Petersburg, Florida. Contract no. NA81-BA-P00006. Publication no. LSU-CEFI-83–18

T. Baker Smith & Son, Inc. (2002) Houma navigation canal secondary impacts study. Prepared for Terrebonne Parish Consolidated Government. Houma, LA, 50 pp

Taylor NC, Day JW, Neusaenger GE (1989) Ecological characterization of Jean Lafitte National Historical Park, Louisiana: basis for a management plan. In: Marsh management in coastal Louisiana: effects and issues—proceedings of a symposium

Taylor NC (1988) Ecological characterization of Jean Lafitte National Historical Park. Louisiana: Basis for a management plan. M.S. Thesis. Louisiana State University. Baton Rouge, LA

Teal JM, Best R, Caffrey J, Hopkinson CS, McKee KL, Morris JT, Newman S, Orem B (2012) Mississippi River freshwater diversions in Southern Louisiana: effects on wetland vegetation, soils, and elevation. In: Lewitus AJ, Croom M, Davison T, Kidwell DM, Kleiss BA, Pahl JW, Swarzenski CM (eds) Final report to the state of Louisiana and the U.S. Army corps of engineers through the Louisiana Coastal Area Science & Technology Program; coordinated by the National Oceanic and Atmospheric Administration, 49 pp

Totten S, Matthew W, Hanan MA, Simpson S (2007) Natural remediation of marsh soil contaminated by oil-field brine containing elevated radium levels, southern Louisiana. Environ Geosci 14:113–122

Troxler TG, Childers DL, Madden CJ (2014) Drivers of decadalscale change in southern Everglades wetland macrophyte communities of the coastal ecotone. Wetlands 34:S81–S90. https://doi.org/10.1007/s13157-013-0446-5

Turner RE (1987) Relationship between canal and levee density and coastal land loss in Louisiana. US Fish Wildlife Service Biological Report 85(14). U.S. Department of Interior, Washington, D.C., 58 pp

Turner RE (1997) Wetland loss in the northern Gulf of Mexico: Multiple working hypotheses. Estuaries 20:1–13

Turner RE, Cahoon DR (eds) (1987) Causes of wetland loss in the coastal Central Gulf of Mexico, volume II: technical narrative. Prepared for Minerals Management Service, New Orleans, LA. Contract No. 14–12–0001–30252. OCS Study/MMS 87–0120, 400 pp

Turner RE, Cahoon DR (1988) Causes of wetland loss in the coastal Central Gulf of Mexico. Prepared under MMS Contract 14–12–0001–30252, New Orleans, LA. USDI, Minerals Management Service, Gulf of Mexico OCS Region

Turner RE, McClenachan G (2018) Reversing wetland death from 35,000 cuts: Opportunities to restore Louisiana's dredged canals. PLoS One 13(12):e0207717

Turner RE, Rao YS (1990) Relationships between wetland fragmentation and recent hydrologic changes in a deltaic coast. Estuaries 13:72–281

Turner RE, Streever B (2002) Approaches to coastal wetland restoration: Northern Gulf of Mexico. Kugler Publications, 147 pp

Turner RE, Swenson E, Lee J (1994a) A rationale for coastal wetland restoration through spoil bank management in Louisiana, USA. Environ Manage 18:271–282

Turner RE, Lee JM, Neill C (1994b) Backfilling canals to restore wetlands: empirical results in coastal Louisiana. Wetlands Ecol Manage 3:63–78

Turner RE, Swenson EM, Milan CS, Lee JM (2007) Hurricane signals in salt marsh sediments: inorganic sources and soil volume. Limnol Oceanogr 52:1231–1238

Twilley RR, Bentley SJ, Chen Q et al (2016) Co-evolution of wetland landscapes, flooding, and human settlement in the Mississippi River Delta plain. Sustain Sci 11:711–731

U.S. Dept of the Army Corps of Engineers (USACE) (2004) Louisiana Coastal Area (LCA), Louisiana Ecosystem Restoration Study, Final vol 1—LCA Study, Main Report. http://www.lca.gov/final_report.aspx

USDA (1951) A report on the relationship of agricultural use of wetlands to the conservation of wildlife in Cameron Parish, Louisiana. U.S. Department of Agriculture, Soil Conservation Service, Fort Worth, Tex

USDA (1994) Calcasieu-Sabine cooperative river basin study report. U.S. Department of Agriculture, Natural Resources Conservation Service, Alexandria, Louisiana, 151 pp

Valentine JM (1976) Plant succession after sawgrass mortality in southwestern Louisiana. In: Proceedings, 30th annual conference, Southeast association of game and fish commissioners, pp 634–640

Veil JA, Puder GM, Elcock D, Redweik RJ (2004) A white paper describing produced water from production of crude oil, natural gas, and coal bed methane. No. ANL/EA/RP-112631. Argonne National Lab., IL, US

Veil JA, Kimmell TA, Rechner AC (2005) Characteristics of produced water discharged to the gulf of Mexico Hypoxic Zone. Report to the U.S. Dept. of Energy, National Energy Technology Laboratory, Argonne National Laboratory, Washington, DC

Wang FC (1988) Dynamics of saltwater intrusion in coastal channels. J Geophys Res 93(C6):6937–6946

Wang J, Xu K, Restreppo GA, Bentley SJ, Meng X, Zhang X (2018) The coupling of bay hydrodynamics to sediment transport and its implication in micro-tidal wetland sustainability. Mar Geol 405:68–76

Welder FA (1959) Processes of deltaic sedimentation in the lower Mississippi River. Louisiana State University, Coastal Studies Institute Technical Report 12, pp 1–90

Wicker K (1981) Chenier Plain region ecological characterization: a habitat mapping study. Prepared for the Louisiana Coastal resources program, Louisiana Department of Natural Resources, Baton Rouge, LA and Office of Coastal Zone Management, National Oceanic and Atmospheric Administration, Department of Commerce, Washington, D.C. 105 pp

Wilson C, Allison MA (2008) An equilibrium profile model for retreating marsh shorelines in southeast Louisiana. Est Coast Shelf Sci 80:483–494

Williams EK, Rosenheim BE (2015) What happens to soil organic carbon as coastal marsh ecosystems change in response to increasing salinity? An exploration using ramped pyrolysis. Geochem Geophys Geosyst 16:2322–2335

Williams SJ, Penland S, Roberts HH (1994) Processes affecting coastal wetland loss in the Louisiana deltaic plain. In: Williams SJ, Cichon HA (eds) Processes of coastal wetlands loss in Louisiana.

Presented at Coastal Zone '93, New Orleans, Louisiana: USGS Open-File Report 94-0275, pp 21–29

Wilson BJ, Servais S, Charles SP, Davis SE, Gaiser EE, Kominoski JS, Richards JH, Troxler TG (2018a) Declines in plant productivity drive carbon loss from brackish coastal wetland mesocosms exposed to saltwater intrusion. Estuaries Coasts 41:2147–2158

Wilson BJ, Servais S, Mazzei V, Kominoski JS, Hu M, Davis SE, Gaiser E, Sklar F, Bauman L, Kelly S, Madden C, Richards J, Rudnick D, Stachelek J, Troxler TG (2018b) Salinity pulses interact with seasonal dry-down to increase ecosystem carbon loss in marshes of the Florida Everglades. Ecol Appl 28(8):2092–2108

Wilson BJ, Servais S, Charles SP, Mazzei V, Kominoski JS, Gaiser E, Richards J, Troxler T (2019) Phosphorus alleviation of salinity stress: effects of saltwater intrusion on an Everglades freshwater peat marsh. Ecology 100(5):e02672. https://doi.org/10.1002/ecy.2672

Wray RD, Leatherman SP, Nicholls RJ (1995) Historic and future land loss for upland and marsh islands in the Chesapeake bay, Maryland, U.S.A. J Coastal Res 11:1195–1203

Xu K, Bentley S, Day J, Freeman A (2018) A review of sediment diversions in the Mississippi deltaic plain. Estuarine, Coastal and Shelf Science. In review

Yuill B, Lavoie D, Reed DJ (2009) Understanding subsidence processes in coastal Louisiana. J Coastal Res Special Issue 54:23–36

Chapter 3
The Geology of the Mississippi River Delta and Interactions with Oil and Gas Activities

H. C. Clark and Charles Norman

3.1 Introduction

To understand the rapid changes happening along the Louisiana coast today, it's important to know about the geologic factors involved, the processes that built, sustain and challenge the delta complex. Beneath the coast lie all of the rivers and deltas that came before, depositing sediment and building the Louisiana coast along the edge of an ocean basin, the Gulf of Mexico, that opened about 165 million years ago—not that long ago in geologic time. The broad Louisiana coast is almost at sea level, it's delicate and vulnerable, and this situation is compounded by global sea level rise together with coastwide subsidence of the almost 10 mile thick (about 15 km) bowl of sediment, that is, the layers of geology beneath the coast that made up all the coasts that have gone before (Fig. 3.1).

The geology here is not hard to understand since the geologic processes we see happening on the surface today are the same that have always shaped the geology here in the past, and the layers of sediment that make up the geology beneath are a record of that geologic history. The geology of the Mississippi River Delta is fascinating but sometimes forgotten because much of it is unseen, yet it is an essential part of the engine that built the delta. This chapter begins with a brief recount of geologic history, followed by a discussion of the almost unique geology that is the basis of oil and gas development on the Louisiana coast, and then applies both to a review of

---

The original version of this chapter was revised: The author "Charles Norman" affiliation has been updated. The correction to this chapter is available at
https://doi.org/10.1007/978-3-030-94526-8_10

H. C. Clark (✉)
Department of Earth, Enviromental and Planetary Sciences, Rice University, Houston, TX 77005, USA
e-mail: hcclark@rice.edu

C. Norman
Charles Norman & Associates, Lake Charles, LA, USA

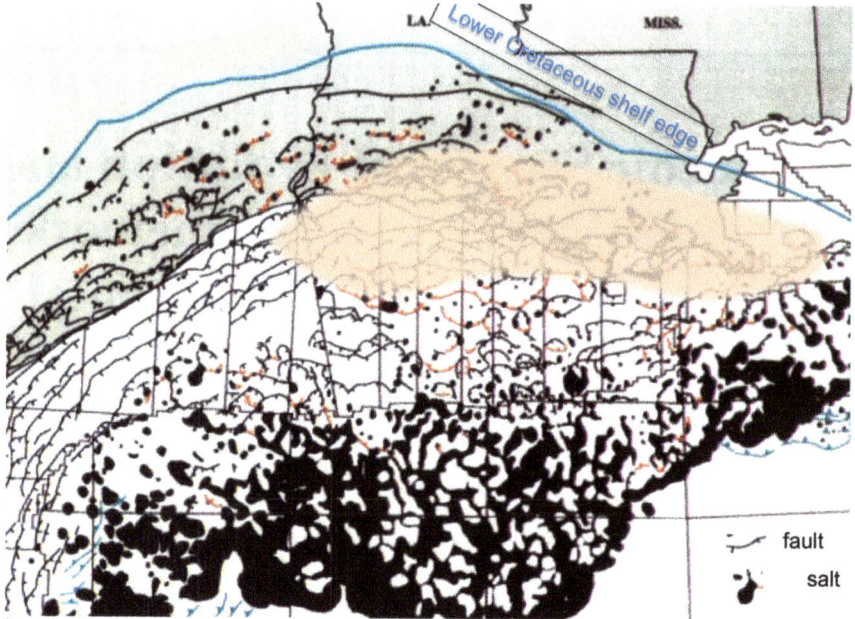

Fig. 3.1 The northern Gulf of Mexico showing growth faults adjusting the geologic section basinward; and malleable salt as domes and extruded features (modified from Diegel et al. 1995). Lines are typical growth faults with tick marks on downthrown side (black basinward dipping, red coastward dipping, blue (south area) are thrust faults. The geologic section is about 15 km thick at the depocenter (peach) paralleling and approximately at the coast

vertical displacement or subsidence along the Louisiana coast and the role that oil and gas activities play in inducing subsidence.

3.2 Gulf of Mexico—Opening and Building the Louisiana Coast

The underpinnings of the Mississippi deltaic plain and the Chenier Plain developed with the opening of the Gulf and beginning of its geologic history. Large scale north–south tears stretched and faulted the existing earth crust across the floor of the new basin, as Mexico pulled away and the Gulf opened, leaving early faulted basins floating like toy blocks on newly formed oceanic crust. This early Gulf edge is shown as the base or foundation of the schematic geologic cross-section shown in Fig. 3.2 (modified from Moffett 2014).

That base, once an ocean floor, is now buried about 15 km deep, and over geologic time the mass has continuously subsided to accommodate new sediment pouring off

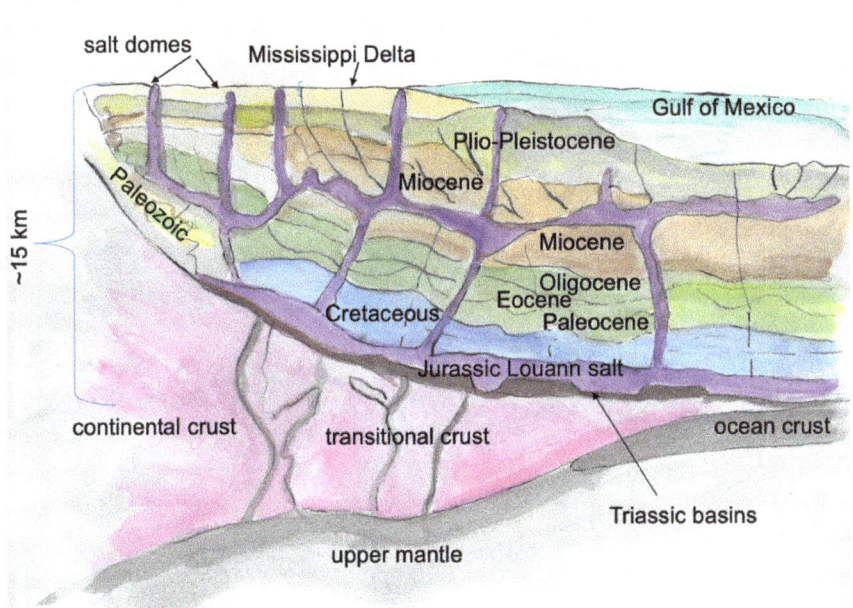

Fig. 3.2 Schematic geologic cross-section, north to south, through the Mississippi Delta and northern Gulf of Mexico basin margin (after Moffett 2014). Approximate location is in the center of previous figure. Based on data from regional seismic sections (Radovich et al. 2007, 2011), potential fields (Filina 2019) and drilling (Moffett 2014)

the continent. The new Gulf of Mexico (GOM) was often not well connected to either the Atlantic or Pacific oceans, impeded by ephemeral land bridges from Yucatan to Florida and the Isthmus of Panama, isolating and saturating the water body, then precipitating the Jurrasic Louann salt in a layer up to hundreds of feet thick that covers the basin floor. This salt forms the first layer of the section as shown in Fig. 3.1 and the stratigraphic section outlined in Fig. 3.3 (modified from Galloway 2009).

It is this salt, with its ability to flow like squeezed toothpaste, rise buoyantly, and be involved in tectonism, or geologic deformation, thousands of meters above the base of the section that makes the Gulf's geology extraordinary. Section, in geologic terms, refers to the total sediment layer that was deposited after the opening of the Gulf of Mexico.

From its beginning, the Louisiana coast received a large share of the sediment pouring off the North American continent as mountains on the east and west rose and eroded, and rivers switched course and drained to the south. These ancestral Mississippi Rivers delivered sediment, built deltas, and much more, loading the existing Louisiana coast that in turn subsided to receive more sediment, with the surface always staying near sea level and building out as the asthenosphere, or the

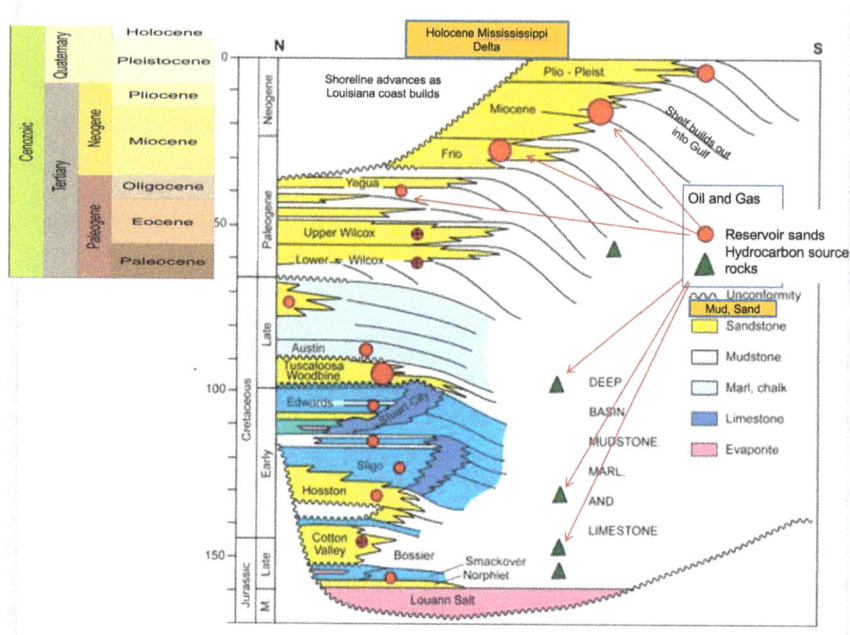

Fig. 3.3 Regional geostratigraphic section (modified from Galloway 2009) with geologic timescale and relative depositional position, boundaries, geology, character, and composition across the diagram. The section is from north to south, through the Mississippi Delta. Sands and sediments (yellow) coming off the continent migrate basinward in time. Earlier (Jurassic and Cretaceous) times were dominated by marine deposits, including reefs (blue). Warm climate and the often limited basin input resulted in times of ideal conditions for hydrocarbon development (triangles), then migration to reservoir sands (red circles)

earth's molasses-like upper mantle just below the crust, stretched and bent to accommodate the load (Blum and Roberts 2012). Since this sedimentation along the Gulf margin was the product of geologic events on the continent when mountain building happened, sediments were coarse and deposition rapid, and during quiet times the sedimentation process itself was less and quiet, often dominated by marine precipitates and skeletal remains. As a layer was buried, water was squeezed out, particles rearranged and chemistry changed, progressively compacting the individual layers within the section as the weight above increased. Mesozoic sediments (more than 100 million years old) are mature rocks, moving down as the lithosphere beneath moves, but no longer moving within the section, that is, not compacting anymore. Older Tertiary rocks (about 70–35 million years old) are less mature, also moving down with the sedimentary section and if compacting, doing so imperceptibly. Younger Tertiary sediments (about 35 million years old and younger) are often unsorted, loosely packed, and minimally cemented, if at all; but these sediments also reflect the compaction effect of burial beneath thousands of meters of similar sands and clays above.

Pleistocene ice took up much of the world's water budget over the most recent 3 million years during recurring ice ages. At each Pleistocene glacial maximum, the gently sloping Louisiana shelf was left dry miles offshore, and cut by ancestral streams like the Sabine, Calcasieu, Mississippi, and Pearl rivers and leaving channels and deposits dominated by these streams rather than processes along the shoreline (Anderson et al. 1996). With the last glacial melting beginning about 12,000 years ago, sea level again rose over the continental shelf and the shoreline returned over the geologic section beneath, shaping the present-day Louisiana coast, with the Mississippi delta and Chenier Plain evolving to the surface we know today.

The end result of this deposition process, or geologic history, is the elongate bowl of sediment about 15 km thick (almost 10 miles or way more than 150 football fields deep) at its maximum, stretching along the northern Gulf of Mexico coast, and cut by faults and punctured by salt domes as shown in the map of Fig. 3.1 and schematic cross-section, or slice through this layer cake of Fig. 3.2. These two diagrams characterize the coast and summarize the geologic history outlined in the previous paragraphs. Sediment sources and deposition rates have varied over time, faults and fault zones developed to accommodate new loads, and salt moved upward to take a variety of forms and create the salt domes distributed along the coast. And all the while, the layer of the asthenosphere below gave way to maintain balance and keep the land surface near sea level. The map of Fig. 3.1 shows the thickened sediment wedge (peach) and faults parallel to the coast, and salt (black) enveloped within and then extruded down slope toward the deep Gulf. Except for the exaggerated fanciful distribution of salt, Fig. 3.2 follows closely the data developed from regional seismic reflection synthesis (Gulfspan) (Radovich et al. 2007, 2011) and refraction, and gravity and magnetic models (Gumbo 2) (Christeson et al. 2014). The schematic cross-section is from north to south, across the mid-Louisana coast (see cross-section line on Fig. 3.1), and shows the geologic section, starting at the bottom with the asthenosphere, then the stretched crust, then early salt and sediments of the beginning Triassic, Jurassic, and Cretaceous, or the Mesozoic (see timescale and stratigraphy of Fig. 3.3 for time perspective), following upward, the major section beneath the coast beginning with the Eocene, Oligocene, and Miocene, then more Miocene up through the Pleistocene, the major oil and gas objectives of the Louisiana Gulf coast, and finally on top the post-glacial Holocene deposits that make up the current delta. The cross-section illustrates the complex deformation of originally horizontal sediments by faulting and salt movement, along with representative oil and gas wells tapping oil and gas reserves that developed as a result, while the stratigraphic column (Fig. 3.3) summarizes the Louisiana coast in time and the changing position of the coastline, along with the character of the sediments along this coastline, and the timing and relative location of source rock, or organic-rich sediments that have led to the prolific coastal and Gulf oil and gas fields. But the geologic section along the Louisiana coast is more than a stack of sediment layers, it's a dynamic operation at several scales in time and space, as discussed in the following sections.

3.3 Overprints—Faults and Salt Movement

Faults and salt movement or tectonism affect the sedimentary section at depth and over geologic history and are active at present in the form of things like salt domes and prominent fault breaks between land and water. A fault is simply a break with displacement of sediment layers. A fault happens when something is stressed to the point where it must yield or break, and with geology, that often happens soon after sediment deposition—and then the process continues as deposition continues, and the fault is called a growth fault, with evidence of the growth a thicker set of layers on the downthrown side of the fault. The underlying reason is adjustment, compensation—a way to accommodate the kilometers thick mass that blankets the continental edge too rapidly to simply stack as a set of layers.

Salt responds to pressure or squeezing, and Louann salt movement began as sediments from the continent poured in, covered the salt layer, and began to load what was to become the Mississippi Delta; this Louann salt underlying the mass began to flow outward like a squeezed tube of toothpaste, and upward also, given an opportunity, because it was lighter than the sediments above. Where a salt mass stopped, usually temporarily in terms of geologic velocity, it formed shapes like salt domes and new layers that both affected deposition and deformed the sediment layers involved. Remnants of the original Louann often intruded as tabular sheets, and once there, provided a mobile base for sediments piled on above to detach from the section, or "skate downhill", even on virtually no slope at all. And when conditions changed, salt that once served to transport the section above re-extruded and moved (another squeezed toothpaste tube analogy) to another level, leaving the detached sheets welded in place. Salt's ability to move and move again, deform and migrate further Gulfward is illustrated in the seismic section of Fig. 3.4 (Diegel et al. 1995).

This north–south profile across the onshore, mid-delta area shows the result of a sediment layer detachment (pink) where a tabular salt layer that once intruded between it and the white layer below, served as a Teflon™-like lubricator, mobilized the section above, and is now gone, leaving this detachment above (pink) welded to the layer (white) below. In this seismic section, while the detachment was actively moving gulfward, a number of faults developed and cut the moving section like large-scale landslides, bottoming out at the detachment (former salt) surface. These faults continued to move over a significant amount of geologic time while deposition across the region and over the faults continued, building a thicker section on the downthrown side of each fault—typical growth faults. The salt has since re-extruded and moved away and most of the growth faults have ended, but offsets in the seismic data show that a few continue upward, perhaps to the surface. Salt tectonism and fault movement are episodic and an episode of salt or fault movement happens locally and over a limited time—then it might begin again later in geologic time. Events like these are a critical part of the Louisiana coastal section geologic history, repeated countless times across the Gulf of Mexico basin and intimately involved with the development of oil and gas here. The pervasive nature of faulting and salt movement is illustrated by the regional seismic section of Fig. 3.5 (Radovich et al. 2011) from

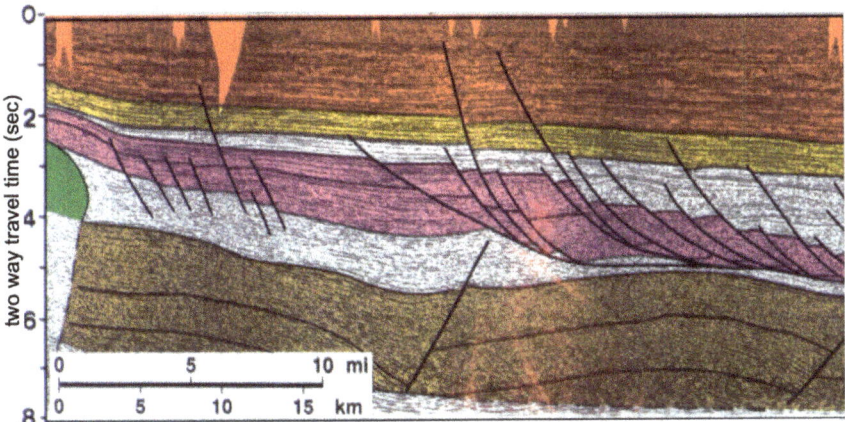

Fig. 3.4 Typical Louisiana coast seismic section, dip aligned, north to south, approximately mid-Delta (from Diegel et al. 1995). Listric (curved) normal, down to the coast, growth faults, coalescing on a detachment (rafting or sliding) surface. Role of episodic growth faulting shown by thickened sections on downthrown blocks. Counter regional faults cutting the Eocene and older beds at the base were possibly feeders for the salt that likely once existed as tabular bodies in the Miocene, mobilizing the faults as detachments moved basinward, then extruded to another location, leaving the Miocene "welded" to the beds below. Growth faults continue to offset younger beds at the top of the section

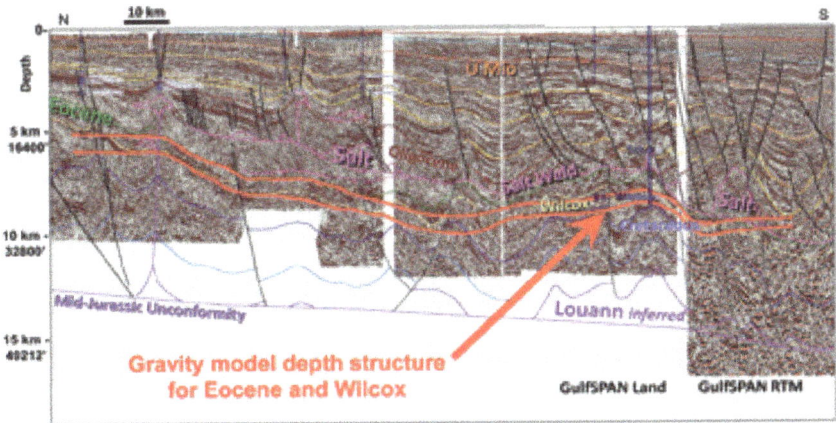

Fig. 3.5 Composite regional seismic section from north to south crossing the western Delta and into the deep basin (from Radovich et al. 2011). The interpretation shows numerous growth faults (Oligocene-Miocene) above a regional salt weld along with gravity modeled location of the Wilcox (Paleocene-Eocene) and Louann salt (Jurassic). Basinward dipping faults give way to north dipping faults further offshore

well onshore to the deep Gulf near the western limit of the Deltaic Plain complex. The down to the coast faults, counter faults, salt structures, and salt domes shown in this section are typical of the Louisiana coast.

In dealing with present-day land loss, it is important to realize that geologic factors at depth are often involved either directly or indirectly and present-day fault activation and movement and salt tectonism are critical parts of the picture. These will be discussed further as this section moves to integrate the deep geologic section and near surface processes.

3.4 The Age of Delta Building—Holocene Geology

While the Pleistocene ice age was a time of limited sedimentation along the present Louisiana coast, the massive Holocene Delta and Chenier Plain are the geology we walk on today. Holocene geology and geologic processes are the most familiar, and so the easier to fit into an understanding of factors involved in land loss and decisions about restoration. The Louisiana Holocene coast is dominated by the Mississippi Delta complex and its related Chenier Plain, but the framework of delta geology was in place virtually from the Gulf opening beginning millions of years ago. The continent between the Appalachian and Rocky mountain belts to the east and west drained to the Gulf from the beginning, and over time, the Mississippi river captured more streams and its fairway dominated the central Gulf Coast. The Mississippi, from its origin down to the modern delta, was mapped by Fisk and remains a classic both as a map and graphic art, and a portion of this map is shown in Fig. 3.6 (Fisk 1944, sheet 11).

Fisk illustrated that over its geologic history, the Mississippi constantly changed its meandering course across a broad river valley, and the map shows by color change, the overlapping meanders through time. With glacial melt, subsequent sea level rise, and a new shoreline, the Holocene Mississippi deposited its sediment load as it met the new coast, concentrated in a new delta, or set of deltas, switching back and forth across the new coastline, but never straying far from the geologic fairway origin. This Delta building history is summarized in the Environmental Setting chapter in Figs. 2.1, 2.2, and 2.3, a distillation and expansion of the original field coring and interpretation by Frazier (1967) where delta lobes or switching positions of the delta are shown by the colors related to the timeline below the map. More than 500 cores and descriptions were combined to develop cross-sections, then were outlined, age dated and mapped originally by Frazier.

Frazier developed cross-sections that described both depositional environments and the age-delta lobe relationships of the Delta components. Figure 3.7 illustrates the relationships between the sands, clays, peat and water bodies that built the several Delta lobes interpreted in the dated cores. As Holocene history unfolded, these deltaic lobes shed sediment that was then carried along the coast and deposited as linear features, barrier islands, or cheniers that paralleled the coast and built the present

3 The Geology of the Mississippi River Delta and Interactions with Oil...

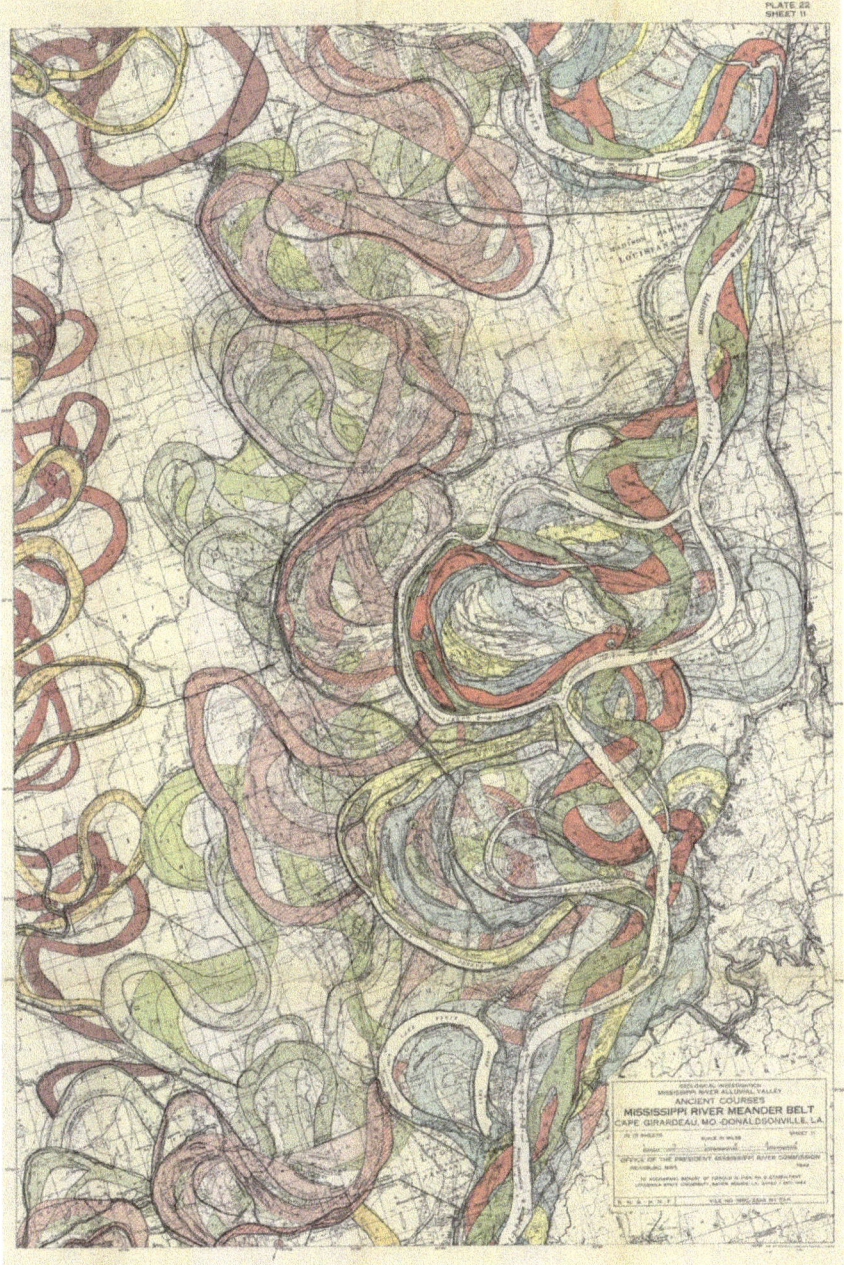

Fig. 3.6 The Mississippi River's moving and switching meanders, mapped by Fisk (1944, sheet 11). This part of the Minnesota to the Gulf mapping project is at the junction of the present day Mississippi, Red, and Atchafalaya rivers and shows earlier river courses across the valley on their way to the several sets of distributary channels that feed the Mississippi River Delta

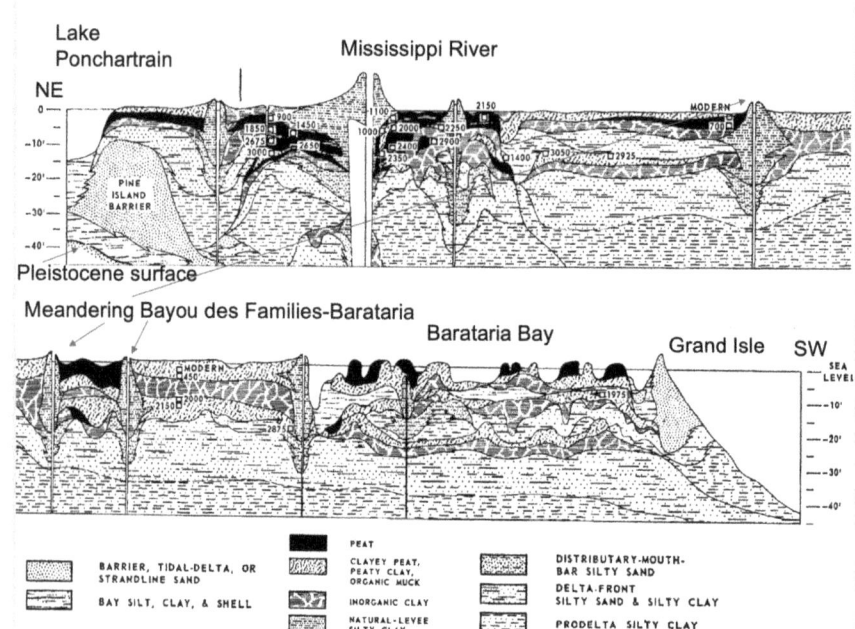

Fig. 3.7 Typical cross section developed by Frazier (1967) assembled from dated core data. This north south cross section crosses Lake Pontchartrain then through the Mississippi river then across Barataria Bay and Grand Isle. The sediment data led to interpretation of depositional environments, and the carbon dates placed those into delta lobe designations

southwestern Louisiana landscape. As each delta lobe developed, it became a sediment source for more cheniers laying out sand-rich, parallel, oak covered sand bars downcurrent and expanding Gulfward. Seaward growth of the Chenier Plain was greatest when the river was flowing to the western part of the deltaic plain and lowest when it was flowing to the east. Holocene geology of the Mississippi Delta section is a history of delta building and switching following the same mechanisms active at the surface today. That geology is superposed on the Pleistocene land surface that lay exposed at epoch's end and formed a weathered typically stiff, oxidized, overconsolidated surface that became the base of the Holocene. Over the continent, the years passed, spring thaws brought Mississippi floods and massive sediment influx, hurricane season drove a lot of sediment back to and along the coast, the systems along the coast thrived and geologic input became a part. Peat layers formed as swamps marked the sea level surface as it moved up and inland with glacial melt phases. All of this was recorded in the geologic history tapped by Frazier's coring (1967) and others (for examples Kulp et al. 2005; Hijma et al. 2017), and illustrates the rising sea level over Holocene time.

Dating this basal peat along the northern Gulf of Mexico has led to interpretations of Holocene history along the coast and development of Holocene sea level curves that, while they may be the subject of vigorous discussion, show a general consistency

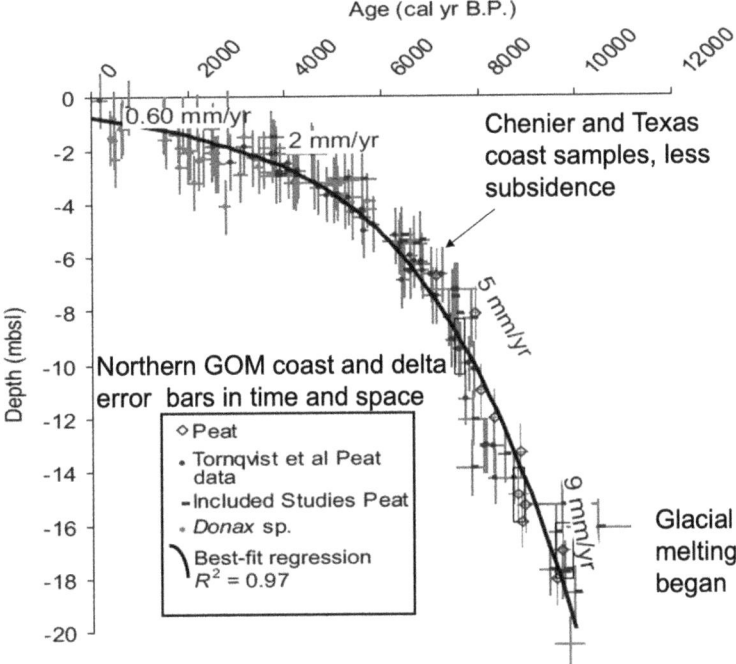

Fig. 3.8 Composite sea level curve (modified from Milliken et al. 2008) assembled from basal peat (on or near the Pleistocene surface) and shoreline (swash zone Donax shell) sampling at sites along the Gulf of Mexico Coast from Texas through Louisiana to Alabama. The plot is of age dates and sample depths fitted by nonlinear regression (after Törnqvist et al., 2004a, b, 2006), where the slope of the curve represents rate of sea level rise over Holocene time

(Fig. 3.8, Milliken et al. 2008) and exponential form, beginning with a rapid rise, then trending to lower measured values, with variations along the way (Anderson et al., 2010, 2014).

3.5 Geology and Oil and Gas Along the Louisiana Coast

There is a fundamental reason the Mississippi Delta has so many oil and gas fields. It has all the elements needed for successful formation of reservoirs and exploration—a thick sedimentary section full of sands for reservoirs, and shales between to seal them, abundant source rocks near and offshore and through the geologic section, temperatures and pressures to mature hydrocarbons from these rocks, fluid mechanisms, and pathways to migrate the oil and gas, and finally, many mechanisms that built structural and stratigraphic reservoir traps to hold the oil and gas for present-day discovery. These elements are illustrated in the enhanced stratigraphic section of Fig. 3.3 where the reservoir candidate sands are shown in yellow, carbonates, sometimes as reefs and often fractured, in blues, and their reservoir levels as red dots proportional to the volume of oil and gas. Sources and their ages are shown as the green triangles offshore. These elements have been combined with structural and stratigraphic migration, trap, and reservoir opportunities over geologic time, typical of the regional cross-section of Fig. 3.2, and spread along the Louisiana coast, well inland, and well out into the Gulf. This generalized section, beginning with the Louann salt layer deposition about 165 million years ago and miles deep, shows that some times were dominated by vigorous sand deposition coming off the continent, others by more quiet seas evidenced by limestones and fine muds. This geology didn't happen as a set of simple uniform layers to be discovered by random drilling, but because each part of the section all along the coast and far offshore is part of a system that is knowable, some parts more readily than others. This is also why there are geologists, lots of them, the Gulf Coast Association of Geological Societies is the largest focused group of geologists in the world, and much of their region has not been studied or explored, at least not as thoroughly as they would like. First, the record of geologic history is fundamentally incomplete—the sediments coming off the continent are a product of its geologic history, one punctuated by pulses of mountain building followed by erosion and subsequent deposition in much the same way as today—transport by rivers and streams, deposited as deltas and other forms along the shoreline at the time, recast and redeposited by advancing seas and storms (more time scale) again and again. The geologic history of a sand grain, or the layer it's a part of or basin where this layer is deposited may be long and complicated. The section (see Figs. 3.2 and 3.3) shows significant periods of geologic history, recorded by the sediments, where the seas advanced across the shoreline leaving behind a record of marine life, evolving over time, but detailed examination would reveal periods of non-deposition or absence of history as well; all this makes it difficult for a geologist to piece together a complete history with information leading to the discovery of oil and gas. Another is that remaining reserve potential is often very deep, may require drilling through salt, and in a hostile temperature and pressure environment, where geophysical information is limited by masking effects of that salt. The difficulty that drilling into these ultra-deep layers creates is exemplified by the Deepwater Horizon disaster.

On the Louisiana coast, the difficulties inherent in geologic exploration have been offset by the almost unique combination of geology leading to an extended life cycle of oil and gas production. The thick section of sands and clays at, beyond, and beneath the Louisiana coast described in earlier sections is one part—plenty of volume of material to serve as accumulator, pathway, and reservoir. Hydrocarbon source environment has been another Gulf of Mexico attribute: always at low latitude, hot, the Gulf periodically restricted—ideal conditions where vegetation and photosynthetic algae thrived, often died in anoxic events, and was rapidly buried and effectively preserved by continental muds and turned to mobile hydrocarbon by fortuitous temperature and pressure conditions. Also, as the sands and muds were buried in the rapidly subsiding section, impermeable clays often sealed sands before they could dewater, leaving zones of overpressure that then drove fluids (containing buoyant oil and gas) updip to intermediate or final traps—the reservoirs.

Oil and gas exploration is an inverse problem—the problem solver must assemble and interpret relatively little data about something that can't be seen directly. The inverse problem has intrigued, not deterred, geologists. The first oil and gas exploration was based on direct evidence—oil and gas seeps; then exploration became more difficult. Oil and gas exploration on the Louisiana coast has progressed, developing a more thorough picture of the geologic section, yet one that is never complete. Information comes from boreholes in the form of geophysical logs of all sorts, well cuttings and cores with micropaleontology sampling, and measurements about the well fluids, encounters including pressures and temperatures, and the progress of the well itself. Geophysical measurements take advantage of almost any contrast in physical properties and integrates that with the drilled framework. Salt, for example, contrasts in density, seismic velocity, and electrical and thermal conductivity and many early salt dome discoveries were the result of torsion balance, gravimeter, refraction, and reflection surveys. Tools, techniques, concepts, and knowledge bases of all sorts have come a long way since the initial Jennings well, and while a great deal is known, much remains to be learned.

The geology of the Mississippi Delta and offshore has been ideal for the creation of oil and gas traps in many ways. For one, the deluge of sediments coming off the continent was partly accommodated, or adjusted for, by growth faults. These growth faults often include extensional and rotational components resulting in "rollover" anticlines, and this inverted bowl, if it involves sands sandwiched between impermeable clay layers, is a common and effective oil and gas trap. A typical exploration plan involves identifying a fault and examining sands along its downthrown side for a positive structural closure, or anticline. Faults along the Louisiana coast are often these growth faults and often more complex than described in this example—and still more often the reason new oil and gas production is discovered where previous geologic interpretations have been abandoned.

The map of Fig. 3.9 (Wallace 1966) is a regional map of subsurface faulting compiled from oil and gas well information and illustrates the density of faulting along the Louisiana coast and their segmented nature as well.

Another "trap" happens when a sand is limited, cut off, "pinched out" or changes character (facies) so that it becomes a reservoir by depositional change—a stratigraphic "trap", one that is often an element of a faulted or structurally controlled reservoir. Moving salt (moving very slowly, but moving just the same) is another component of the Louisiana coast's oil and gas development and entrapment. Salt is lighter than the stack of sediments that buried it; it's ductile and resilient. Initially precipitated on the newly opened basin floor of the Gulf of Mexico, the Louann layer began to rise whenever a discontinuity offered a density difference (2.2–2.67 gm/cc) that drove a salt mass upward, much like a volleyball rises in a swimming pool. Salt intruded the section in sheets, spires, and into horizontal tablets forming new layers or sources of additional salt movement, finally peeling off as domes and penetrating near, and uplifting the surface that existed at the time. These salt domes dragged up and involved the stratigraphy they rammed through, developing new radial and tangential faulting and leaving behind new oil and gas "traps" sealed against their edges. Salt domes are seldom just cylinders and often mushroom with

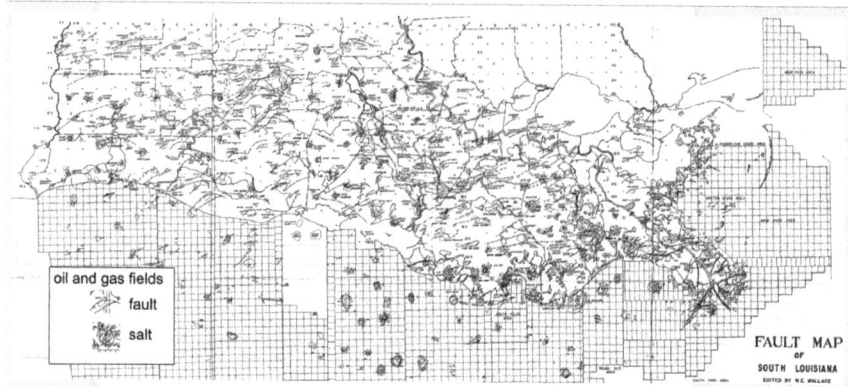

Fig. 3.9 Oil and gas fields and their reservoir structure as shown on the fault and salt map of the Louisiana coast (Wallace 1966). The map illustrates the incredible density of producing fault and salt structures related the development of the Mississippi River Delta and the Louisiana coast

overhang concealing additional reserves. Where salt moved into a new area or elevation between horizontal layers, it often has served to limit faulting above, detach, and mobilize the section (as in the example of Fig. 3.4). Moreover, salt is a strong seismic reflector, complicating imaging of the section around and hiding that below a salt structure. In summary, salt is both a reason for many oil and gas traps on the Louisiana coast and a confounding factor in the complexity of Mississippi Delta and Gulf of Mexico petroleum geology and remains a challenge. Here are two examples of typical Louisiana coast oil and gas fields, one on a fault, the Lirette field, the other around a large salt dome, the Lake Washington field. These are typical of the Louisiana coast and serve as a close-up defining the lines and circles of the Wallace map.

The Lirette field (Fig. 3.10, Silvernail 1967) incorporates a down to the coast fault (downdropped side toward the coast) and a rollover anticline (upside down bowl shape) trapping lighter than water oil and gas against the fault and under the anticline. The field produced initially from the source of a gas seep at about 2,500 feet, then from a thick section of Miocene sands. The map (a) and cross-section (b) combination is the usual way to describe a field where (a) is a plan or map view that shows the faults interpreted cutting a mid-Miocene sand layer and an anticline defined by the closed contours circling a group of producing oil and gas wells. Figure 3.10b is a south to north cross-section through the field connecting the wells shown by the location line on the map, and illustrating the faults and anticline. A more dramatic example of Louisiana coastal geology is the Lake Washington salt dome and the oil and gas field surrounding it (Fig. 3.11, Clem 2010).

Here, the salt dome has risen through the sediment layers above like the hydraulic ram that lifts your car at the service center. The map shows the white salt dome top in 3D perspective coming up through a prominent shallow sand shown with the contour lines and with colors, red down to green, then blue, along with radial faults

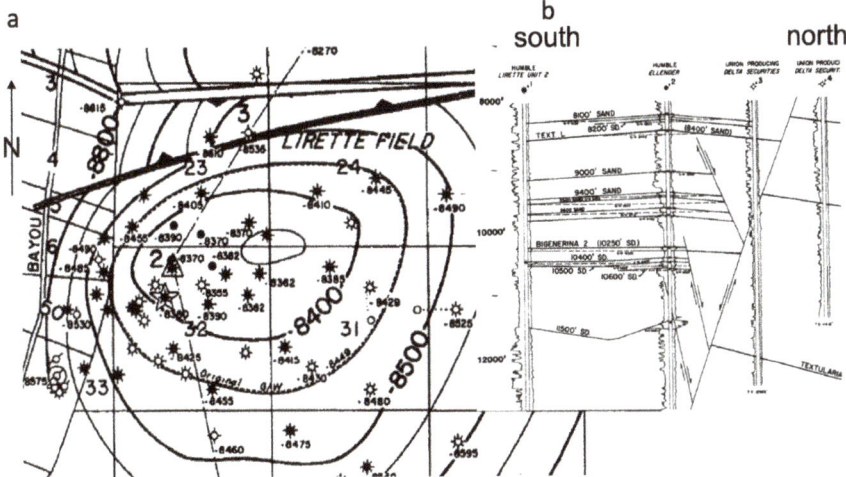

Fig. 3.10 Typical oil and gas field structural contour map (**a**) and geophysical log cross-section (**b**) of Lirette oil and gas field (from Silvernail 1967). The map shows the down to the coast (and an antithetic) fault that formed the rollover anticline (closed contours) and oil and gas trap of the reservoir sands. The structural contour map and cross section depict one of the several productive Miocene sands at about the 8500 foot depth. The cross-section shows the dominant fault, several antithetic faults, and the gentle inverted bowl anticlinal structure

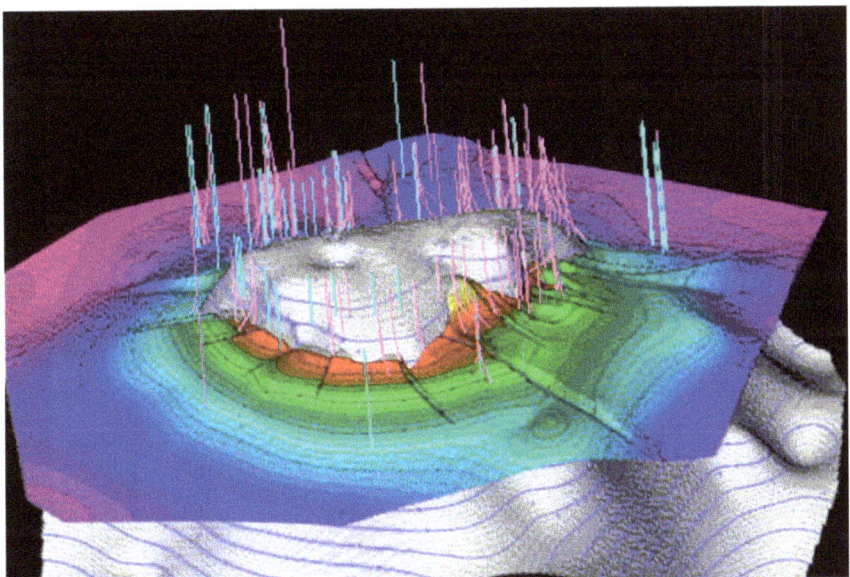

Fig. 3.11 Three dimensional view of Lake Washington field (vertical lines are wells, from Clem 2010) with salt dome mass (white) rising through sands (colors indicate depths) dragged up alongside and radially faulted. Wellbores shown produce from numerous sand reservoirs trapped against both the dome and faults

cutting that surface. All around the dome, oil and gas are trapped in sands like this, against the dome and along the radial fault offsets. The Louisiana coast geologic section remains complex, but its framework provides a valuable perspective about the changes taking place on the coast today.

3.6 Present State of the Geologic Section: Compaction and Subsidence

Much of the land loss in the Mississippi Delta involves subsidence of one form or another, and it's important to know something about the geologic contributions to this subsidence. Surface subsidence on the Louisiana coast is the sum of the displacement of the whole geologic section plus displacement within the section—that movement happens at different rates at different places, depending on compaction processes and things like faulting and salt tectonism within and through the section. The section is moving down to isostatically accommodate, or balance and adjust for the sediment load (load isostatic adjustment, LIA), as it has since deposition began with the Gulf opening. The whole coast is also moving down as the last effects of the ice age die away, the forebulge effect fronting the melted continental glacial ice mass relaxes to achieve another isostatic balance (glacial isostatic adjustment, GIA). Both are very deep processes and happen as the viscous part of the asthenosphere moves down and away beneath the Moho, or the interface between crust and mantle (see Fig. 3.2). Isostatic adjustments and their contribution to Mississippi Delta subsidence have been studied by several researchers (Love et al. 2016; Kuchar et al. 2018; Dokka 2011) and because of different measurements, scale, and conceptualizations, the range of displacement rates is broad. The present rate of isostatic adjustment from the combination of the Mississippi Delta isostatic adjustment and glacial isostatic adjustment is -2 mm/a (Wolstencroft et al. 2014), is a long term contribution to subsidence and rate of subsidence, and affects the broad Mississippi Delta and Chenier Plain. This adjustment by upper mantle displacement is the deepest form, or component, of subsidence and it affects the whole coast—slowly and broadly and may be separated from more shallow and regional subsidence. The processes, ages and time spans, character, rates, and how or if these are presently active are summarized in Table 3.1 and outlined in the discussion that follows.

The Louisiana coast's 15-km-thick sedimentary section above the isostatically adjusting crust and asthenosphere is the result of a complex geologic history, a history that varies across the delta and along the coast. Subsidence due to sediment compaction, compaction rates, and factors affecting compaction are important subjects to consider. At the time of deposition, sand grains are loosely packed with a lot of open space between (~40% porosity), and clay platelets drift (~ >50% porosity) like so many Frisbees™. As these sediments are buried, particles are rearranged, packed closer together and water is squeezed out. This compaction process is summarized by Fig. 3.12, a sediment column illustrating trends in typical pressure

Table 3.1 A summary of factors that affect displacement in the Mississippi Delta, where the processes or geologic age are evaluated in terms of active time, displacement character, whether the process is active now, and the related rate of compaction, displacement and subsidence. Note that through geologic time, sedimentary processes were active at and after deposition but reached a limit after; while some processes like growth faulting, salt deformation and migration, and isostatic adjustment transcend time boundaries

Time or process in geology	Time scale of process	Displacement character	Is process active now?	Displacement rate
Holocene	Began ~8,000 years ago	Compacting with organic and load	Yes, Chapters on environment and impacts	~up to 5 mm/a total, delta load center; ~10^{-2}–~10^{1} mm/a if faulting, marsh processes
Plio-Pleistocene	Began ~5 ma ago ended ~2.5 ma ago	Compacted with load	Nearing asymptotic, re-activated with faulting	~0.22 mm/a compaction α delta load center
Growth faulting	Began after deposition, continues episodically	Mega landslide or radial with salt piercement	Yes, episodic	~up to cm/a over episode of 10s to 1,000s of years, α to center of fault segment ~ kms length
Salt movement	Began with load over Louann, continues locally, slowly	Extruded, "toothpaste", tabular, ridges, columns as piercement salt domes	Yes, very slow, most activity ended with Miocene salt welds	Some salt domes deform the Holocene surface, some dissolution, slow salt withdrawal
Mio-Pliocene	Exponential decrease with depth, ~15 ma	Sediment load, compaction related subsidence	No, but fault affected	Compaction ~ 10^{-2} mm/a, Subsidence ~ 10^{-1} mm/a averaged to present, not now
Cretaceous through Eocene	At time and ended tens of millions ago	Sediment load compaction	No, exponential asymptotic	Moves with total geologic section
Louann salt	Deformed since Jurassic precipitation	Viscous, plastic masses tabular, piercement dome	Approaches limit	Dome uplift, limited withdrawal basins
Ice sheet bulge relaxation subsidence	Reversed with ice melt at end Pleistocene	Isostatic balance	Yes, coastwide, inexorable, episodic w/glacial history	~10^{-1} to ~2 mm/a subsidence component of RSLR

(continued)

Table 3.1 (continued)

Time or process in geology	Time scale of process	Displacement character	Is process active now?	Displacement rate
Sediment load subsidence	Continued depression since Gulf opening	Isostatic balance	Yes, also proportional to delta load	~1–8 mm/a subsidence component of RSLR

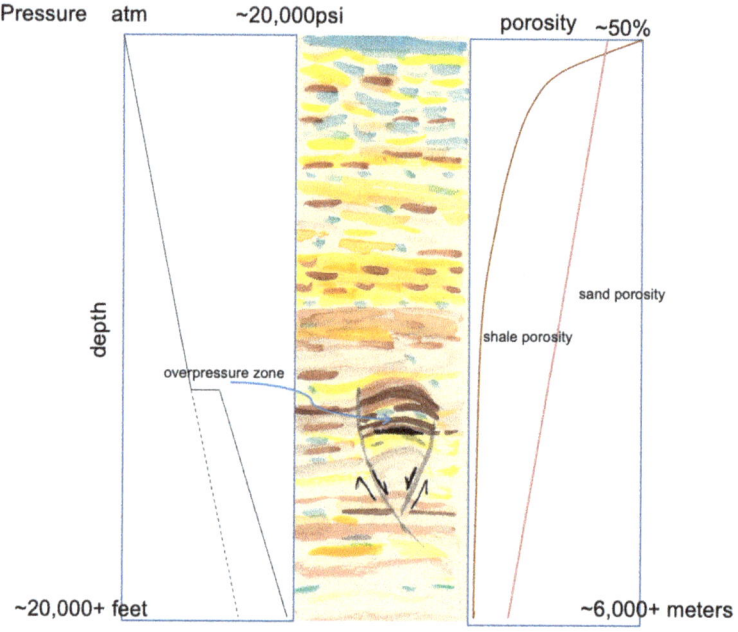

Fig. 3.12 Compaction process in Gulf coast sediments where the increasing load of overburden extrudes water and re-arranges sand and clay particles over time and with increasing depth. The graph on the left is the pressure increase with depth, and where anomalous pressure increases when fluid is trapped by conditions at depth. The middle illustration depicts loose, newly deposited sand and clay progressing to defined layers that continue to compact if not prevented by a structural block, the fault with a rollover anticline (and an oil reservoir) as shown in the lower portion of the column—and leading to the pressure increase shown on the left. The graph on the right is the result of compaction, as clays and sands reach asymptotic maxima for sand and clay (shale)

on the left, and therefore compaction of sands and clay (porosity) on the right, both with increasing depth.

The older the rock in the sedimentary section, the greater the depth of burial under the increasing load above, and the more time the rock has had to rearrange sand and clay particles and extrude water incorporated at the time of burial. The compaction rate and its contribution to present subsidence is nonexistent for the oldest, deepest rocks in the section; that is, while older sediments of Eocene, Oligocene, and Miocene time compacted at their beginnings, they are effectively not compacting now. The diagram of Fig. 3.12 begins with loosely packed sand and newly deposited clay at the surface, saturated with water. Pressure increases linearly with depth, but changes from hydrostatic (a column of water) to lithostatic (a column of sediment) if a permeability barrier (e.g. fault or clay) traps water, oil and gas as shown by the rollover anticline fault. Oil and gas well core measurements show that sand porosity decreases almost linearly with depth to less than half that at deposition at deeper levels as age increases from the present to say, Miocene time; that is, sand grains rearrange and compact with depth and time to an optimal packing. Core measurements and lab experiments show that clay porosity decreases exponentially with depth, reaching a limit, and becoming shale. So, future compaction within the deeper sedimentary section is expected to be near zero for contribution to present and future compaction rates. The exception to this orderly process happens where layers of sands were sealed by thick shales and faults before the entrapped water could escape; and as burial continued, pressure built and these overpressure zones became permanent features, usually at deep Miocene levels in the geologic section. Overpressure is both a problem in oil and gas drilling, but also an important part of the geochemical work at depth.

It is thus important to understand that there are a number of factors that affect vertical movement of the Delta surface. Much of its history is millions of years old and the compaction effective at and after deposition has long since run it course, while very new deposits are rapidly compacting and contribute to surface subsidence. Also, the entire section moves in response to the mass or load posed by the kilometers of sediments—and it also moves because the ice sheet was suddenly (geologically) taken off the continent (much like getting off a water bed).

Older sedimentary layers, say Paleocene up through Miocene are mostly finished with compacting and not moving except by means of fault and salt displacement, younger sediments are progressively subject to a variety of compaction mechanisms. Younger sediments in the Mississippi Delta geologic section, Miocene (about 15 ma) and younger, have been loaded less, and over less time than older sediments, and compaction and related surface subsidence effects may be incomplete, for example in overpressure situations or with reservoir depletion. Moving up in the geologic section to the very young Pliocene and Pleistocene, and almost to the surface, compaction and subsidence reflect loading by the younger features of the Louisiana coast. That is, the near coastline and Delta morphology influence the activity within the younger column. Compaction and subsidence from the late Pliocene (3.85 ma) to the middle Pleistocene (0.53 ma) along the Louisiana coast and the mid Delta region was the subject of a micropaleontologic study of oil and gas well samples by Frederick et al. (2019). The authors created a series of biostratigraphic maps from a proprietary industry library of 80,000 wells, where the data points contoured were the youngest, or topmost, occurrence of several key chronostratigraphic microfossils.

A cross-section across the coast and Delta summarizes the results as shown in Fig. 3.13. This section illustrates both the long-term, broad mid-Louisiana depositional fairway and the isostatic depression of the depocenter as sediment from the continent accumulated; that is, middle of the delta mass, greatest load, deepest isostatic load adjustment. This depression history reflects that a major part of the North American continent from the Rockies to the Appalachians drained to the Gulf

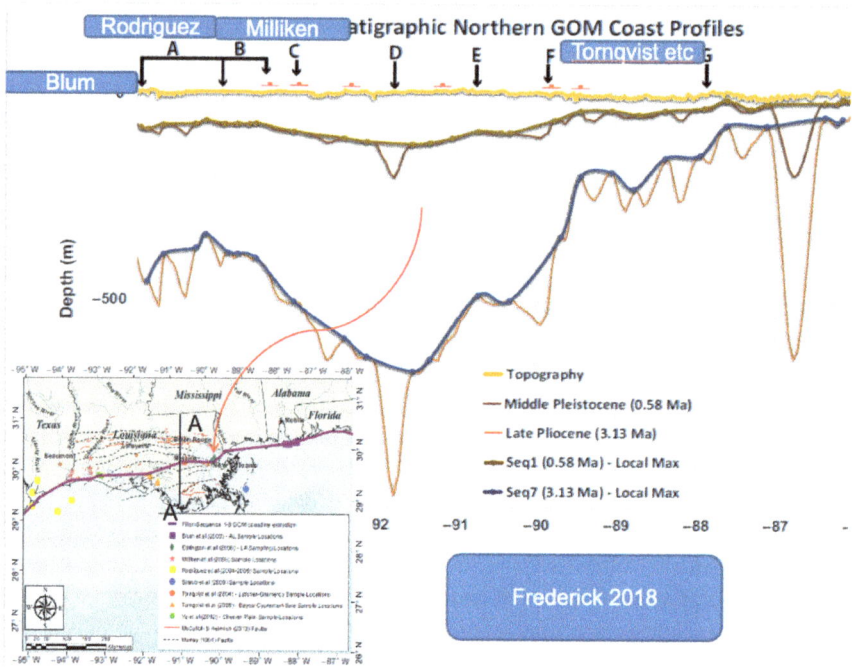

Fig. 3.13 Age date (chronostratigraphic) profiles through the Mississippi Delta (modified from Frederick et al. 2019) paralleling the coast as shown on the inset map. Present topography is shown at the top (yellow) along with the Holocene sampling location and authors as well as fault locations projected from Murray (1961) (red). The next profile represents the mid-Pleistocene sequence surface (0.58 ma) from the authors' dated, gridded and smoothed map, with known incised stream evidence superposed as the more detailed dark brown. The lower profile is the late Pliocene sequence surface (black) with the interpreted incised stream channels superposed

with rivers emptying into the Gulf across Texas, through Louisiana to Alabama (see Fig. 3.13 inset map) carrying a massive sediment load. Thus, ancient Mississippi Rivers flowed further to both the east and west than in the current Holocene delta. This study illustrates the influence of the broad continent edge depocenter from middle Pliocene through middle Pleistocene time and the expected variability of coastal subsidence across this depocenter. It also illustrates the lesson once more that measurements of compaction and subsidence at one location along the coast should not be applied to another without qualification of position, geologic time, and consideration of local geology. The Frederick et al. study determined a middle Pliocene through middle Pleistocene time averaged subsidence rate across broad depocenter shown in Fig. 3.13 is ~0.22 mm/y. The study also shows that subsidence rates with depth measured here decrease as the time considered increases. This interpretation is reasonable for two reasons, at least; the deeper portion of a compacting section has done so early in the history of that part of the section and the compaction profile becomes exponential; and also, much of the geologic record is no record at all, a hiatus, since there are long periods where nothing is deposited and long periods where the main activity is erosion or erasure of the record—averaging over longer time periods must consider these contributions.

The same mechanisms of deposition, compaction, salt and fault movement, isostatic adjustment, and consequent subsidence of the land surface that controlled the earlier geologic history of the coastal section have been and are still active in the Holocene—with significant variability within the young section and across the present Louisiana coast (Bomer et al. 2019). Compaction and subsidence and Holocene sediment behavior variability are demonstrated by several studies; for example, Törnqvist et al. (2008) determined compaction rates on a series of ^{14}C dated cores taken along a transect perpendicular to Bayou Lafourche and plotted these values against depth of overburden above a dated horizon (Fig. 3.14) to show an increasing compaction rate with greater depth of overburden above the marker peat layer.

Also, cores that included sand layers (red dots in Fig. 3.14) compacted at a lower rate, and while a linear relationship between overburden pressure and compaction rate is reasonable, the distribution of the shallow points suggests a threshold compaction effect as well. Another comparison of measured compaction rates using ^{14}C dated samples, with those expected from engineering properties showed a range of compaction rates, only generally falling within that predicted. This variation in compaction behavior is expected in the young Holocene section; at places, it is dominated by organic materials, at others by sands and everywhere

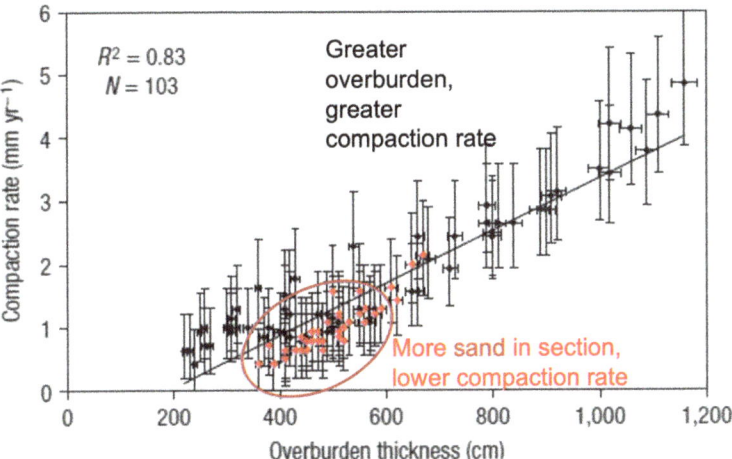

Fig. 3.14 Holocene compaction rate variation with deposited overburden load and stratigraphy (modified from Törnqvist et al. 2008). The Delta cores were taken at locations on Bayou Lafourche. Compaction rate was determined by present displacement of basal peat below a paleo surface, and overburden load (including sand facies) was measured from more than 100 cores. Presence of a significant sand component was determined by core description

fluxed through with shallow groundwater, all still in the process of compacting—leaving the present geology a dynamic situation. Holocene compaction is variable and on the order of 5 mm/yr as shown by these data; and Holocene compaction is ongoing. Vertical displacement of the Mississippi delta section beyond sediment compaction was examined by Yu et al. (2012) in a comparison of basal peat samples (top of the Pleistocene surface, base of the Holocene section) taken from the western Louisiana Chenier Plain with basal peat from the delta. Here, the Chenier basal peat sample data all plotted above the sea level rise curve determined primarily from Delta samples. The interpretation of the difference is that during the Holocene, the Pleistocene surface beneath the delta has been and is being displaced differentially by the greater delta sediment load like the Mio-Pliocene isostatic adjustment discussed earlier.

In summary, sediments of the Louisiana coast geologic section compact with depth, and for sediments of the deeper Mio-Pliocene to Paleocene depths, they compacted to their limit long ago, and so, while they are moving with the section, they are no longer significantly involved in movement within the section unless impacted by other elements like abnormal pressure. The Miocene and younger portion of the section dominate the geology of oil and gas production and so dominate the trapping mechanisms responsible. Earlier, we discussed shallow subsidence in the upper few meters due to surface alteration to hydrology due to such things as canals and spoil banks.

3.7 Faults, Subsidence, and Oil and Gas Fields

Faults and subsidence: subsidence on the downdropped side of a fault and subsidence over collapsed oil and gas reservoirs are a significant part of Louisiana coastal land loss. While the preceding sections have outlined the concepts of faulting and salt movement as a part of the Mississippi Delta geologic section, and how this faulting and salt movement are involved in the development of oil and gas fields, the following section is about how some of those faults reach the land surface today and play a role in Louisiana coast land loss. Faults in the Mississippi Delta geologic section do not move continuously over geologic time, but in spurts or episodes; and move along individual segments, not the whole fault or fault zone at once (Straub and Mohrig 2010; George 2008; Armstrong 2012). Salt diapirism, or piercement, or dome formation, with radial faults disrupting beds surrounding a dome as it rises from depth adds the complexity of salt movement and this too happens over brief times (geologically) and short distances. So, a fault that began long ago as sediments were deposited, with its beginnings now deep in the geologic section, may reach upward with time as faulting conditions continue, a growth fault, and that growth fault, if activated, may offset the land surface.

There is a coincidence of faults, oil and gas fields, and coastal land loss. A number of faults along the Louisiana coast have reached the surface, the faults are active, and this active faulting is often evidenced as a recently developed, arcuate break

3 The Geology of the Mississippi River Delta and Interactions with Oil...

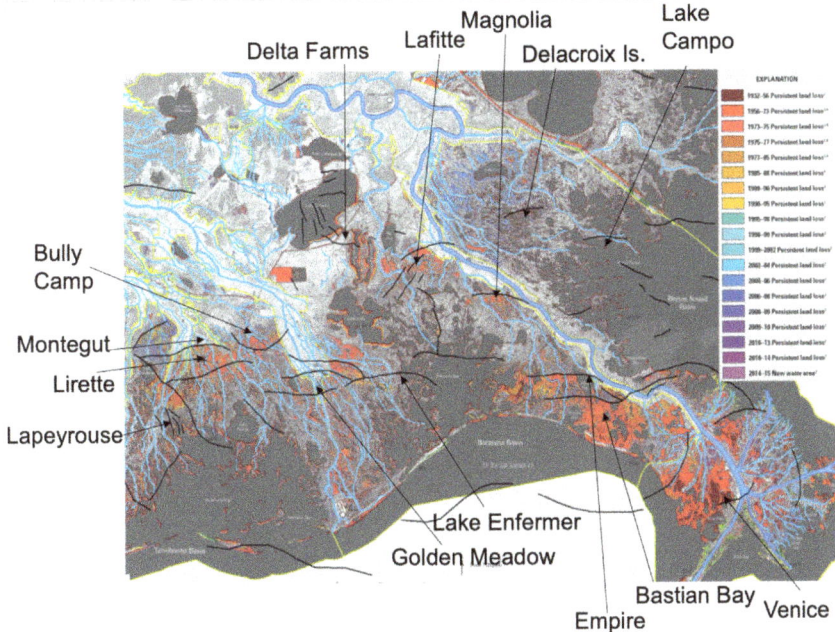

Fig. 3.15 Map of land loss, active faults and related oil and gas fields over the Mississippi Delta. Land area change between 1932 and 2016, where red represents maximum loss (hot spot) is from USGS map SIM3381 (Couvillion et al. 2017), faults superposed on USGS map (from McLindon 2020a, b), and oil and gas fields (from LDNR SONRIS 2020)

separating land from water, or land from lost land. This land to water fault expression is illustrated across the delta by Fig. 3.15, a map of surface faults (arcuate black lines with black squares toward fault downthrown blocks) superposed on the map of delta region land loss between 1932 and 2016 (Couvillion et al. 2017; Culpepper et al. 2019), together with the names of several of the oil and gas fields and fault names related to these faults. On the map, the deeper red represents maximum land loss, or "hot spots", and these areas of historic wetland loss are concentrated along the downthrown sides of the faults, and many of these faults, in turn, are related to oil and gas fields.

The map shows a spatial connection between active faults, oil and gas fields, and land loss across the broad delta area. There is a temporal connection as well. The land loss shown on the map of Fig. 3.15 has happened over the life cycle of Louisiana coastal oil and gas production, with the peak land loss following the oil and gas cycle's peak production, as illustrated by Fig. 3.16, a composite graph of land loss and oil and gas production in coastal Louisiana.

These historical data show close temporal and spatial correlations between rates of wetland loss and rates and locations of hydrocarbon production. This compilation of several land loss studies shows similar trends for different studies, each study done using different methods and each considering slightly different conceptualizations of

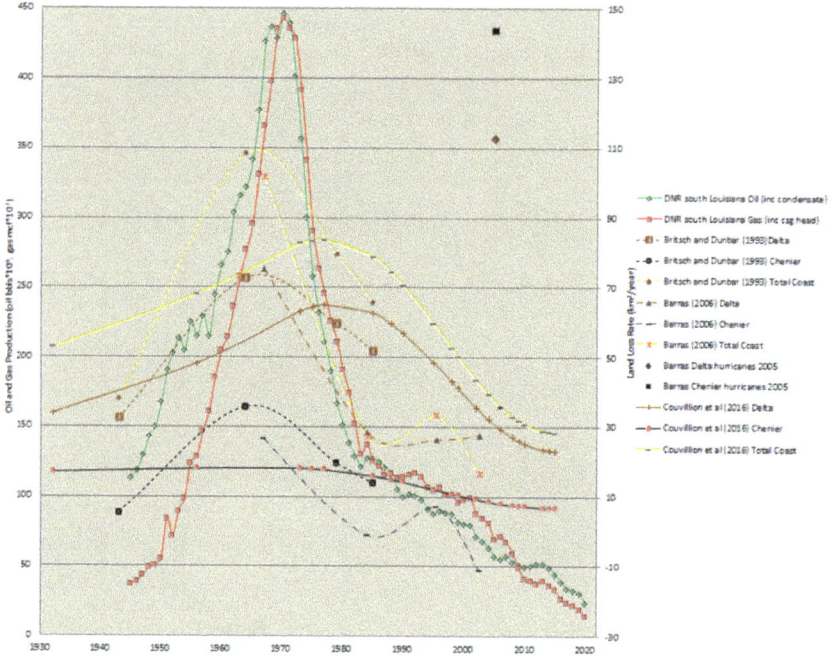

Fig. 3.16 Historical summary of oil and gas production in south Louisiana (LDNR data updated April, 2021) compared to rate of land loss as measured by Britsch and Dunbar (1993), Barras et al. (2008), and Couvillion et al. (2017). Each study (listed with date of project) divided the Louisiana coast into the Delta area (browns) and Chenier Plain (blacks) and the integrated whole (yellows). The studies illustrate the relationship between the life cycle of oil and gas development and rate of land loss

the delta and chenier areas evaluated. The patterns of land loss are generally similar, with peak loss between 1960 and 1980. The curve for total oil and gas production peaks before land loss likely because of delayed impacts of oil and gas as well as other contributing causes of land loss, for example, isolation from riverine input, edge erosion, hurricane effects, longer-term water level variability, and surface impacts of oil and gas canals. Regardless of the method or area considered, each land loss study reflects the cycle of land loss and correlates with the life cycle of south Louisiana oil and gas production rise, peak, and decline. And regardless of the proxy used to describe the oil and gas life-cycle, the cycle's temporal relation to land loss is unmistakable.

While coincidence of oil and gas production, faulting, and land loss is not causality, the connections are clear and the mechanisms are generally understood. These connections along the Louisiana coast have been the subject of a number of studies ranging from surface feature correlations, projections to depth, causal mechanisms of faulting and resulting subsidence, and collapse of oil and gas reservoirs translated to surface subsidence. Historically, the role of faulting in shaping water bodies such as Lakes Salvador, Pontchartrain, and Borgne was noted by Fisk as part of his Mississippi folio (1945), active faulting and geomorphic features across the delta were mapped by Gagliano (2003a, b, c), seismic sections were used by Kuecher et al. (2001) to trace active faulting from depth to surface and from the Terrebonne area to Plaquemines, and to link fluid movement and causality, Gagliano and Haggar (2010) described the characteristic "D" shape signature of water bodies created by fault activation, and Armstrong et al. (2014) interpreted a 3D seismic survey over the Breton area to trace a number of growth faults from Miocene depth to the surface and determine their segmented and episodic behavior. Individual fault studies of the Delacroix (Levesh 2018), Golden Meadow (Johnston 2019; Scates 2020; Akintomide 2021), Magnolia (Bullock 2020), Ironton (Bridgeman 2019), Little Chenier (O'Leary 2018), Bastian Bay/Empire (Martin 2006) are examples of recent fault studies using multiple data sources including 3D seismic surveys, and several of these are shown on the fault map of Fig. 3.15.

3.8 Reservoir Depletion, Fault Activation, and Subsidence Mechanisms

The Louisiana coast is not unique with regard to fault activation. A causal relationship between fluid production from sediment reservoirs and subsequent subsidence and faulting is studied worldwide in many geologic situations, often dramatic and always with consequences, for examples: the Central Valley of California agricultural groundwater withdrawal with subsidence of up to 9 m (Poland et al. 1975), Goose Creek oil and gas field where onshore became offshore in Galveston Bay, Texas (Pratt and Johnson 1926), Ekofisk oil field in the North Sea with structural damage to the reservoir and production facilities (Zoback and Zinke 2002), and Wilmington oil field at Long Beach, California, with more than 9 m of subsidence and associated faulting (Mayuga 1970). The production of oil, gas, and water from a fault involved reservoir leads to subsidence and consequent land loss both from fault activation and from compaction or collapse of the reservoir itself as illustrated in Fig. 3.17 where both the characteristic "D" shaped water body created by the fault displacement and the subsidence depression over the depleted and compacted reservoir come together as a composite land loss.

Faults related to Louisiana coast oil and gas fields that reach up to and break the surface today are mostly, if not entirely, reactivated growth faults including salt dome radial faults, and these faults are involved with the reservoirs below—reservoirs that these faults created in the first place. The mechanism of reactivation along these growth faults is typically related to the differential stress created as the pressure depleted reservoir is juxtaposed against originally pressured side of the fault, that is called poroelastic reduction of horizontal confining stress, which occurs as a result of the fluid withdrawal and consequent pore pressure decrease (Hillis 2001; Zoback and Zinke 2002) that creates a dislocation nucleation that activates the fault segment.

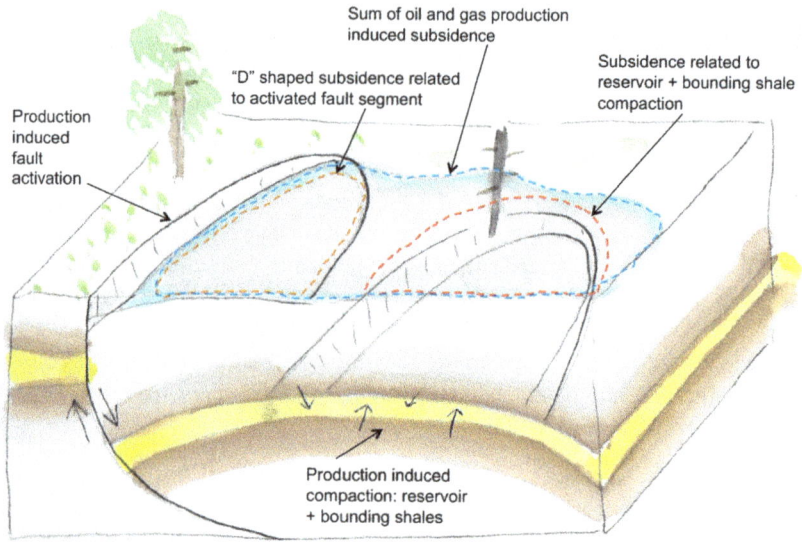

Fig. 3.17 Typical Mississippi River Delta oil and gas reservoir, a rollover anticline on the downthrown side of a down-to-the-coast normal fault. Oil, gas and water production affects the subsurface in and around the producing reservoir. The depletion process leads to production induced fault activation, reservoir compaction with additional compaction of bounding shales as shown in the subsurface. The surface manifestation of these subsurface changes is shown as a composite of land loss on the downthrown side the activated fault plus subsidence over the compacted oil, gas and water producing reservoir (from Day et al. 2020)

That is, when pressure is decreased on one side of the fault by producing oil and gas from the reservoir sand layer, the imbalance triggers the fault re-activation, and displacement spreads across a fault segment. In each case, even a small change in the overall stress situation may move the metastable geologic situation over the threshold, and faulting follows. Other mechanisms that may be involved in fault activation are fluid and gas migration along a fault relieving overpressure at depth (Kuecher et al. 2001), geochemical alteration along a fault plane (Diao and Espinosa-Marzal 2018), lubrication by fluids along a fault plane (Gagliano 2003a, b, c), and combinations of these mechanisms.

Subsidence related to the reservoir itself happens when oil, gas, and brine production depletes the hydrocarbon reservoir, often precipitously, resulting in pressure drop then compaction; and this change at depth translates upward, manifest at the surface as subsidence and faulting (Pratt and Johnson 1926). This is shown by the depression over the reservoir in Fig. 3.17. Geertsma (1973) modeled this surface subsidence using a simple disc, and Mallman and Zoback (2007) and Chan and Zoback (2007) applied more comprehensive compaction models including faulting to match surface subsidence data. More often, the situation at the surface results from a combination of subsidence and faulting mechanisms; for example, without pressure support the depleted reservoir collapses, and the pressure difference at the nearby associated fault plane nucleates slippage there (Ko et al. 2004; Morton et al. 2005, 2006; Chan and Zoback 2007), leading to a subtle, and sometimes dramatic, overprint integrated with other oil and gas related processes such as shallow subsidence due to surface hydrologic alteration.

3.9 Subsidence and the Louisiana Coast

Observations generally confirm the relationship between fluid production and subsidence in coastal Louisiana: for examples, wetland loss is typically higher in the vicinity of oil and gas fields than in marsh away from these fields (Bondesanf et al. 1995; Day et al. 2000; Morton et al. 2002, 2006; Morton and Bernier 2010; Mallman and Zoback 2007; Dokka 2011; Caro Guenca et al. 2011; Kolker et al. 2011; Yu et al. 2012; Chang et al. 2014; Day et al. 2020). On the coastal Cameron Parish Chenier Plain, Yu et al. (2012) suggested that paleo-sea level elevations, vertically offset by 0.5–1 m on a transect near the area of maximum oil and gas production, were influenced by production of the oil and gas nearby. Local rates of measured subsidence in oil and gas fields in south-central Louisiana (often more than 20 mm/year) were much higher than regional rates in the Mississippi River Delta (about 10 mm/yr or less; Morton et al. 2002). In a detailed study, Morton et al. (2005) analyzed releveling surveys, remote images, subsurface maps, stratigraphic sections, and hydrocarbon production data in relation to wetland loss for the Terrebonne-Lafourche basins and found that the highest rates of subsidence coincided with the location of oil and gas fields (Fig. 3.18, from Morton 2005).

Analysis of survey data along Louisiana Highway 1 between Raceland and Leeville, Louisiana showed much greater subsidence rates near and over producing oil and gas fields and related faults than between these fields during two periods of measurement (Morton et al. 2002, 2005, 2006) and concluded that these rapid changes were caused by induced subsidence and fault reactivation resulting from oil and gas production. Faulting and subsidence related to oil and gas fields along the Louisiana coast are more complicated than disc compaction or fault slip approximation models might explain. For example, Morton et al. (2005, 2006) noted that subsidence over oil and gas fields they studied continued well after peak production and with increasing rates. Chang et al. 2014) explained this continued subsidence by first measuring dewatering of deep clay core samples in the laboratory, then modeling the surveyed subsidence behavior over the oil and gas fields with reservoir depletion followed by continuing bounding clay compaction dewatering into the depleted reservoir. The results of the modeling are shown in Fig. 3.19 where the oil fields and releveling line along Highway 1 were matched by the model disc reservoir representation of the Valentine field. Here, Chang and Zoback envisioned subsidence as a two-step process where the initial component involved the compacting reservoir sand followed by an extended period of dewatering of the shales that bounded the

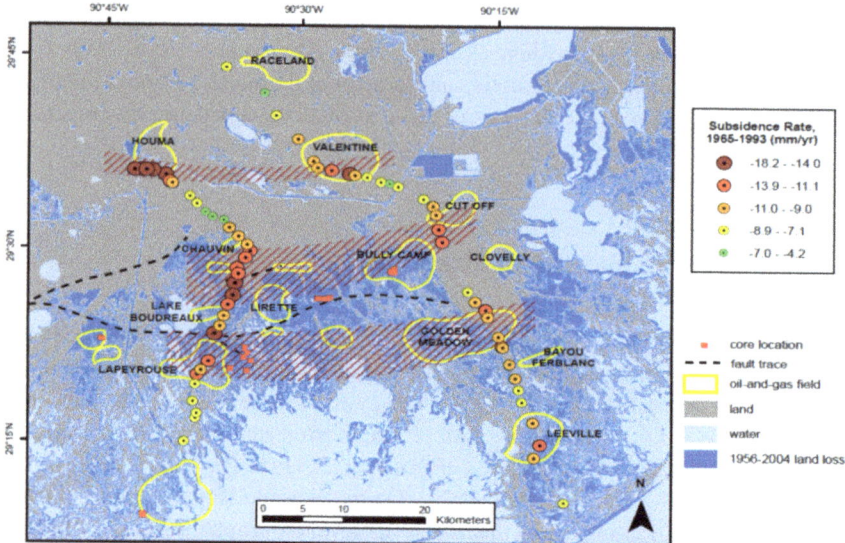

Fig. 3.18 Map showing the relationship between oil and gas fields and average subsidence rates for the epoch between 1965 and 1993 over a portion of the the Mississippi Delta (modified from Morton et al. 2005). Areas of highest subsidence rates (>12 mm/yr; hatched pattern) correlate closely with locations of oil and gas fields. Two regional fault areas shown by dashed lines. Lowest average subsidence rates are located between major producing fields. Several of these fields are shown on the map of Fig. 3.15

reservoir. The shale dewatering and related compaction was based on their earlier laboratory results characterizing the process.

While their modeling of reservoir compaction alone accounted for early subsidence, the addition of the bounding shale dewatering phase resulted in additional compaction and subsidence that more closely approximated the observed releveling survey results. A combination of factors likely creates faulting here, and a reservoir is not simply a uniform sand disc, but a complexly faulted set of sands bound by shales, so modeling is an approximation, but one that improves with the addition of geologically reasonable features. One such case of the combined effect of faulting and reservoir depletion was where Chan and Zoback (2007) modeled subsidence in the Lapeyrouse field, a faulted anticlinal structure related to the Golden Meadow fault. The Lapeyrouse field location is shown on Figs. 3.15 and 3.18. The area has suffered a great deal of land loss regionally and Chan and Zoback sought to replicate the measured subsidence over the Lapeyrouse field portion, first with a series of compacting discs to approximate the individual reservoir sand compartments, then with the addition of offset on the Golden Meadow fault as shown in Fig. 3.20.

This combination of faulting, reservoir compaction, fluid movement, and other factors is likely the typical case for faulting and subsidence along the Louisiana coast, and there are other combinations as well that activate faulting and induce subsidence in metastable fault related oil and gas fields. In summary, the connection

3 The Geology of the Mississippi River Delta and Interactions with Oil...   67

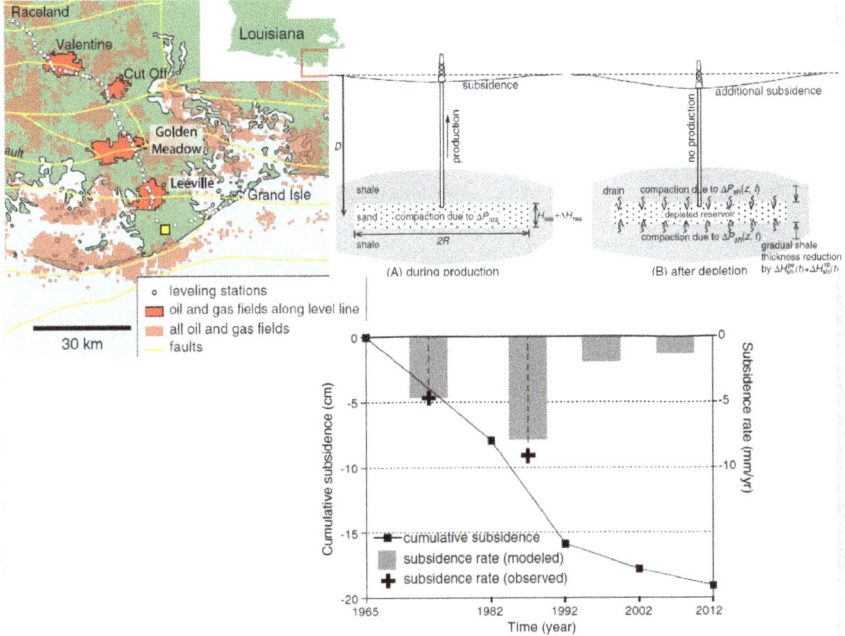

Fig. 3.19 Relationship between reservoir compaction, bounding shale drainage and surface subsidence (from Chang et al. 2014). Study area map showing oil and gas fields, subsidence leveling lines and faults (upper left), conceptual model of oil and gas reservoir sand compaction followed by additional compaction with shale fluid drainage (upper right), and results of model comparison (center)

between faults, oil and gas fields, subsidence, and land loss is evident along the Louisiana coast. The faults move along limited segments over limited time spans or episodes. The relationship of faulting and subsidence to the life cycle of oil and gas development indicates that as the cycle comes to a close, expected fault activity may end as well. The lesson from this is that it is important to understand faulting and subsidence and to include measurements about faulting and subsidence on the Louisiana coast in long term planning.

3.10 Getting Oil and Gas Out of the Ground—Framing the Impacts

The Mississippi Delta is probably the most studied subsurface geology in the world, and all this knowledge came about because of oil and gas. But getting oil and gas out of the ground here involves a number of operations, each with a significant impact. Succeeding chapters in this book deal with those impacts, so it's appropriate to close this chapter on geology with a discussion of the history of getting oil out of the ground

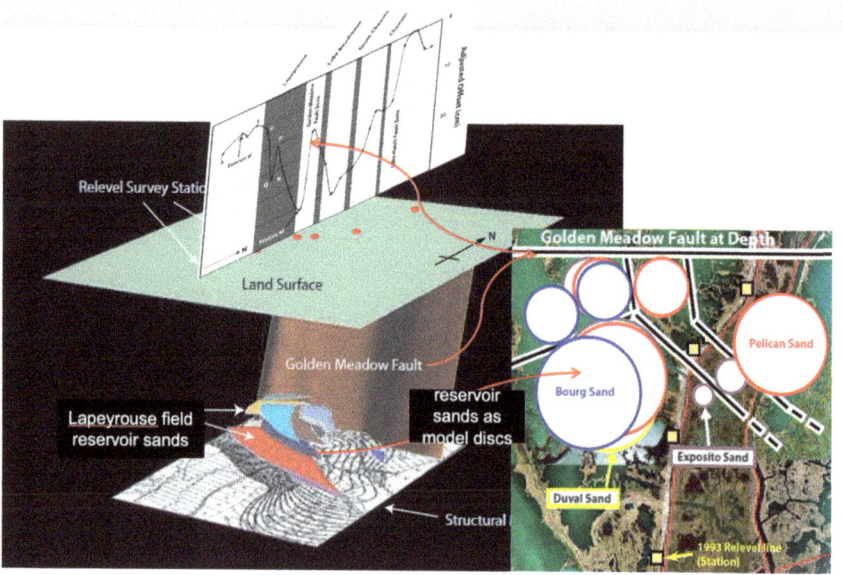

Fig. 3.20 Subsidence and the Lapeyrouse oil and gas field (modified from Chan and Zoback 2007). Subsidence releveling data curve (upper left) along bayou that crosses field (red dots, yellow dots right), multiple Lapeyrouse field reservoir sands (lower left) and Golden Meadow fault (middle left) modeled by disc approximations and faults (right). Subsidence could not be completely modeled by reservoir compaction alone, thus the addition of fault re-activation for a full explanation

in coastal Louisiana, in order to better understand how these impacts happened over the life cycle of oil and gas development.

At the dawn of the twentieth century, none of the oil fields shown on the maps in this book were known, coastal Louisiana was a blank slate, largely unaltered. But there were a number of surface indications of oil and gas—bubbling ponds, oil seeps, paraffin bogs, and subsequently, circular (think salt domes like Avery Island) topographic rises or depressions and linear features like stream course changes that hinted of oil reservoirs below. So, when Spindletop came in as a gusher outside of nearby Beaumont, Texas, these telltale features inspired the first exploratory drilling in Louisiana. In September 1901, the first successful oil well was completed in a rice field at Jennings, Louisiana (Spencer and Miller 2001). It was called the Heywood #1 Jules Clement well, drilled to a depth of 1,700 feet deep (Fig. 3.21). This event marked the beginning of oil and gas production in Louisiana. Drillers traced out the top and flanks of the Jennings dome, and soon, the countryside was covered with derricks (Fig. 3.21).

Oil companies fresh off of their successes in Pennsylvania were eager to begin production in this newly discovered region. The oil and gas companies were welcomed to the region of coastal Louisiana. The relationship between oil and gas companies and local populations and the state is a profitable one as oil and gas exploration and production brought with it jobs and a thriving economy. Many of coastal

3 The Geology of the Mississippi River Delta and Interactions with Oil...

Fig. 3.21 The Jennings field with the discovery well, the Heywood #1 Jules Clement, on the left and the developed field on the right. Note the horse and buggy are the mode of transportation in both (Louisiana Department of Natural Resources)

Louisiana's citizens depended, and continue to depend, either directly or indirectly on the success of oil and gas companies, either through jobs associated with the oil and gas fields or merchants selling their goods to the growing population or the petrochemical industry, which uses oil and gas as feedstocks. The production of oil and gas in the wetlands of Louisiana benefited the oil and gas companies, local populations, as well as local, state, and federal governments. By the end of 1905, more than 6,000,000 bbl. of petroleum was produced. It was priced at $0.72 cents/bbl. which created a huge sum of money for the time totaling $4.3 million dollars.

Adventurers drilled a lot of wells in the early days, most of them failures, and a lot based on less than honest promotion. It's important to realize at the outset that a great deal of oil and gas information comes from drilling that makes measurements and observations, but does not find oil and gas—dry holes, and in some circumstances that rate approaches 60%. This dry hole rate was the case as entrepreneurs covered the Jennings and other domes with discovery (first well with production), development, and wildcat (little or no knowledge of the subsurface) wells. Early exploration was "loosely" controlled in terms of well spacing, safety and environmental consideration. Scenes like Jennings (Fig. 3.21) were repeated over Louisiana, and spills, fires and general damage were common. Little did the oil and gas companies know the obstacles they would face in attempting to develop an oil and gas field in coastal Louisiana. Early drilling was generally on the coastal prairie where sites were accessed by board roads. Scenes of early exploration and drilling are shown in Fig. 3.22.

The early wells, and for a long time after, were drilled with constructed wood derricks built on pads and left in place after drilling. Subsurface information came from the cuttings from nearby wells. By the 1920s, additional methods of looking at the subsurface indirectly with geophysical instruments had been developed in

Fig. 3.22 Scenes from early Louisiana oil and gas exploration, from upper left: wooden derrick and crew, cable tool rig drawing, wooden derricks covering a field left in place after drilling, a torsion balance survey measurement, explorationists wading through marsh (State Library of Louisiana)

Europe. The salt domes and fault controlled oil reservoirs of the Louisiana coast (Fig. 1.1 coastal oil and gas fields, Fig. 3.1 structure of the coast, Fig. 3.9 faults and salt map of south Louisiana) provided an ideal application of these new geophysical methods: salt domes are much less dense than the surrounding sediments, salt transmits seismic energy at very high velocity and changes the path of a seismic wave, and fault offset is an ideal target of reflected seismic waves. Many of the early discoveries were developed through torsion balance (gravity gradient and salt density contrast), seismic refraction (seismic waves altered by salt), and seismic reflection (seismic waves returned to the surface by contrasting layers). Each of these surface geophysical surveys required extensive clearing along the survey and service paths, and involved topographic survey teams, measurement crews, and for seismic energy, and setting off significant explosive charges at depth. All of this clearly took an incredible amount of work and left behind a changed landscape—but not the dramatically changed landscape that was to come.

By the 1920s oil and gas explorers realized that much of Louisiana's potentially productive dome and fault structures were located beneath the wetlands of the Delta and Chenier area. The first wells in this more difficult setting were drilled from wood derricks set on driven piers, and served by boats. The early wells were drilled by cable tool rigs, where the bit (like a very large chisel suspended on a cable) was dropped by a rotating drum then pulled up several feet, tripped, and dropped again.

3 The Geology of the Mississippi River Delta and Interactions with Oil...

The tool was periodically pulled out of the hole and cuttings were retrieved with a bailer (a bucket with a hinged base). These gave way to rotary drilling with mud circulating continuously through the drill pipe, and the drill bit cuttings carried to the surface by the circulating mud. Wooden derricks gave way to steel, and those derricks often remained in place after wells were developed.

The wetland environment is not favorable for building roads and bridges. Accessing and operating oil and gas fields in this wetland environment comes with its share of difficulties. Exploration and production in coastal Louisiana necessitated the dredging of canals through wetlands and water bodies for navigation, drainage, pipelines, and oil and gas extraction. Barges brought in materials to build a wood platform on driven piers, then timbers or steel to build a derrick, followed by more barges with equipment and drillpipe, then another to power the works. If successful, tanks and more pipe followed, and the process continued. Photos showing some of these early canal based operation illustrate some of the challenges (Fig. 3.23).

Fig. 3.23 Early canals and exploration, from upper left: barge dredge excavation; clamshell bucket placing spoil; driven pier drilling platform on canal, steel derrick, pumping well; canal, well and tanks at Raceland field (State Library of Louisiana)

Texaco built the first drilling barge to move the entire process intact from drillsite to drillsite. It was constructed by connecting two barges, leaving a keyway between and mounting a steel derrick across the platform. Though a floating, portable drilling rig was a good idea, Texaco realized early on that their engineers were the second to think of it. The patent (US1681533, Submarine Drill, issued 1928) holder, Louis Giliasso, located tending bar in the jungles of Panama, resolved the problem, and the first prototype drilling barge, The Giliasso, was put in service in Lake Pelto in 1933 (Fig. 3.24) and a new era began. Soon, oil and gas exploration in south Louisiana dominated the landscape as exploration and production advanced and adapted to the wetlands. Figure 3.25 shows several scenes of drilling during the 30s and 40s where canals and drilling operations ranged from excavation through relatively dry land to open water.

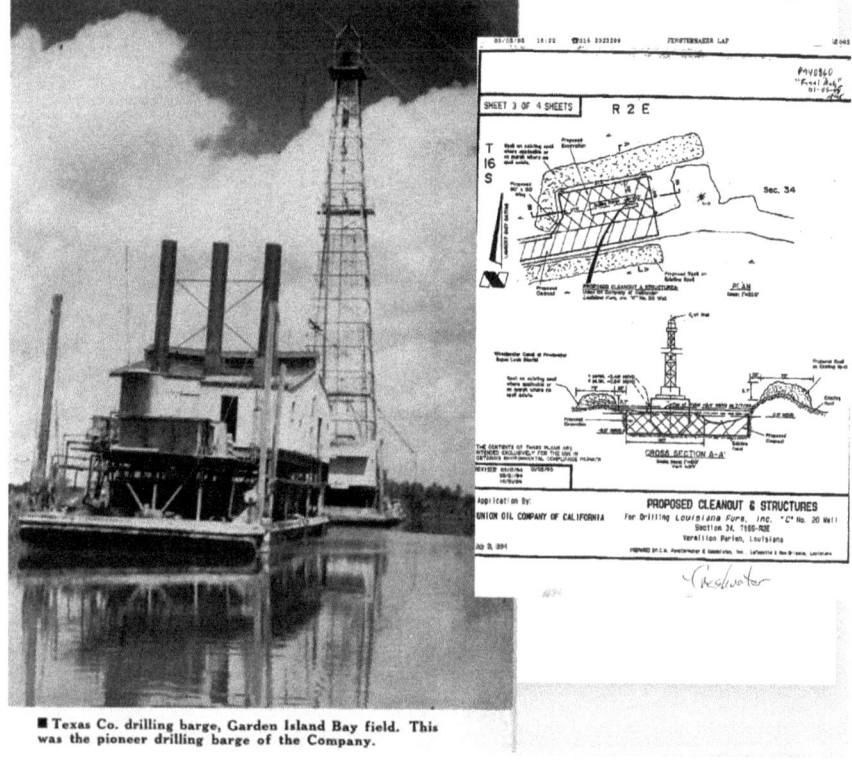

■ Texas Co. drilling barge, Garden Island Bay field. This was the pioneer drilling barge of the Company.

Fig. 3.24 First barge mounted drilling rig, the Giliasso, with barge for power in foreground (left). The barge was a combination of two standard 24 foot wide barges with an 8 foot keyhole between, and a 137 foot high mast straddling the set. The barge was floated into place and temporarily sunk for stability. The barge in front provided power. Also shown is a typical canal keyhole dredging plan (right), where the excavation accommodated a Giliasso class drilling system (State Library of Louisiana)

3 The Geology of the Mississippi River Delta and Interactions with Oil...

Fig. 3.25 Wells drilled using canals (circa 1930s, 40s). Access canals were widened with slips, or modified with keyholes to accommodate the barges. In many cases, the barges transported the materials and the constructed derricks and equipment were left in place (State Library of Louisiana)

The drilling operation involves the rig itself, crew and shop space, a source of power (sometimes on a second barge), an excavated pit or barge for drilling mud used to ensure the integrity of the borehole and retrieve cuttings, drill pipe, casing pipe, pumps and cement to seal the well, fuel and additional supplies. As the well is drilled, each step may require specialized equipment and people—there is a lot of traffic as a well is drilled in the wetlands.

In the process of developing these fields extensive permitting was required, even in the early days. These permitting activities included surface and mineral leases, dredging access canals, drilling wells, constructing infrastructure such as facilities, pipelines, and compressors, and oftentimes construction of production pits.

At the same time that oil and gas drilling technology was evolving, geophysical exploration methods that had proven successful in finding salt domes and faults inland, were adapted to the wetlands—if swimming, slogging, and pirogues for workers and geophysical measuring equipment can be considered adaptation. Some salt domes in the coastal environment were known at the outset, but geophysical surveys became essential elements of exploration. A seismic survey involved a sound source, historically explosive, and now often a vibration sequenced pad, accompanied by sound receivers, or geophones—a few feet apart, laid out in lines. The survey required transport of a small drilling rig and pipe, drilling mud, explosives, geophones and recording equipment—and then moving it all to the next shot and receiver locations. The seismic lines and the accompanying elevation measuring required extensive clearing that is often still visible on aerial photos. The explosive charges were usually placed in shotholes drilled to the base of the weathered zone using backpack carried drilling equipment. All this equipment and associated cabling had to be carried through the marsh by man, boat and sledge—and that was after a crew cut a path. It's not difficult to envision the motivation to mechanize things, and this was accomplished with the marsh buggy, actually first developed by trappers modifying Ford Model A trucks with paddle-wheels, in the 1930s. Figure 3.26 is a montage of photos of geophysical transport through the marsh over time.

Over the years, geophysical exploration survey footprints have reduced with things like lighter equipment and transportation, and cable-eliminating data communication; but then expanded with nets, or orchards, of geophones and shotpoints to do the 3D seismic surveys that are the technology of today. The subject geophysical survey impact remains an environmental concern addressed by survey planning and monitoring (Ensminger et al. 1997).

Drilling a well after geologists assemble the geophysical work and other data and pick a location, still involves the same excavation elements and impacts as in the days of the Giliasso, though these days, much of the drilling uses existing canals and begins through existing wellbores, or sidetracks. From the 1930s on, drilling involved a canal deep enough to float, then sink and rest the drilling barge on the

Fig. 3.26 Geophysical survey "progress" from the first marsh buggies to the 1950s. The buggy went from Model A driven to large floating wheeled vehicles, capable of moving an operation, but setting up and retrieving equipment remained labor intensive (State Library of Louisiana)

excavated bottom—typically about 8 feet; and wide enough to accommodate the barge, and power, mud and miscellaneous barges and boats. Canal excavation was again an adaptation of existing equipment; Louisiana has a history of dealing with canal excavation for agriculture, flood protection and other projects. Canals were dug first with a bucket "steam shovel", and then and to the present, with a clamshell dragline, both often barge-mounted versions of land based equipment. A typical barge is 56 feet wide and canals are excavated about 60–65 feet, until reaching the drillsite, which is dredged as a canal keyhole or sideslip about twice that width to allow maneuvering. The excavated marsh, of varying composition, was and is still dropped as a spoil bank paralleling the canal, set back from the excavation, but limited by the dragline boom length. Several examples of barge drilling are shown in Fig. 3.27.

Fig. 3.27 Examples of barge drilling in Louisiana wetlands. Note the combinations that make moving and drilling more efficient, yet the basic components, their width and supply components remain the same. Much of the drilling now takes place on existing canals with excavation for the drill site

If the initial well at a prospect is successful, succeeding additional drillsites are chosen and a canal system laid out as the extent of the structure and reservoir are defined. For examples, canals at the Leeville and the Venice field outline the salt structure, while canals at the Lapeyrouse field area were expanded parallel to the fault blocks involved. Producing wells usually involve a wellhead platform, a pump if necessary, and a flowline to the production facility.

Once production is established, a production facility is constructed to process the oil and gas before barging or pipelining it to a central operation. It is fed by flowlines from the wells, and in turn, feeds pipelines or barges. The production facility is typically near the center of the field and adjacent to a canal or bayou leading inland that can be used for transport and boat access. At the field level, the production facility typically separates oil, gas and produced water, sending the produced water to a nearby injection well. Prior to the early 1990s, final oil and water separation was done in an earthen pit with a center berm, the oil that rose to the surface behind the berm was skimmed and produced brine discharged into the adjacent water body. From the 1990s on, produced water was injected into a relatively shallow formation by means of an injection well, usually a well nearby the production platform, re-completed for the purpose.

Like any active operation, an oil and gas field requires maintenance, monitoring, testing and repairs, so the canal network continues over the life cycle of the project. As the field produces, a reservoir sand may reduce pressure, be affected by a rising water column as the oil and gas above are produced, or wells blocked or filled with sand at reservoir depth. These and other changes may require treatment that could include bringing in a barge to work on the well. The barge may be small for surface mechanical or coiled tubing cleanout or treatment or plugging, or larger if pulling tubing or downhole components is necessary. If a major workover is necessary or the well is re-completed at a greater depth, then a full sized barge set and rig returns. Each of these functions involves traffic, ranging from small but frequent boats, to crew and product transport vessels, to full scale well drilling with all the attendant supplies and waste.

The development of these fields and their access canals greatly altered the ecosystem of the wetlands of coastal Louisiana. Several examples of oil and gas canals of various ages are shown in Fig. 3.28. As canals were dredged, spoil material was typically placed on both sides of canals partially impounding wetlands; altering natural hydrology and salinity; decreasing nutrient, organic matter, and sediment exchange; changing vegetation composition and reducing vegetation productivity (e.g., Bass and Turner 1997; Cahoon and Turner 1989; Ko et al. 2004; Day et al. 2019, 2020). Canal and drillsite access today is often accomplished by using the force of a boat's turning propellers, "prop washing", to move accumulated sediment out of the existing canal and spread it over the subsided water bottom. The canals themselves are a dynamic system, generally widening with time, partially filled with sediment, and the spoil banks subside, and serve as an anchor for vegetation. The interaction of these canals and oil and gas operations with the natural system is the subject of chapters that follow.

3 The Geology of the Mississippi River Delta and Interactions with Oil...

Fig. 3.28 Oil and gas exploration and development canals in the Mississippi River Delta. Note the various canal components, drilling rigs, production facilities, relic spoil banks and marsh impacts (United States Geological Survey and Southeastern Louisiana University)

3.11 Summary

Geology forms the supporting structure of the ecological framework of the Delta. Much of geology is about differences from the norm, things like faults, sands and peat. Geology can be measured, and a clear picture of the subsurface should be part of any restoration project. Geologic things going on today, storms and sediment transport for example, are the same as those that happened in the past—and came together to build the geologic section. Land loss along the Louisiana coast is a critical issue that involves geology—how the geology beneath the coast got there in the first place; faults, salt movement, compaction, and subsidence of the geologic section; geology in oil and gas exploration and development; and how all this is involved with changes we see taking place today. This is not a matter of geochauvinism, but a realization that geology happens and building with it is always better than attempting work against it. The processes and consequences are many and complex, and they interact with shallow subsidence due to surface hydrologic alteration, but general ideas about them are straightforward and easily understood.

References

Akintomide A (2021) Neogene to quaternary fault activity and salt tectonics within the terrebonne salt withdrawal basin: effect of sediment loading on subsidence and salt-fault interaction. Doctor of Philosophy thesis, Tulane university, 178 p. https://digitallibrary.tulane.edu/islandora/object/tulane%3A122064

Anderson JB, Abdulah K, Sarzalejo S, Siringan F, Thomas MA (1996) Late quaternary sedimentation and high-resolution sequence stratigraphy of the East Texas Shelf. Geolog Soc, London, Special publications 117:95–124. https://doi.org/10.1144/GSL.SP.1996.117.01.06

Anderson J, Milliken K, Wallace D, Rodriguez A, Simms A (2010) Coastal impact underestimated from rapid sea level rise. EOS Trans Am Geophys Union 91:205–206. https://doi.org/10.1029/2010EO230001

Anderson J, Wallace D, Simms A, Rodriguez A, Milliken K (2014) Variable response of coastal environments of the Northwestern Gulf of Mexico to sea-level rise and climate change: implications for future change. Mar Geol 352:348–366. https://doi.org/10.1016/j.margeo.2013.12.008

Armstrong CP (2012) 3D seismic geomorphology and stratigraphy of the late Miocene to Pliocene Mississippi River Delta: fluvial systems and dynamics (thesis)

Armstrong C, Mohrig D, Hess T, George T, Straub KM (2014) Influence of growth faults on coastal fluvial systems: examples from the late Miocene to recent Mississippi River Delta. Sed Geol 301:120–132. https://doi.org/10.1016/j.sedgeo.2013.06.010

Barras JA, Bernier JC, Morton RA (2008) Land area change in Coastal Louisiana: a multidecadal perspective (from 1956 to 2006) (Scientific investigations map no. USGS SIM 3019)

Bass AS, Turner RE (1997) Relationships between salt marsh loss and dredged canals in three Louisiana Estuaries. J Coastal Res 13:895–903

Blum M, Roberts H (2012) The Mississippi Delta region: past, present, and future. Annu Rev Earth Planet Sci 40:655–683. https://doi.org/10.1146/annurev-earth-042711-105248

Blum MD, Misner TJ, Collins ES, Scott DB, Morton RA, Aslan A (2001) Middle Holocene sea-level rise and highstand at +2 M, Central Texas Coast. J Sediment Res 71:581–588

Bomer EJ, Bentley SJ, Hughes JET, Wilson CA, Crawford F, Xu K (2019) Deltaic morphodynamics and stratigraphic evolution of Middle Barataria Bay and Middle Breton Sound regions, Louisiana, USA: implications for river-sediment diversions. Estuar Coast Shelf Sci 224:20–33. https://doi.org/10.1016/j.ecss.2019.03.017

Bondesanf M, Castiglioni GB, Elmis C, Gabbianellis G, Marocco R, Pirazzolift PA, Tomasin A (1995) Coastal areas at risk from storm surges and sea-level rise in Northeastern Italy. J Coastal Res 11:1354–1379

Bridgeman J (2019) Understanding Mississippi Delta subsidence through stratigraphic and geotechnical analysis of a continuous holocene core at a subsidence superstation. Master of Science thesis, Tulane university, 96 p

Britsch LD, Dunbar JB (1993) Land loss rates: Louisiana Coastal Plain. J Coastal Res 9:16

Bullock J (2020) Evolution of the Magnolia growth fault: an evaluation of cenozoic activity, Plaquemines Parish, Southeastern Louisiana. Master of Science thesis, university of New Orleans

Cahoon D, Turner R (1989) Accretion and canal impacts in a rapidly subsiding wetland II. Feldspar Marker horizon technique. Estuaries Coasts 12:260–268. https://doi.org/10.2307/1351905

Caro Cuenca M, Hanssen R, Hooper A, Arikan M (2011) Surface deformation of the whole Netherlands after PSI analysis. FRINGE 2011 Workshop, ESA SP-697 1-8

Chan A, Zoback M (2007) The role of hydrocarbon production on land subsidence and fault reactivation in the Louisiana Coastal Zone. J Coastal Res 233:771–786. https://doi.org/10.2112/05-0553

Chang C, Zoback M (2009) Viscous creep in room-dried unconsolidated Gulf of Mexico Shale (I): experimental results. J Petrol Sci Eng 69:239–246. https://doi.org/10.1016/j.petrol.2009.08.018

Chang C, Mallman E, Zoback M (2014) Time-dependent subsidence associated with drainage-induced compaction in Gulf of Mexico shales bounding a severely depleted gas reservoir. AAPG Bull 98(6), 1145–1159. https://doi.org/10.1306/11111313009

Christeson G, Van Avendonk H, Norton I, Snedden J, Eddy D, Karner G, Johnson C (2014) Deep crustal structure in the eastern Gulf of Mexico. J Geophys Res: Solid Earth 119(9) 6782–6801. https://doi.org/10.1002/2014JB011045

Clem RK (2010) Lake Washington field Plaquemines Parish, Louisiana. In: Oil and gas fields of South Louisiana. New Orleans Geological Society, p 12

Couvillion B, Beck H, Schoolhammer D, Fischer M (2017) Land area change in Coastal Louisiana (1932 to 2016), Scientific investigations map 3381 (Scientific investigations map no. SIM 381), Scientific investigations map. U.S. Geological Survey

Cuenca M, Hanssen R, Hooper A, Arakan M (2011) Surface deformation of the whole Netherlands after PSI analysis

Culpepper D, McDade E, Dawers N, Kulp M, Zhang R (2019) LSU Digital commons synthesis of fault traces in SE Louisiana relative to infrastructure recommended citation transportation consortium of south central states synthesis of fault traces in SE Louisiana relative to infrastructure

Day JW et al (eds) Energy production in the Mississippi River Delta, 978-3-030-94525-1

Day JW, Britsch LD, Hawes SR, Shaffer GP, Reed DJ, Cahoon D (2000) Pattern and process of land loss in the Mississippi Delta: a spatial and temporal analysis of wetland habitat change. Estuaries 23:425–438. https://doi.org/10.2307/1353136

Day JW, Shaffer GP, Cahoon DR, DeLaune RD (2019) Canals, backfilling and wetland loss in the Mississippi Delta [WWW Document]. https://pubs.er.usgs.gov/publication/70207538. Accessed 17 July 21

Day J, Clark H, Chang C, Hunter R, Norman C (2020) Life cycle of oil and gas fields in the Mississippi River Delta: a review. Water 12:30

Diao Y, Espinosa-Marzal R (2018) The role of water in fault lubrication. Nat Commun 9. https://doi.org/10.1038/s41467-018-04782-9

Dickinson G (1953) Geological aspects of abnormal reservoir pressures in Gulf Coast Louisiana. Bulletin 37. https://doi.org/10.1306/5CEADC6B-16BB-11D7-8645000102C1865D

Diegel F, Karlo J, Schuster D, Shoup R, Tauvers P (1995) Cenozoic structural evolution and tectono-stratigraphic framework of the Northern Gulf Coast continental margin. In: Salt tectonics, vol 65. AAPG Memoir, pp 109–151

Dokka RK (nd) Subsidence of South Louisiana: measurement, causes, and human implications 34

Dokka RK (2006) Modern-day tectonic subsidence in Coastal Louisiana. Geol 34:281. https://doi.org/10.1130/G22264.1

Dokka R (2011) The role of deep processes in late 20th century subsidence of New Orleans and coastal areas of Southern Louisiana and Mississippi. J Geophys Res 116. https://doi.org/10.1029/2010JB008008

Dokka RK, Sella GF, Dixon TH (2006) Tectonic control of subsidence and southward displacement of Southeast Louisiana with respect to Stable North America. Geophys Res Lett

Ensminger A, Fossier R, Gagliano MH, Gagliano SM, Mouton E, Windham M (1997) Lake sand: a reduction of environmental impacts during a 3-D seismic survey in the Louisiana Coastal Wetlands 47

Filina I (2019) Crustal architecture of the Northwestern and Central Gulf of Mexico from integrated geophysical analysis. Interpretation 7:T899–T910. https://doi.org/10.1190/INT-2018-0258.1

Fisk H (1944) Geological investigation of the alluvial valley of the Lower Mississippi River (atlas). Mississippi River Commission, Vicksburg, MS

Frazier D (1967) Recent deltaic deposits of the Mississippi River: their development and chronology. Trans Gulf Coast Assoc Geol Soc 27:287–315

Frederick BC, Blum M, Fillon R, Roberts H (2019) Resolving the contributing factors to Mississippi Delta subsidence: past and present. Basin Res 31:171–190. https://doi.org/10.1111/bre.12314

Gagliano S (2003a) Active geological faults and land change in SE LA Part 1 (Research Study for Mississippi River Commission). USACAE Contract No. DACW 29-00-C-0034

Gagliano S (2003b) Active geological faults and land change in SE LA Part 2

Gagliano S (2003c) Active geological faults and land change in Southeastern Louisiana: executive summary

Gagliano S, Haggar K (2010) Effects of D-shaped fault deformation on South Louisiana Landscape. AAPG annual meeting abstracts

Galloway W (2009) Gulf of Mexico. GEO ExPro 6, 11

Geertsma J (1973) Land subsidence above compacting oil and gas reservoirs. J Petrol Technol 25:734–744. https://doi.org/10.2118/3730-PA

George T (2008) 3-D seismic evaluation of fault control on quaternary subsidence patterns, rates, and related. University of Texas

Hijma MP, Shen Z, Törnqvist TE, Mauz B (2017) Late Holocene evolution of a coupled, mud-dominated delta plain–Chenier plain system, Coastal Louisiana, USA. Earth Surf Dynam 5:689–710. https://doi.org/10.5194/esurf-5-689-2017

Hillis R (2001) Coupled changes in pore pressure and stress in oil fields and sedimentary basins. Pet Geosci 7:419–425. https://doi.org/10.1144/petgeo.7.4.419

Johnston A (2019) Investigating the relationship between subsurface geology and land loss near Golden Meadow, Louisiana by utilizing 3-D seismic and well log data. Master of Science thesis, University of Louisiana at Lafayette, 102 p

Johnson A, Zhang R, Gottardi R, Dawers N (nd) Investigating the relationship between structural geology and wetland loss near Golden Meadow, Louisiana by utilizing 3D seismic reflection and well log data [WWW Document]. AGU—2017 AGU Fall Meeting. https://agu.confex.com/agu/fm17/meetingapp.cgi/Paper/287572. Accessed 20 May 18

Ko J-Y, Day J, Barras J, Morton R, Johnston J, Steyer G, Kemp P, Clairain E, Theriot R (2004) Impacts of oil and gas activities on coastal wetland loss in the Mississippi Delta. In: Winters K, Nipper M (eds) Environmental analysis of the Gulf of Mexico, vol 1, Special publication series no. 1. Corpus Christi, Texas Harte Research Institute for Gulf of Mexico Studies, pp 608–621. Harte Research Center, Corpus Christi, TX, p 14

Kolker AS, Allison MA, Hameed S (2011) An evaluation of subsidence rates and sea-level variability in the northern Gulf of Mexico. Geophys Res Lett 38. https://doi.org/10.1029/2011GL049458

Kuchar J, Milne G, Wolstencroft M, Love R, Tarasov L, Hijma M (2018) The influence of sediment isostatic adjustment on sea level change and land motion along the U.S. Gulf Coast: SIA and land motion along U.S. Gulf Coast. J Geophys Res Solid Earth 123:780–796. https://doi.org/10.1002/2017JB014695

Kuecher GJ, Roberts HH, Thompson MD, Matthews I (2001) Evidence for active growth faulting in the Terrebonne Delta Plain, South Louisiana: implications for wetland loss and the vertical migration of petroleum. Environ Geosci 8:77–94. https://doi.org/10.1046/j.1526-0984.2001.82001.x

Kulp M, Penland S, Williams SJ, Jenkins C, Flocks J, Kindinger J (2005) Geologic framework, evolution, and sediment resources for restoration of the Louisiana Coastal Zone. J Coast Res 56–71

Levesh J (2018) Middle Miocene through present fault history of the Delacroix Island fault system

Louisiana Department of Natural Resources (nd) Department of Natural Resources—State of Louisiana, Oil and Gas Production Values [WWW Document]. Louisiana Energy Facts and Figures. http://www.dnr.louisiana.gov/index.cfm/page/209. Accessed 12 July 21

Love R, Milne G, Tarasov L, Engelhart S, Hijma M, Abdulah K, Horton B, Törnqvist T (2016) The contribution of glacial isostatic adjustment to projections of sea-level change along the Atlantic and Gulf Coasts of North America. Earth's Future 4:440–464. https://doi.org/10.1002/2016EF 000363

Mallman EP, Zoback MD (2007) Subsidence in the Louisiana Coastal zone due to hydrocarbon production. J Coast Res 6

Martin E (2006) Fault induced subsidence near Empire and Bastian Bay, Louisiana. Tulane

Mayuga MN (1970) Geology and development of California's Giant—Wilmington Oil Field. Geol Giant Petrol Fields AAPG Memoir 14:158–184

McLindon C (2020a) Louisiana's oil and gas industry—the missing link in coastal sustainability [WWW Document]. McLindon Geosciences, LLC. https://www.mcgeo.me/blog/louisianas-oil-and-gas-industry-the-missing-link-in-coastal-sustainability. Accessed 3 Jan 21

McLindon C (2020b) The Montegut fault—a conduit for fluid migration [WWW Document]. McLindon Geosciences, LLC. https://www.mcgeo.me/blog/the-montegut-fault-a-conduit-for-fluid-migration. Accessed 17 July 21

Milliken KT, Anderson JB, Rodriguez AB (2008) A new composite Holocene sea-level curve for the northern Gulf of Mexico. In: Special paper 443: response of upper Gulf Coast estuaries to Holocene climate change and sea-level rise. Geological Society of America, pp 1–11. https://doi.org/10.1130/2008.2443(01)

Moffett JB (2014) Deep gas play in Gulf of Mexico, presentation at American association of petroleum geologists playmaker 2.0, Houston. https://www.aapg.org/videos/playmaker/Articleid/37745/jim-bob-moffett-deep-gas-play-in-gulf-of-mexico

Morton RA, Bernier JC (2010) Recent subsidence-rate reductions in the Mississippi Delta and their geological implications. J Coast Res 555–561. https://doi.org/10.2112/JCOASTRES-D-09-00014R1.1

Morton R, Bernier J, Barras J, Ferina F (2005a) Historical subsidence and wetland loss in the Mississippi Delta plain. GCAGS Trans 55:17

Morton R, Bernier J, Barras J, Ferina N (2005b) Rapid subsidence and historical wetland loss in the Mississippi Delta Plain: likely causes and future implications. USGS open file report 05–1216, 116 p

Morton R, Bernier J, Barras J (2006) Evidence of regional subsidence and associated interior wetland loss induced by hydrocarbon production, Gulf Coast Region, USA. Environ Geol 50:261–274

Morton RA, Buster NA, Krohn MD (2002) Subsurface controls on historical subsidence rates and associated wetland loss in Southcentral Louisiana. Trans Gulf Coast Assoc Geol Soc 52:767–778

Murray G (1961) Geology of the atlantic and Gulf coastal province of North America. Harper Geoscience Series. New York, 692 p

O'Leary M (2018) Relationship between growth faults and subsidence: impact on coastal erosion, an example from Cameron Parish, Southwestern Louisiana, USA. Master of Science thesis, University of Louisiana at Lafayette, 227 p

Poland JF, Lofgren BE, Ireland RL, Pugh RG (1975) Land subsidence in the San Joaquin Valley, California, As of 1972. USGS Professional Paper 437-H 87

Pratt WE, Johnson DW (1926) Local subsidence of the Goose Creek Oil Field. J Geol 34:577–590. https://doi.org/10.1086/623352

Radovich BJ, Moon J, Connors CD, Bird D (2007) Insights into structure and stratigraphy of the Northern Gulf of Mexico from 2D pre-stack depth migration imaging of mega-regional onshore to deep water, long-offset seismic data. In: GCAGS transactions. Presented at the GCAGS Annual Meeting, p 7

Radovich B, Horn B, Nuttall P, McGrail A (2011) The only complete regional perspective. GEO Ex Pro 8(2)

Rodriguez A, Anderson J, Siringan F, Taviani M (2004) Holocene evolution of the East Texas Coast and inner continental shelf: along-strike variability in coastal retreat rates. J Sediment Res 74(3):405–421

Scates AR (2020) Geomorphic and shallow subsurface expression of growth faults in Mississippi River Delta holocene sediment; Golden Meadow, Louisiana. Master of Science thesis, University of Louisiana at Lafayette, 133 p

Silvernail JD (1967) Lirette field: Terrebonne Parish. Louisiana pp 78–79
Spencer JA, Miller B (2001) 100 years of exploration and production at Jennings Field 8
Straub K, Mohrig D (2010) Subsidence associated with active growth faulting on the Mississippi Delta: displacement rates and steering of the Mississippi River, long-term estuary assessment group ann mtg presentation, 27 p. http://leag.tulane.edu/PDFs/Straub-LEAG-4.28.10.pdf
Törnqvist T, Bick S, Gonzalez J, van der Borg K, Jong A (2004a) Tracking the sea-level signature of the 8.2 ka cooling event: new constraints from the Mississippi Delta. Geophys Res Lett 31:L23309. https://doi.org/10.1029/2004GL021429
Törnqvist TE, González JL, Newsom LA, van der Borg K, de Jong AFM, Kurnik C W (2004b) Deciphering Holocene sea-level history on the U.S. Gulf Coast: a high-resolution record from the Mississippi Delta. Geol Soc America Bull 116:1026. https://doi.org/10.1130/B2525478.1
Törnqvist T, Bick S, van der Borg K, de Jong A (2006) How stable is the Mississippi Delta? Geol 34:697. https://doi.org/10.1130/G22624.1
Törnqvist TE, Wallace DJ, Storms JEA, Wallinga J, van Dam RL, Blaauw M, Derksen MS, Klerks CJW, Meijneken C, Snijders EMA (2008) Mississippi Delta subsidence primarily caused by compaction of Holocene Strata. Nat Geosci 1:173–176. https://doi.org/10.1038/ngeo129
Wallace WE (1966) Fault and salt map of South Louisiana. GCAGS Trans 16:2
Wolstencroft M, Shen Z, Törnqvist TE, Milne GA, Kulp M (2014) Understanding subsidence in the Mississippi Delta region due to sediment, ice, and ocean loading: insights from geophysical modeling. J Geophys Res Solid Earth 119:3838–3856. https://doi.org/10.1002/2013JB010928
Yu S-Y, Törnqvist TE, Hu P (2012) Quantifying Holocene lithospheric subsidence rates underneath the Mississippi Delta. Earth Planet Sci Lett 331–332:21–30. https://doi.org/10.1016/j.epsl.2012.02.021
Yuill B, Lavoie D, Reed D (2009) Understanding subsidence processes in Coastal Louisiana. J Coast Res 23–36. https://doi.org/10.2112/SI54-012.1
Zoback M, Zinke J (2002) Production-induced normal faulting in the Valhall and Ekofisk oil fields. Pure Appl Geophys 159(1), 403–420. https://doi.org/10.1007/PL00001258

Chapter 4
The Regulatory and Legal Framework—Oil and Gas Influence Over Environmental Management in Louisiana

Paul H. Templet

4.1 Introduction

The oil and gas industry has a long history of using their political and financial clout to kill, modify, hinder and otherwise weaken environmental management in Louisiana, and elsewhere, to benefit the industry at the expense of the public (Houck 2015). After the passage of the Coastal Zone Management Act (CZMA) of 1972, I was in charge of developing Louisiana's coastal zone management program (CZM) from 1975 to 1979 and I ultimately resigned after legislation establishing the program was passed. I resigned because of oil and gas industry interference in our promulgation of regulations to implement the CZM program. Oil and gas industry lobbyists had killed the first CZM legislation we proposed in 1976. I know that because the two lead lobbyists for oil and gas approached me in the legislative committee during the hearing on our bill and told me they were going to kill our bill and substitute their bill under our title. This is a legislative trick used to kill legislation that is not in a special interest group's favor. Apparently, there is an unwritten rule in the Louisiana Legislature that if you want to kill someone's bill you must notify them shortly beforehand, in my case it was five minutes beforehand. They did just that but their legislation did little and was rejected by the federal agency (NOAA) in charge of federal CZM a year or so later. One of the oil lobbyists that warned me ultimately became the chief legal counsel for the regulatory agency overseeing oilfield wastes, an example of the indirect influence oil and gas wields in oil states like Louisiana.

Two years later (1978) we passed our original legislation, the State and Local Coastal Resources Management Act of 1978 (Act 361) and it was approved by NOAA

The original version of this chapter was revised: The author "Paul H. Templet" affiliation has been updated. The correction to this chapter is available at
https://doi.org/10.1007/978-3-030-94526-8_10

P. H. Templet (✉)
Templet Resources, Inc., Rancho de Taos, NM, USA
e-mail: ptemplet1@gmail.com

© The Author(s), under exclusive license to Springer Nature Switzerland AG 2022, corrected publication 2022
J. W. Day et al. (eds.), *Energy Production in the Mississippi River Delta*, Lecture Notes in Energy 43, https://doi.org/10.1007/978-3-030-94526-8_4

but then oil and gas interfered again. The agency head for whom I worked stated that he had made "a deal" to get favorable treatment of our bill in the Legislature and, in response, he sent the proposed regulations only to oil and gas for their modifications. The oil industry inserted language in the regulations that weakened them but they were unable to kill them or make then useless. I publicly blew the whistle on their secret activities and sent out the original regulations to many interested parties so there was a basis for comparison between the two sets of regulations before resigning in 1979. Then I left the state.

Similarly, I was the Secretary of the Louisiana Department of Environmental Quality (LDEQ) from 1988–1992 and I witnessed firsthand the role oil and gas played in attempting to downgrade environmental laws, regulations and programs. For example, a committee composed of oil and gas representatives wrote the regulations for the Louisiana Department of Natural Resources (DNR) that attempts to control their exploration and production (E&P) activities (Sect. 29B) to this day. The committee hired a consultant to draft their regulations and the regulations were based on agronomy, i.e., impacts to plants rather than impacts to humans or the environment, probably the only set of rules in the US based on plant impacts. Plants are generally more tolerant than humans to chemicals, cancer-causing agents, etc. The regulations were ultimately promulgated into law by the DNR/OC (Department of Natural Resources/Office of Conservation) with few, if any, changes. Oil and gas efforts benefited the industry financially to the detriment of Louisiana's environment by ignoring groundwater impacts, lax enforcement and lax standards.

Whenever a special interest group externalizes their costs to the public, e.g., by creating conditions that make it easier to pollute, they create a subsidy for themselves. I discuss this in more detail in this chapter. Oil and gas interfered in Louisiana's environmental efforts both directly and indirectly but before I get into that I need to set the stage about oilfield wastes and impacts.

4.2 Production Wastes in the Oil and Gas E&P Industry

Day to day activities in the oil and gas E&P industry involve the generation of numerous wastes including, but not limited to, hydrocarbons, drilling fluids, produced water containing mercury, arsenic, barium, and other metals, and other toxic substances. Produced waters are the hypersaline brines that come up from depth and almost invariably accompany oil and gas production. The brines can contain salt (NaCl) in concentrations 7–8 times that of seawater, hydrocarbons, metals, and radioactive materials (Boesch and Rabalais 1989, St. Pe 1990). It is common for these substances to be discharged or spilled onto the ground during oil and gas exploration and production and for many years produced waters were placed in leaking unlined pits. The leaking pits contaminated groundwater even though the Statewide Order No. 29-B passed in 1986 that governed the disposal of oilfield waste prohibited it (Theriot 2016). An earlier version of Rule 29 prohibited groundwater contamination as early as 1941.

It was general oil industry knowledge since at least the mid 1930's that the use of earthen pits for storage and disposal of produced water, and other oilfield wastes, was not a reliable management practice due to the high risk of contamination of soils, surface water and groundwater (Schmidt and Devine 1929, Rima et al. 1971, Silverstein 2013). In addition, it was also oil industry knowledge that alternative technologies such as subsurface injection wells and storage tanks, were available and successfully used to dispose of such wastes and would eliminate pit leakage and related contamination (Rima et al. 1971).

The E&P part of the oil industry knew of the leaking pits from reports issued by the US Bureau of Mines (Schmidt & Devine, 1929; Schmidt and Wilhelm, 1938). In addition, the American Petroleum Institute (API, the national oil and gas trade association) issued reports critical of unlined pits in 1932 (Martin, Chairman of the API Committee on Disposal of Production Wastes), and in 1942, (Elliston, H.). In 1944 an API report (API 1974) advised against the use of pits and promoted injection into the subsurface. They recommended against the pits because "These were constructed to impound the salt water and dispose of it by solar evaporation. However, in many of the fields precipitation almost equaled evaporation; therefore, this method proved of little value in water disposal." "Our records reveal that as early as 1925 salt water incident to the production of oil was injected into subsurface strata… and at this time it is now recognized as the best method of water disposal." Unfortunately, the API recommendations were not heeded in Louisiana and elsewhere for many years. These reports, and there are others that were similar, were ignored by the oil industry as they continued to use unlined pits in Louisiana that contaminated groundwater until the late 1980s.

The 1932 Martin API report is particularly straightforward on the issue of unlined pits. "We are only 'kidding' ourselves when we think we can dispose of salt water by solar evaporation from earthen ponds." "What we have attributed to evaporation was due to seepage, …but with very few exceptions it is impractical. Eventually, such seepage may either follow an impervious stratus to the surface where it may affect vegetation or may find its way to fresh water sources, either surface or subsurface, and in such quantities as to be objectionable." "…We cannot expect to successfully impound salt water without seepage, and that disposal by seepage is not as practical as methods which will confine the water to definite and known channels." Unfortunately, the Louisiana oil and gas E&P industry ignored these reports and continued to use unlined pits for produced water disposal until forced to change by state regulation in the late 1980s, and the E&P industry wrote the new regulations. The E&P legacy is extensive contamination of groundwater even though state regulation (the DNR/OC Rule 29) prohibited the contamination of 'strata water' in 1941 (Theriot 2016). No testing of groundwater was done or available to the public until "Legacy Lawsuits" and plaintiff's attorneys began groundwater testing in the mid 1990s. The vast majority of unlined pits were found to have leaked to groundwater.

Other states had begun to ban unlined pits, e.g., Kansas in the late 1930s, Texas and California in the late 1960s. Unfortunately, Louisiana regulation remained in the sphere of influence of oil and gas and was the last to regulate unlined pits. Legislation in Louisiana was killed in the 1960s that would have addressed the problem with unlined pits. It wasn't until the late 1980s that the problem was partially addressed by the regulations known in Louisiana as 29B, a clear example of one of the indirect effects of oil industry manipulation. Another indirect example is that the head of the primary regulatory agency for oil and gas is usually an oil industry executive or lobbyist. I have heard industry lobbyists tell Louisiana legislative investigating committees that pits don't leak because of extensive clays in spite of the 1929 Bureau of Mines study (1929) that said the opposite. That study pointed out that clays lose their ability to seal pits when exposed to hypersaline brines. There is no evidence that the industry looked for suitable clays when deciding where to locate their pits.

4.3 Oilfield Radiation

The radiation problem in oilfields was a new one for me. The secret that there was radiation in oilfields had not gotten out to the public in 1988 and that's the way the oil industry wanted it. Naturally occurring radioactive material (NORM) is a substance associated with oil and gas production and its toxic wastes can be spilled, concentrated or discharged during operations (Otton 2006). I promulgated a DEQ rule in 1988–89, the first in the US, that prohibited discharge into waters of the state or on the land surface. By the time the rule was implemented in early 1992, approximately 29,000 sites across Louisiana with contamination from radioactive material had been identified (Smith 2015). Radium, and other radionucleids, are brought up from deep in the earth in the hyper saline produced waters from oilfields (Neff et al. 1992). The radioactive metals tend to concentrate in pipe scales and, when cleaned by blowing compressed air and reaming the pipes, the dust becomes a health hazard to workers and the public (Smith 2015). Radioactive wastes from oil and gas field operations presented a threat to public health, safely and the environment in Louisiana and the DEQ, during my administration and over oil company opposition, promulgated the first set of rules regulating such waste in the United States (Rolling Stone 2020, https://www.rollingstone.com/politics/politics-features/oil-gas-fracking-radioactive-investigation-937389/).

4.4 Oil and Gas E&P Exemptions from Regulation

The oil and gas industry's E&P lobbyists and campaign contributions have been effective in removing them from federal regulation, another example of the indirect impacts from the industry. Industry lobbyists in 1980 got the US Congress to exempt them from the Federal solid and hazardous waste law (the Resource Conservation and Recovery Act, RCRA). This congressional action removed E&P operations from EPA authority and regulations under Ronald Reagan. Oil and gas like to call their wastes "Nonhazardous Oilfield Waste" (NOW). It is nonhazardous by law, though not necessarily in fact and differs from hazardous waste only in the manner of regulation and treatment given it. The "nonhazardous" classification was granted by congressional action and thus the waste is not treated as a hazardous waste under the RCRA and is not regulated as are other hazardous wastes. However, if various NOW wastes were to be tested a significant percentage would fit in the "hazardous" classification of RCRA because they contain a number of listed hazardous wastes (e.g. benzene, toluene and polynuclear aromatic compounds, radium, etc.) and the wastes may fail one or more of the "characteristic" tests of flammability, reactivity, corrosivity, leaching or toxicity. Thus NOW is nonhazardous because of its source, that is, because it originated in an oilfield. It must be emphasized that, because of the NOW exemption from RCRA, there is no way to determine if an oilfield waste is hazardous merely by analyzing it and applying the standard tests. One must also know, and trust, the source. If the oilfield operation from which the waste originated mixed NOW wastes with hazardous waste, such as spilled solvents and cleaning agents, then the entire waste would be classified hazardous under RCRA but the treatment facility receiving the waste or the public would not know that.

The congressional exemption of E&P wastes from RCRA provides a telling example of oil and gas influence over regulatory and lawmaking bodies, even over the Congress. Imagine their influence over state and local agencies and lawmakers. The oil and gas industry has been effective in delaying and preventing regulation. For example, Wascom (1990—Louisiana Assistant Director Injection and Mining Division, Office of Conservation (OC), Department of Natural Resources (DNR)) noted "Oil and gas industry lobbying efforts resulted in the US RCRA amendments of 1980 which exempted most oil and gas wastes from the hazardous waste requirements of subtitle C until the outcome of further study by EPA." In a 1989 report on closure and disposal practices in Louisiana under DNR/OC the authors (BP Exploration and consultants) noted (Wells, et al., 1989, p. 2):

> EPA's recommendations were sent in a final Report to Congress in June 1988. Two early drafts of the report had been prepared. The first draft argued in favor of regulating oilfield E&P wastes as hazardous waste under RCRA. The American Petroleum Institute (API) calculated the nationwide cost of compliance with such a program to be over $40 billion. The second report, following widespread industry opposition, represented a considerable retreat from the earlier position, but left the door open to classification of some oilfield wastes as hazardous. The current exemption is defined in the July 6, 1988 Federal Register (53 FR 25446-25459) and represents a compromise between EPA and the American Petroleum Institute (API).

The first draft report mentioned above to regulate E&P wastes under RCRA was from professional EPA staff and the "second report", favorable to the industry position of no federal regulation of E&P wastes and malleable state regulation, was from the EPA political appointees under president Ronald Reagan.

4.5 Externalities and Subsidies

As bad as the direct impacts of E&P operations on Louisiana's environment was/are, these impacts are dwarfed by the indirect impacts on the people and government. However, indirect impacts are difficult to measure. The environmental impacts, and others, created by the oil industry's E&P activities can be measured as the industry externalizes its wastes and other costs in energy and taxes. An externality is a cost imposed by the industry on the public, such as pollution effects on health. The externality also generates a subsidy that is reaped by the industry. For example, an industry can minimize the amount they spend on pollution control by preventing good environmental regulations with the use of campaign contributions and having lobbyists influence state and federal legislatures and agencies. I gave the example above about how the oil industry escaped US regulation under RCRA and the EPA using their political and financial clout, and about how the Louisiana oil and gas industry weakened Louisiana's Coastal Management program and regulations.

In an earlier publication (Templet, 2001) I calculated the subsidies across the 50 states reaped by industry in the 1990s, relative to the US average, when they externalized costs on pollution, energy and taxes. I am not talking about the direct subsidies states grant to corporations to attract them and their jobs, I'm addressing the hidden or indirect subsidies that few know about. The pollution subsidy is defined as the costs that manufacturers avoid by spending less than the national average per pound of toxic pollution released in a state. The pollution subsidy thus calculated amounted to approximately $410/capita annually in Louisiana in the late 1990s, the highest in the US, and correlates well with poorer economic performance in terms of poverty, income disparity, unemployment and personal income. The total subsidy, including the energy and the tax subsidy is in the $900 per person/year range. I found that states with high subsidies have significantly higher pollution levels, use more energy less efficiently, are poorer socio-economically and have weaker prospects for economic development. As industry externalizes more of its pollution costs the public becomes economically poorer, to say nothing of being more polluted with the associated health costs. Industry does better, as an industry externalizes its costs it improves its profit margin, its stock price and its dividends. But people on the lower end of the income scale don't share in these improvements so income, poverty and income disparity increases.

The energy subsidy, calculated from the disparity in energy costs for industry and the public, shows similar effects. The tax subsidy is calculated from the disparity between progressive and regressive taxes and shows similar effects. I put a tax subsidy program in place in Louisiana that linked pollution levels with state tax subsidies (Farber, et al, 1995). The higher the releases the lower the tax subsidy. It lasted only one year because industry opposition and a pliable governor killed it, another indirect impact. States with relatively high hidden subsidies tend to be in the South and to a lesser extent the West. The Northeast and Midwest tend to grant fewer hidden subsidies and are better off because of it.

4.6 Hidden Subsidies and Political Power

A troubling result of high subsidies it that it correlates to low voting rates across all 50 states. An active voting public is critical for a democratic society and is one indicator of the extent of social capital. Among states with above-average total subsidies, participation in federal elections is on average 15 percent lower than the U.S. average. Citizens in high-subsidy states may well feel disenfranchised, perceiving that their elected representatives cater to special interests. They may doubt that voting will change anything. Yet low participation itself contributes to the further concentration of power to corporations in a dismal spiral of decline in high subsidy states.

Those receiving hidden subsidies can use their additional financial capital in a number of ways. One obvious way is to spend more on campaign contributions of friendly candidates, and to hire more lobbyists to protect and augment the subsidies. Industrial corporations are major contributors to political campaigns and hire armies of lobbyists at the federal and state levels. In making contributions, special interests not only help to elect representatives who serve their interests, but likely also influence choices for appointed positions.

In my experience as DEQ Secretary in Louisiana's state government, I found that the quality of public leadership declines as special interests increase their sway. Even federally funded programs tend to languish. State agencies become less responsive to citizens, who in turn withdraw from the political process. The state becomes a less attractive place to live and do business. Being known as the 'northernmost banana republic' or 'cancer alley' is not a good thing for a states' reputation. The end result is institutional failure, a loss of trust in government and the loss of social capital.

Again and again, decision makers hear the refrain that higher environmental standards will drive away firms and jobs and result in lower public welfare. Yet much empirical research supports my own findings to the contrary. For example, Meyer (1992) found that worker productivity, employment, and growth rates are lower in states with poorer environmental rankings. Cannon (1993) likewise found that economic growth rates were lower in states with poorer environmental records. The fact that environmental protection and economic welfare go hand in hand must be impressed upon the public and their elected representatives.

4.7 Leakage of Wealth from States

Large companies are the biggest beneficiaries of subsidies; they use the most energy and other resources, discharge the most waste, and have the largest incomes and property holdings. If all of the profits generated by subsidies were to remain in the state, the recycling of the extra income might counter at least some of the harm that hidden subsidies inflict on public welfare and the political system within a state. Many of the profits are exported, however, to shareholders and corporate managers living in other states. Value added per job by state - a surrogate measure for corporate profits - is positively and significantly related to the share of gross state product that leaves the state (Templet, 1995). The greater the subsidies, the greater the profits, and the greater the rate of leakage. Louisiana again offers a case in point. It has the highest subsidies in the country as well as the highest value added per manufacturing job. Yet only two-thirds of the annual gross state product (GSP) accrues as income within the state (U.S. Bureau of the Census 1997). The remainder, roughly $5,000 per person annually, is exported to other states, or even to other countries since some subsidy-seeking firms are foreign-owned. If Louisiana were instead to retain this wealth, the state's per-capita income would be close to the U.S. average.

Leakage is a major source of income disparities among states. It drains income from states that use the most resources and generate the most pollution, relying heavily on such industries as mining and manufacturing. These states tend to be poorer than average. The hidden subsidy enhances profits that leak to richer states that tend to pay smaller subsidies and to export less income. As leakage declines, income per capita rises; as income per capita declines, leakage rises. A number of the richer states import net income; that is, their total income exceeds their state product. As income leaks from a state, there are rises in unemployment, poverty, and pollution. The states that import net income also benefit from products resulting from mining and manufacturing yet the costs remain behind in the producing states. The situation is analogous to colonialism, in which the mother country draws resources and other wealth from the colony, proffering little compensation in return. In this respect, the United States displays a kind of internal colonialism.

4.8 Summary

As oil and gas E&P secures weak environmental laws and regulations for itself by manipulating government in Louisiana, and elsewhere, it also secures advantages in its income, influence and profits. That's the motivation for their actions. Unfortunately, their actions corrupt governments and reduce citizens environmental protection as they seek to gain more advantages, and to keep them, leading to a downward spiral that negatively affects many indicators of the economic and social health of a state. Some of these indicators are environmental rankings, income/capita, institutional failure, social capital, low voting rates and leakage of state wealth. There are undoubtedly others.

References

API Drilling and Production Practice (1942) HH. Elliston, published by the central committee on drilling and production practice (published 1943) (ind1669). Sinclair-Prairie Oil Co. Tulsa, OK

API, 1944, by H.H. Elliston and W.B. Davis, A Method of Handling Salt-Water Disposal, including Treatment of water, in Drilling and Production Practice, sponsored by the Central Committee on Drilling and Production Practice, Div. Of Production, API (sit0319).

API (1974) Recommended land drilling operating practices for protection of the environment, API RP 52, Feb. 1975, 1st edn. API Washington, DC. Issued by API Prod. Dept., Dallas, TX. And Oct. 74, reissued May 82 (ind1576)

Boesch DF, Rabalais NN (eds) (1989) Produced waters in sensitive coastal habitats: an analysis of impacts, Central Coastal Gulf of Mexico. In: OCS Report/MMS 89-0031, U.S. Department of the interior, minerals management service, Gulf of Mexico OCS regional office, New Orleans, Louisiana, p 157

Boyce JK, Klemer AR, Templet PH, Willis CE (1999) Power distribution, the environment, and public health: a state level analysis. Ecol Econ 29:127–140

Costanza R, D'Arge R, de Groot R, Farber S, Grasso M, Hannon B, Naeem S, Limburg K, Paruelo J, O'Neill RV, Raskin R, Sutton P, van den Belt M (1997) The value of world's ecosystem services and natural capital. Nature 387:253–260

Cannon F (1993) Economic growth and the environment. In: economic and business outlook. Bank of America, Economics-Policy Research Department, San Francisco

Farber S, Moreau R, Templet PH (1995) A tax incentive tool for environmental management: an environmental scorecard. Ecol Econ 12:183–189

Houck OA (2015) The reckoning: oil and gas development in the Louisiana Coastal Zone. Tulane Environ Law J 28:185–296

Martin V (1932) Disposal of production division wastes. In: For presentation at Panhandle chapter meeting of division of production. Martin was Chair of the API committee on disposal of production wastes (sit0320)

Meyer SM (1992) Environmentalism and economic prosperity: testing the environmental impact hypothesis. In: Working paper series no. 1. Project on Environmental Politics and Policy, Massachusetts Institute of Technology, Cambridge, Mass

Otton JK (2006) Environmental aspects of produced-water salt releases in onshore and estuarine petroleum-producing areas of the United States—a bibliography: U.S. Geological survey open-File report 2006-1154, p 223

Neff JM, Sauer TC, Maciolek N (1992) Composition, fate and effects of produced water discharges to nearshore marine waters. In Ray JP, Engelhardt FR (eds) Produced water. environmental science research, vol 46. Springer, Boston, MA. https://doi.org/10.1007/978-1-4615-2902-6_30

Rima DR, Chase EB, Myers BM (1971) Subsurface waste disposal by means of wells—a selective annotated bibliography. In: Geological Survey Water Supply Paper 2020. U.S. Department of the Interior, U.S. Government Printing Office, Washington, DC, p 309

Rolling Stone (2020) Justin Nobel author. https://www.rollingstone.com/politics/politics-features/oil-gas-fracking-radioactive-investigation-937389/

Schmidt L, Devine J (1929) The disposal of oil field brines. Department of Interior, Bureau of Mines (ind1560)

Silverstein K (2013) Secret oil company memos on pollution in Louisiana. HARPER'SMAG. (Oct. 11, 2013, 8:00 AM), http://harpers.org/blog/2013/10/secret-oil-company-memos-onpollution-in-louisiana/

St. Pe KM (1990) An assessment of produced water impacts to low-energy, brackish water systems in Southeast Louisiana. In: Louisiana Department of Environmental Quality, Water Pollution Control Division, Baton Rouge, Louisiana, p 204

Schmidt L, Wilhelm C (1938) Disposal of petroleum wastes on oil producing properties. US Dept. of interior, Bureau of Mines (sit0323)

Smith S (2015) Crude justice: how i fought big oil and won, and what you should know about the new environmental attack on America. Benbella Books, Inc., Dallas, Texas, p 252

Templet PH (1995) Grazing the commons; externalities, subsidies and economic development. Ecolog Econ 12:141–159

Templet PH (2003) Defending the public domain: pollution, subsidies, and poverty. In Boyce JK, Shelley BG (eds) Natural assets; democritizing environmental ownership. Island Press, Washington, DC20009

Templet PH (2001) Energy price disparity and public welfare. Ecol Econ 36:443–460

Theriot JP (2016) Oilfield battleground: Louisiana's legacy lawsuits in historical perspective. La Hist: J La Hist Assoc 57:403–462

Wascom CD (1990) A regulatory history of commercial oilfield waste disposal. In: Proceedings the state of La. 1st international symposium on oil and gas exploration waste management practices. New Orleans. Sponsored by EPA, p 821

Wells SK, Strandberg BM, Evangelisti R (BP Exploration) 11/1/89, closure and disposal practices under La. Statewide Order 29-B. American Filtration Society National Meeting, Houston, TX

Young CA (1932) Letter. (API Production waste issue group secretary) to R.W. Taylor (Simms oil company) re Martin's paper disposal of production wastes and response from Martin to Young (brousa03082–03085)

Chapter 5
Impacts of Oil and Gas Activity in the Mississippi River Delta

John W. Day, Rachael G. Hunter, and H. C. Clark

5.1 Introduction

There are three general types of impacts of oil and gas production including (1) alteration of surface hydrology, (2) induced subsidence due to fluids withdrawal (oil, gas, and produced water), and (3) toxic impacts of produced water and spilled oil. In addition, abandoned infrastructure (e.g., platforms, well heads, pipes) can cause problems such as leaking toxins and navigation hazards. In this chapter, we discuss alteration of surface hydrology and induced subsidence. In the following chapter, toxic impacts are addressed.

Direct impacts from oil and gas exploration and production on surface hydrology and ecosystems include soil and vegetation removal during canal dredging and vegetation burial from the placement of spoil material alongside canals (Stone et al. 1978; Johnston et al. 2009). Gagliano (1973) studied wetland loss in coastal Louisiana between 1936 and 1959 and concluded that direct loss from canal dredging accounted for 39% of wetland loss. If wetland loss from spoil bank construction was included in that estimate, Craig et al. (1979) concluded that total direct loss would be much greater (69% of wetland loss).

The original version of this chapter was revised: The author "Rachael G. Hunter's" affiliation has been updated. The correction to this chapter is available at
https://doi.org/10.1007/978-3-030-94526-8_10

J. W. Day (✉)
Department of Oceanography and Coastal Science, Louisiana State University, 2005 Olive St, Baton Rouge, LA 70806-6660, USA
e-mail: Johnday@lsu.edu; johnwday@bellsouth.net

R. G. Hunter
Comite Resources, Inc., Baton Rouge, LA, USA

H. C. Clark
Department of Earth, Environmental and Planetary Sciences, Rice University, Houston, TX 77005, USA

© The Author(s), under exclusive license to Springer Nature Switzerland AG 2022,
corrected publication 2022
J. W. Day et al. (eds.), *Energy Production in the Mississippi River Delta*,
Lecture Notes in Energy 43, https://doi.org/10.1007/978-3-030-94526-8_5

In addition to direct loss, oil and gas activities cause indirect wetland loss through alterations in surface hydrology, induced subsidence, and production and disposal of produced water and spills of various kinds. Scaife et al. (1983) studied the impacts of oil and gas canals on coastal wetlands in southeastern Louisiana and southern Mississippi and concluded that the indirect effects of canals caused more land loss than the direct conversion of land to open water. They determined that from 1955 to 1978, direct land loss from canal dredging was less than 10% of the actual land loss caused by canal construction with the other 90% of loss caused by indirect impacts. Other scientists have concluded that the indirect impacts of canal dredging may be double or more than that of the direct impacts (McGinnis et al. 1972; Gagliano 1973; Deegan et al. 1984; Turner and Cahoon 1987; Bass and Turner 1997).

Day et al. (2000) reported that wetland loss due to canals varied by hydrologic basin and ranged from 72% in the Barataria Basin to an inverse relationship in the Atchafalaya region where high riverine input many impacts of oil and gas activities on surface hydrology and ecosystem processes. Other researchers have estimated that direct impacts of oil and gas activities have caused between 9 and 69% of wetland loss and indirect impacts have caused between 22 and 80% of wetland loss (Table 5.1). Turner (1997) and Turner and McClenachan (2018) concluded that nearly all wetland loss in the coastal zone was due to canal impacts.

Wetland loss in coastal Louisiana oil and gas fields typically extends from 3 to 150 m on either side of dredged canals, causing between 6 and 31 ha of wetland loss for every 1.0 km of canal constructed (Olea and Coleman 2014). Bass and Turner (1997) measured the relationship between salt marsh loss and dredged canals in 27 salt marshes in three Louisiana estuaries (Barataria, Terrebonne, and St. Bernard) and found that for every 1 ha of canal dredged there was 2.85 ha of open water formed. Scaife et al. (1983) studied Louisiana coastal land loss in major delta complexes and found a significant correlation between percent annual land loss and canal density between 1955 and 1978. Annual land loss rates in areas without canals were 0.8% between 1955 and 1978 while areas with canals had annual land loss rates ranging between 0 and 40%, depending upon canal density (Scaife et al. 1983). However, conclusions that canals are responsible for practically all wetland loss (Turner 1997; Turner and McGlenachan 2018) are not correct because of methodological issues and the complex way that wetland loss occurs (Day et al. 2000, 2019, 2020).

5.2 Alteration of Surface Hydrology

5.2.1 Canal Dredging and Spoil Placement

Oil and gas canals are typically dredged to a depth of 2.5 m (7–8 feet), a width of 20–40 m (60–130 feet), and a length of 100–1000 m (about 300–3000 feet) or more (Turner et al. 1994). When canals are dredged, wetland soil is removed to form the canal and placed alongside the canal to create spoil banks that smothers

5 Impacts of Oil and Gas Activity in the Mississippi River Delta

Table 5.1 Estimates of the percentage of wetland loss caused by all oil and gas activities (overall) or by direct or indirect impacts of oil and gas activities

Location	Time period	Overall impacts (% wetland loss)	Direct impacts (% wetland loss)	Indirect impacts (% wetland loss)	Source
Coastal Louisiana	1931–1967		69		Gagliano and van Beek (1970)
Coastal Louisiana		45			Gagliano (1973)
Coastal Louisiana			10–69		Craig et al. (1979)
Coastal Louisiana			33–67		Deegan et al. (1979)
Coastal Louisiana			10	Up to 80	Turner et al. (1982)
Coastal Louisiana	1955–1978		10	48–97	Scaife et al. (1983)
Mississippi River Deltaic Plain	1955/1956–1978		25–39		Deegan et al. (1984)
Coastal Louisiana	1932–1990	Majority of losses, >90%			Turner (1987, 1997) Turner and McClenachan (2018)
Coastal Louisiana	1955–1978	20–60	14–16	20–60	Turner and Cahoon (1987)
Coastal Louisiana	1955–1978		16		Turner (1990)
Coastal Louisiana	1955–1978		6.6		Baumann and Turner (1990)
Coastal Louisiana	1955–1978		16	30–59	Boesch et al. (1994)
Breton Sound	1933–1990	68			Day et al. (2000)
Barataria Basin	1933–1990	72			Day et al. (2000)
Mermentau Basin	1933–1990	35			Day et al. (2000)
Mississippi River Deltaic Plain	1930–1990	33.2	10.8	22.4	Penland et al. (2001a, b)

vegetation and impacts hydrology and sediment input into adjacent wetlands (Davis 1973; Turner and Cahoon 1987; Gagliano and Wicker 1989; Reed 1995; Day et al. 1999; Ko and Day 2004; Ko et al. 2004). Most often spoil is placed on both sides of the canal with spoil deposition generally placed away from the canal so that there is initially a strip of marsh between the canal and the spoil bank. The construction of these canals causes direct loss of wetlands through vegetation and soil removal during canal dredging and through burial when dredged spoil is placed on top of wetland vegetation alongside canals. Over time, oil and gas canals widen as a result of spoil bank undercutting, erosion, and collapse and cause additional wetland loss (Davis 1973; Doiron and Whitehurst 1974). Turner and Cahoon (1987) found a direct relationship between the percentage of coastal wetland loss in Louisiana and the percentage of wetland loss through canal dredging and spoil bank placement.

Spoil banks alter natural hydrology by blocking overland sheet flow and causing impoundment. Natural levees adjacent to water bodies average up to about 0.5 m high and 2–10 m wide while a typical spoil bank ranges between 2 to 4 m high and 15 to 40 m wide (Bass and Turner 1997; Fig. 5.1). The side of the spoil bank facing the canal is typically steep and unvegetated while the remaining vegetated portion of the spoil is colonized by species such as *Iva frutescens* and *Batis* sp. Because of the bank heights, plant roots often do not penetrate to the water level and thus do little to prevent erosion, as is the case in a natural wetland channel (Doiron and Whitehurst 1974). Monte (1978) studied spoil bank vegetation and size at numerous canals in

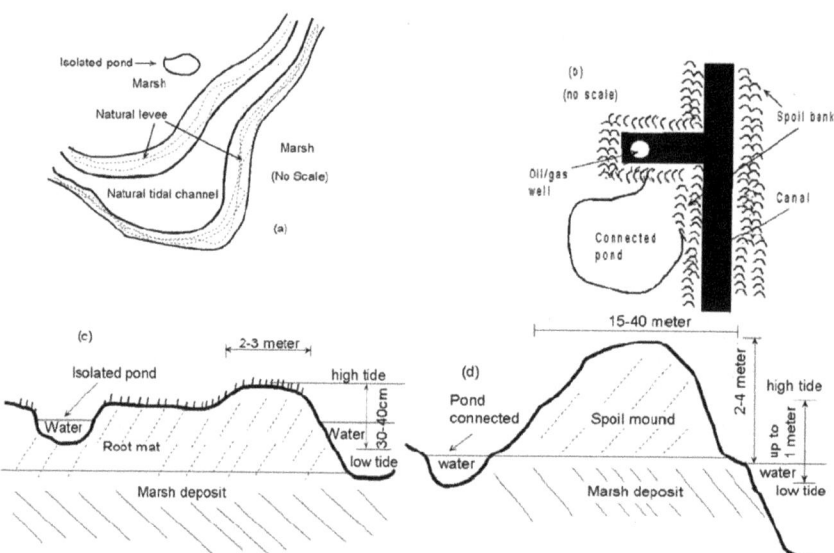

Fig. 5.1 Schematic diagrams of the impact of canal dredging on coastal marshes: **a** top view showing a natural tidal channel and natural levee; **b** top view of canal dredging showing straight canal and spoil bank; **c** cross-sectional view of natural marsh; **d** cross-sectional view of a dredged canal (from Ko et al. 2004)

the Bayou LaFourche Delta and found that spoil bank ridges were, on average, 2.86 times wider than canals (200 ft vs. 70 ft) and the total amount of wetlands altered by spoil bank placement was 5 1/2 times the width of the originally dredged canal.

Localized subsidence along pipeline canals can also occur along the wetland flanks of spoil banks when the weight of the spoil depresses the surface of the marsh and leads to the formation of linear ponds behind the spoil banks. This is due to subsidence from the weight of the spoil, the trapping of overland flow of water at the base of the spoil banks and the blocking of sediment input due to the spoil banks (CEI 2017; Day et al. 2020). This type of wetland loss can be seen along the Tennessee Natural Gas pipeline canal in the Breton Sound basin (Figs. 5.2 and 5.3).

Fig. 5.2 Linear pond formation behind spoil banks along the Tennessee Natural Gas pipeline, Breton Sound, Louisiana (photo by senior author)

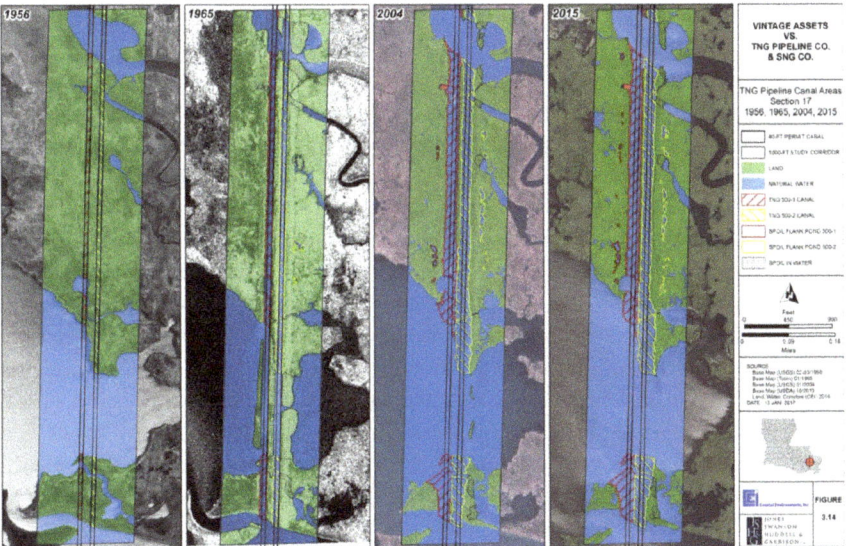

Fig. 5.3 Widening of the Tennessee Natural Gas pipeline canals and formation of spoil flank ponds (CEI 2017)

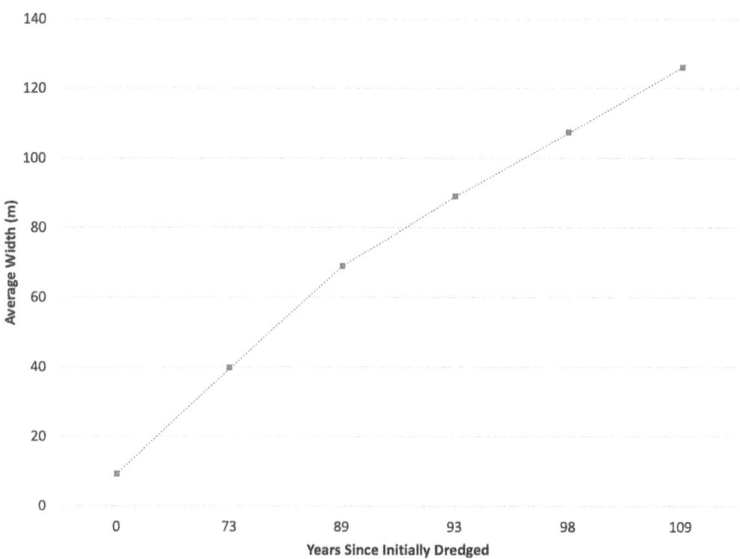

Fig. 5.4 Relationship between canal width and age in the Southwestern Louisiana Canal, LaFourche Parish, LA (data from Johnson and Gosselink 1982 and Google Earth)

Construction of canals, which are typically straight channels, increases tidal flows in coastal marshes that can cause erosion and lead to canal widening over time (Fig. 5.4; Davis 1973; Doiron and Whitehurst 1974; Johnson and Gosselink 1982; Ko and Day 2004; Ko et al. 2004). Increases in canal widths range between 2 and 14% per year, for a doubling time of 5–60 years (Craig et al. 1979). Johnson and Gosselink (1982) examined canal widening at the Leeville oil field in coastal Louisiana and found that canals in areas of greatest boat activity widened at a rate of 2.58 m yr^{-1}, while those in areas of minimal boat activity widened at a rate of 0.95 m yr^{-1}. A review of canal widening rates measured in navigation and oil and gas canals throughout coastal Louisiana between 1926 and 1998 (Supplementary Material Table 1) found that the average widening rate was 6.1% per year. The amount and type of boat traffic greatly influences the rate of widening (Davis 1973). As a canal becomes wider and deeper, a greater volume of water can flow through it, which increases erosion rates over time (Doiron and Whitehurst 1974), so canal width in excess of original cut dimensions is directly correlated with age. Johnson and Gosselink (1982) also suggested that canal widening rates were greatly increased once spoil banks were eroded away. Once dredged, spoil banks are not permanent features and have a life cycle of their own. They disappear over time due to compaction, subsidence, sea-level rise, and erosion. Spoil banks protect remnant marsh from wave erosion and as spoil banks disappear remnant marsh can be lost due to wave attack, leading to further wetlands loss. In Fig. 5.5, subaerial land that disappeared between the two mapping dates (center panel in Fig. 5.5) in the Leeville Field includes both spoil banks and marsh.

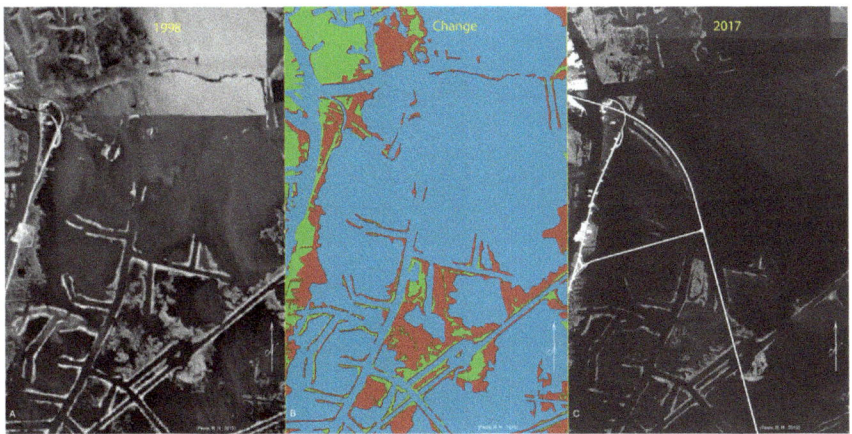

Fig. 5.5 Spoil banks along oil and gas canals near Leeville, Louisiana in 1989 and in 2015 (From Day et al. 2019). Red in the central panel shows loss between 1989 and 2015

Over time as the interior, impounded marshes disappear, often the only features remaining in an oil and gas field are the spoil banks and remanent marshes. But as they compact and subside, these features disappear as well (Fig. 5.5).

As stated previously, most oil and gas canals are deep and straight channels while natural waterways in a marsh are primarily shallow and sinuous tidal channels (Ko and Day 2004). Swenson and Turner (1987) and Craig et al. (1979) determined that as canal density increased, the density of natural channels decreased because dredged canals tend to preferentially capture water flow from natural channels. This process, termed 'channel theft', disrupts hydrology within the affected marsh and may increase the rate of sheet flow drainage, saltwater intrusion, and sediment export because deep, straight canals transport water more efficiently than natural shallow and sinuous channels (Johnston et al. 2009). Oil and gas canals can also connect water bodies that previously had no hydrologic connection or interfere with existing waterways through construction of an intersecting canal network (Johnston et al. 2009).

Land loss associated with canals in coastal Louisiana typically extends from 3 to 150 m on either side of pipeline canals, causing 25–125 acres of wetland loss for every mile of canal construction (Olea and Coleman 2014). Bass and Turner (1997) measured the relationship between salt marsh loss and dredged canals in 27 salt marshes in three Louisiana estuaries (Barataria, Terrebonne, and St. Bernard) and found that for every 1 ha of canal dredged there was 2.85 ha of open water formed and an additional 1 ha of wetland converted to spoil bank vegetation. Scaife et al. (1983) studied Louisiana coastal land loss in major delta complexes and found a significant correlation between percent annual land loss and canal density between 1955 and 1978. Annual land loss rates in areas without canals was 0.8% between 1955 and 1978 while areas with canals had annual land loss rates up to 40%, depending upon canal density (Scaife et al. 1983).

Fig. 5.6 Interior marsh with spoil bank in the background (upper left, lower left) and spoil bank with woody vegetation (upper right, lower right)

Although placement of spoil material onto the wetland surface causes direct loss of vegetation through burial, the elevated spoil banks create a new, human-made terrestrial habitat that is generally colonized with upland vegetation that is different from the adjacent wetlands (Fig. 5.6; Monte 1978). In coastal Louisiana, small changes in elevation often define the difference between two vegetation species and the species that occur along the width of the spoil bank are related to elevation differences. Monte (1978) studied succession of vegetation on spoil banks of different ages in areas with varying salinities along Bayou LaFourche using the line-transect method. She found that once a new spoil bank had dried and compacted, invasion of new species began. Plant succession on spoil banks went through several stages including grass, grass/herb, shrub, shrub/tree, and tree, ultimately resulting in predominantly terrestrial forest climax vegetation. After 30 years of plant succession, 71–81% of the spoil bank plants were terrestrial in nature (Monte 1978).

After spoil is placed on the marsh surface there is a significant loss of soil volume due to consolidation and compaction and loss of organic matter due to oxidation in the aerobic soils. Thus, backfilling in almost all cases cannot completely refill a dredged canal.

Because spoil banks are significantly higher in elevation than surrounding wetlands, they impound or semi-impound adjacent coastal marshes. Impoundment reduces or eliminates surface and subsurface water flow, tidal influence, and exchange of materials (e.g., nutrients, sediments, organisms) between the impounded areas and the surrounding marsh (Turner 1987; Ko et al. 2004; Johnston et al. 2009). Impoundment also traps salt water that overtops spoil banks during high water events or by entry via existing channels, increases length of inundation, decreases drainage (both surface and subsurface) and accretion, reduces vegetation productivity, and increases vegetation mortality and conversion to open water (Davis 1973; Swenson and Turner 1987; Turner and Cahoon 1987; Mendelssohn and McKee 1988; Day et al. 1990, 2020; Taylor et al. 1989; Cahoon and Groat 1990; Turner and Rao 1990; Reed 1992, 1995; Reed and Cahoon 1992; Rogers et al. 1992; Boumans and Day 1994; Cahoon 1994; Asano 1995; Gascuel-Odux et al. 1996; Reed et al. 1997; Ko and Day 2004; Reed and Wilson 2004; Mitsch and Gosselink 2015; Olea and Coleman 2014). Increased inundation can also lead to sulfide toxicity, anoxia, and salt stress that will negatively impact vegetation (Mendelssohn and McKee 1988; McKee and Mendelssohn 1989; Mendelssohn and Morris 2000).

The foregoing discussion generally addressed non-floating marshes growing on a firm substrate. In rooted marshes, overland flow leads to the exchange of water, sediment, nutrients and other constituents over the surface of the marsh. In floating marshes, there is little flooding over the surface of the marsh because the marshes move up and down with fluctuating water levels. In floating marshes, most water exchange is below the floating mat. Subsurface water exchange is extremely important for floating marshes. Previous discussion showed how important this sub-surface water exchange is. In general, impoundment impacts water exchange in a manner generally similar to attached marshes, except that subsurface water exchange is impacted. The following discussion compares spoil bank impacts in attached versus floating marshes.

Turner and Swenson (2020) described the subsidence and demise of spoil banks deposited from dredging canals in an attached Mississippi delta salt marsh over a 30 year period. They showed that the spoil bank height decreased by 1.9 cm yr^{-1} (Fig. 5.7) and that the weight of spoil bank depressed the original marsh surface by a similar amount (Fig. 5.8). The bulk density of the soils ranged from 0.14 to 0.27 g cm^{-3}, which is considerably higher than soils of floating marshes (generally less than 0.1 g cm^{-3}, refs).

Deposition of spoil along canals dredged in floating marsh depress the soils below the bank so that most horizontal flow of water below the mat is interrupted. Thus, the canal network in floating marshes act to impound the marsh and disrupt water flow in a manner similar to attached marshes, except that it occurs below the marsh rather that as overland flow.

In attached marshes, there is a natural streamside levee that is about 10 cm higher than interior marshes. Natural tidal channels bordering the streamside marshes are generally up to a meter deep and tens of meters wide. Tidal channels in interior marshes are smaller and a few tens of cm deep. At high tide, water exchange occurs over the surface of the marsh as sheet flow. In a natural state, floating marshes occur in broad expanses with little tidal channel development. The floating mat or floatant is made up of a surface of living roots with a peat layer attached to the living root mat. Water exchange is mostly subsurface below the mat. Spoil banks impact attached and floating marshes in different ways because of the differences in hydrology and structure (Fig. 5.9).

Placement of the spoil on an attached marsh results in a high spoil bank generally 2–3 m in elevation and the marsh is compressed below the spoil bank. The spoil bank results in an elimination of most surface water exchange. Only during strong

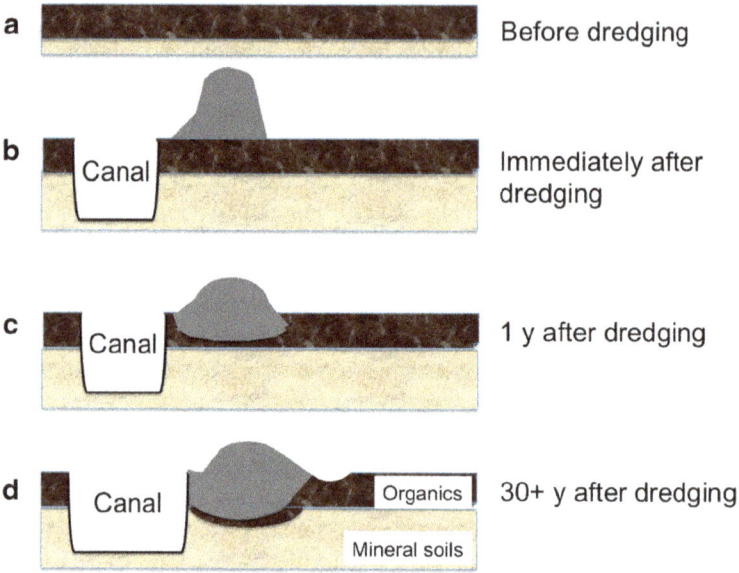

Fig. 5.7 A schematic summary of spoil bank morphology before dredging (**a**), within weeks after dredging (**b**), and then 1 year (**c**) and 30 years later (**d**). Not shown: the spoil bank trees and shrubs on the spoil bank whither, creeks cross through small openings and widen, the marsh becomes open water, and the spoil bank disappears (From Turner and Swenson 2020)

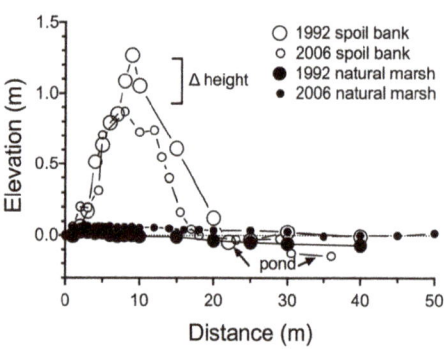

Fig. 5.8 An elevation profile of the natural marsh and the marsh with the spoil bank in 1993 and 2006 (from Turner and Swenson 2020)

climatic events such as hurricanes are spoil banks overtopped. By contrast in floating marshes, placement of spoil results in a rapid compression of the floatant on which the spoil is placed. This eliminates subsurface water exchange. In addition, because much of the spoil is organic floating marsh with very low bulk density, there is little surface expression of the spoil bank with elevations generally less than 20–30 cm.

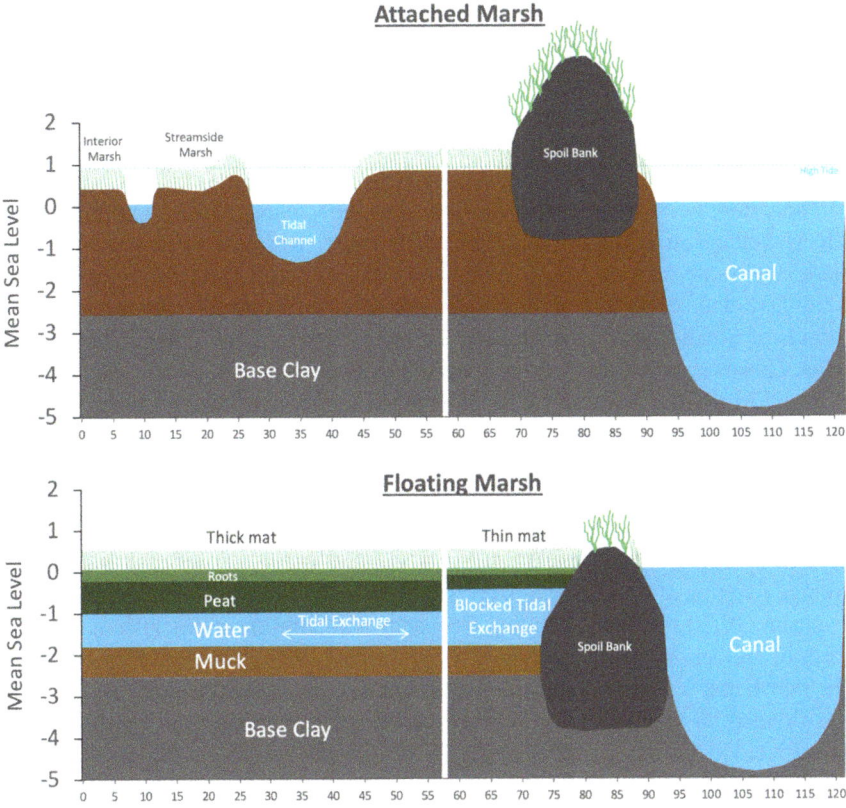

Fig. 5.9 Impacts of spoil banks on attached and floating marshes

5.2.2 Impoundment

Louisiana coastal wetlands are strongly dependent on water exchange between wetlands and water bodies to remain healthy and sustainable. For attached marshes, overland sheet flow is important for inputs of water, sediments, and nutrients, export of organic matter, and soil drainage to maintain wetland health. Because spoil banks are significantly higher in elevation than surrounding wetlands, they impound or semi-impound adjacent coastal marshes. Impoundment reduces or eliminates surface water flow, tidal influence, and exchange of materials (e.g., nutrients, sediments, organisms) between the impounded areas and the surrounding marsh (Turner 1987; Ko et al. 2004; Johnston et al. 2009). Impoundment also traps salt water that overtops spoil banks during storms, increases duration of inundation, decreases drainage and accretion, reduces vegetation productivity, and increases vegetation mortality and conversion to open water (Davis 1973; Swenson and Turner 1987; Turner and Cahoon 1987; Mendelssohn and McKee 1988; Day et al. 1990, 2020; Taylor et al. 1989; Cahoon and Groat 1990; Turner and Rao 1990; Reed 1992, 1995; Reed and

Cahoon 1992; Rogers et al. 1992; Boumans and Day 1994; Cahoon 1994; Asano 1995; Gascuel-Odux et al. 1996; Reed et al. 1997; Ko and Day 2004; Reed and Wilson 2004; Mitsch and Gosselink 2015; Olea and Coleman 2014). Increased inundation can also lead to sulfide toxicity, anoxia, and salt stress that negatively impacts vegetation (Mendelssohn and McKee 1988; McKee and Mendelssohn 1989; Mendelssohn and Morris 2000).

There have been numerous studies that show how managed marsh impoundment affects wetland structure and function and, because spoil banks impound wetlands, these studies are relevant to how oil and gas activities affect wetlands (Ko and Day 2004; Day et al. 2019, 2020). Impounded and semi-impounded wetlands have fewer periods of flooding and reduced water exchange compared to unimpounded marsh areas. However, when water does enter impounded areas (e.g., during the passage of a cold front), the wetlands are flooded much longer than unimpounded wetlands because water is essentially trapped by the spoil banks and water drains from the area more slowly than in unimpounded areas (Swenson 1983; Swenson and Turner 1987; Boumans and Day 1994; Cahoon 1994; Day et al. 2000). Swenson and Turner (1987) found that semi-impounded marshes had fewer flooding events compared to marshes without hydrologic alterations (4.5 events vs. 12.9 events per month), but that the average duration of flooding was significantly longer (149.9 h vs. 29.7 h per month). Swenson (1983) measured flooding frequency and duration at brackish marsh sites north of Golden Meadow, Louisiana, from August 1982 through September 1983. The marsh that was semi-impounded due to placement of spoil from dredged canals was flooded an additional 141 h more per month compared to a marsh that had unrestricted hydrology. Snedden et al. (2015) studied the impact of flooding on health and productivity of *Spartina patens* and *Spartina alterniflora* in Breton Sound using marsh organs (Morris et al. 2013), including areas of the Gentilly oil field. They found that increased flooding duration reduced above- and belowground wetland plant growth. Thus, oil and gas activities increase the duration of flooding due to lowered suspended sediment input, increased subsidence rates, and entrapment of water by spoil banks.

As part of a study of structural marsh management, Boumans and Day (1994) and Cahoon (1994) studied semi-impounded and natural sites at the Rockefeller State Wildlife Refuge and at the Fina-La Terre marsh management area. Structural marsh management is used to control water levels and reduce water exchange by using spoil banks and water control structures. Boumans and Day (1994) measured water level variability during three 48 h periods in the canal that supplied water to two sites and in natural and managed marshes at the refuge (Fig. 5.10). The tide range in the canal for the three measurement periods was 60–70 cm. The tide range over the 48-h measurement period was between 25 and 30 cm in the natural marsh while water level variability in the impounded marsh was less than 5 cm. Water flux was dramatically different for semi-impounded and natural marshes with a free tidal connection (Fig. 5.10).

Boumans and Day (1994) also quantified materials exchange (e.g., nitrogen, sediment, salt) at impounded and unimpounded marshes at Rockefeller Refuge and Fina la Terre, Louisiana. The impoundment of marshes at both Fina and Rockefeller led

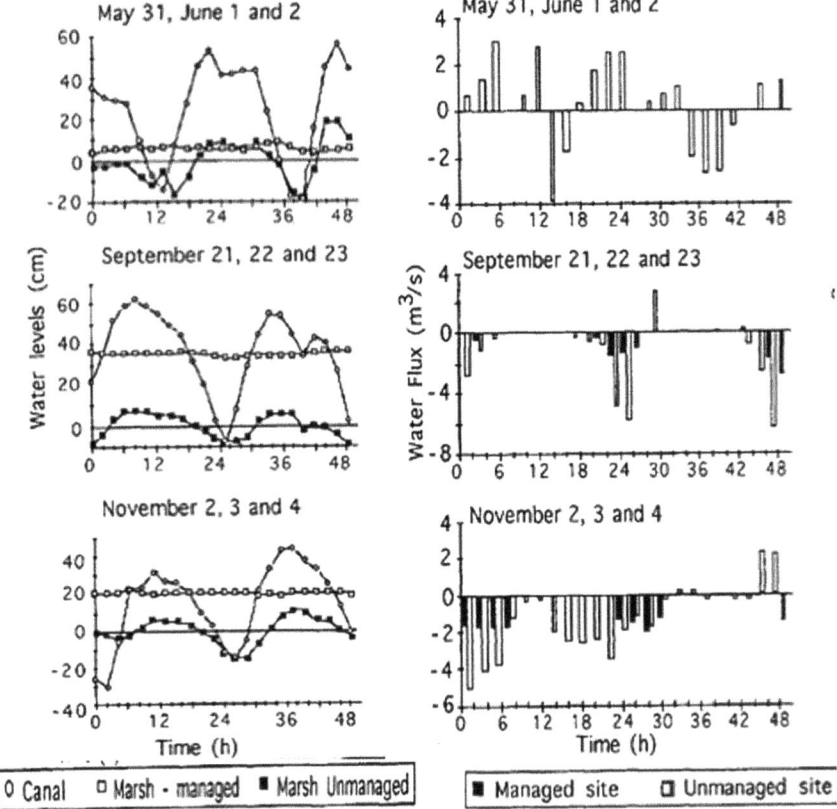

Fig. 5.10 Water levels measured in semi-impounded and unimpounded marshes and the canal that fed both marshes in the Rockefeller Refuge (left). Instantaneous water fluxes per m² of drainage area measured every 2 h for three 48-h measurement periods. Positive values indicate flux into the area, and negative values are flux out of the area (right) (Boumans and Day 1994). Replace this with Fina data

to a dramatic reduction in material exchange (Fig. 5.11). This shows that semi-impoundment leads to a loss of sediment and nutrient inputs. Soil phosphorus was lower and soil organic matter was higher at the managed areas.

5.2.3 Sedimentation and Accretion

Canals and associated spoil banks decrease sediment delivery into adjacent marshes and lead to reduced soil accretion, enhanced subsidence, increased inundation, and vegetation water logging stress (Day et al. 2000, 2011; Mendelssohn and Morris 2000). Accretion of both mineral and organic matter helps offset relative sea level

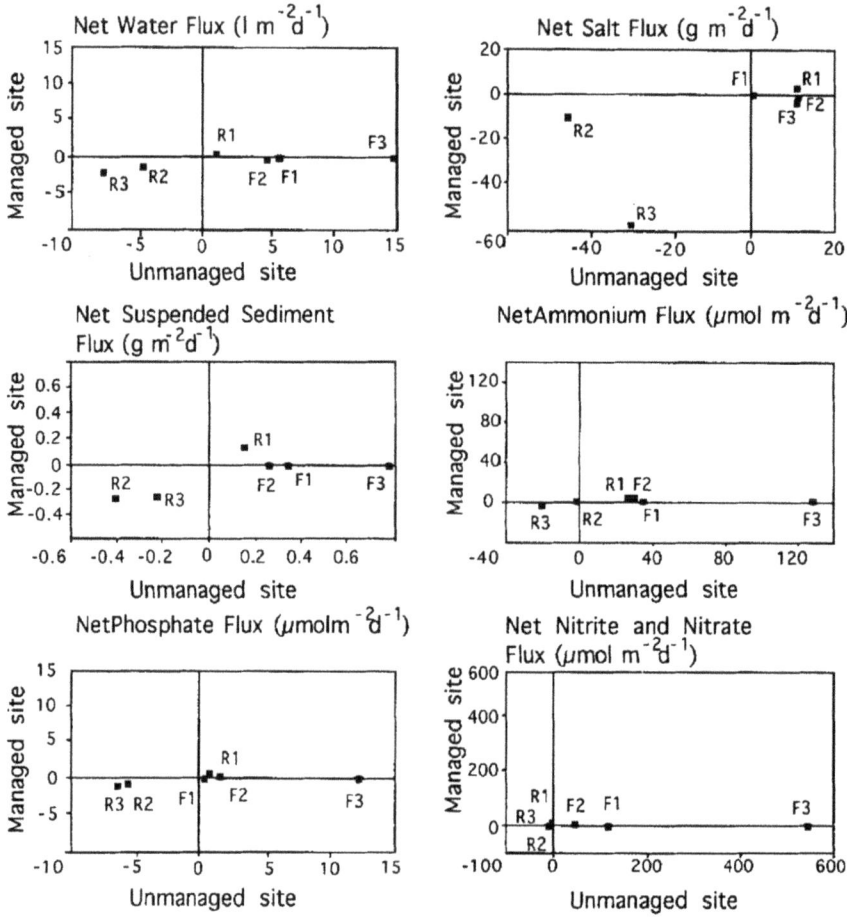

Fig. 5.11 A comparison of net material fluxes per m^2 of the study area for the semi-impoundment managed and unmanaged areas for each 48 h sampling period. The net fluxes for the unmanaged areas are on the horizontal axis and for the managed areas are on the vertical axis. Each circle compares the net flux between managed and unmanaged areas for a particular constituent (water, salt, suspended sediments, ammonium, phosphate, and nitrate plus nitrite) for an individual sampling trip. Positive values indicate flux into the area and negative values are flux out of the area. R and F stand for Rockefeller and Fina, respectively, and 1, 2, and 3 are for the first, second, and third trips (Boumans and Day 1994)

rise (RSLR) while healthy plant roots hold soil in place to minimize soil erosion and lead to organic soil formation.

The regional rate of geologic subsidence in the Mississippi River Delta averages about 10 mm per year, primarily due to compaction, dewatering, and consolidation of Holocene sediments with a historically small part due to eustatic sea-level rise (Day et al. 2000; Ko et al. 2004). This gives rise to a high rate of relative sea-level rise (RSLR) due to subsidence plus eustatic sea-level rise. In sustainable wetlands, RSLR is offset by elevation gain due to mineral sediment and organic matter accretion but in marshes impounded by canal spoil banks accretion is often disrupted because sediment input is restricted and vegetation productivity is decreased (Reed 1992; Jarvis 2010). In a coastal marsh in southwest Louisiana, Cahoon and Turner (1989) measured accretion rates of 0.99 ± 0.20, 0.66 ± 0.25, and 0.60 ± 0.12 cm yr^{-1} in natural marshes, marshes with a continuous spoil bank, and marshes with a discontinuous spoil bank, respectively.

A review of scientific studies on sedimentation in impounded and non-impounded coastal Louisiana wetlands between 1994 and 1999 showed that mean sedimentation in impounded areas averaged 0.75 g m^{-2} yr^{-1} compared to 2.28 g m^{-2} yr^{-1} in non-impounded wetlands (Nichols 1959; Craig et al. 1979; Doiron and Whitehurst 1974; Johnson and Gosselink 1982). Similarly, a review of studies on accretion rates in impounded and non-impounded coastal Louisiana wetlands showed that accretion rates in impounded wetlands averaged 3.40 mm yr^{-1} compared to 9.15 mm yr^{-1} in non-impounded wetlands (Cahoon and Turner 1989; Taylor et al. 1989; Cahoon 1994; Bryant and Chabreck 1998). These data demonstrate that spoil banks significantly reduce sediment inflow and decrease processes important for soil accretion.

Neill and Turner (1987) also found lower sedimentation rates in marshes impounded by spoil banks than marshes without impoundments. Taylor (1988) measured accretion rates that were 1.5 times higher in marshes along natural waterways when compared to marshes impounded by spoil banks in coastal Louisiana. Cahoon (1994) reported that accretion rates and organic matter accumulation in managed, impounded marshes in coastal Louisiana were about five time lower than in unmanaged marshes without impoundments (Fig. 5.12). Similarly, Reed et al. (1997) measured sediment deposition at impounded marshes and unimpounded reference marshes in the Barataria and Terrebonne Basins, Louisiana, and found that deposition was two to five times higher at reference marshes than those that were impounded (Fig. 5.13). Boumans and Day (1994) found that short-term sedimentation rates were significantly higher in unmanaged marshes (1 ± 0.7 g m^{-2}d^{-1}) compared to managed

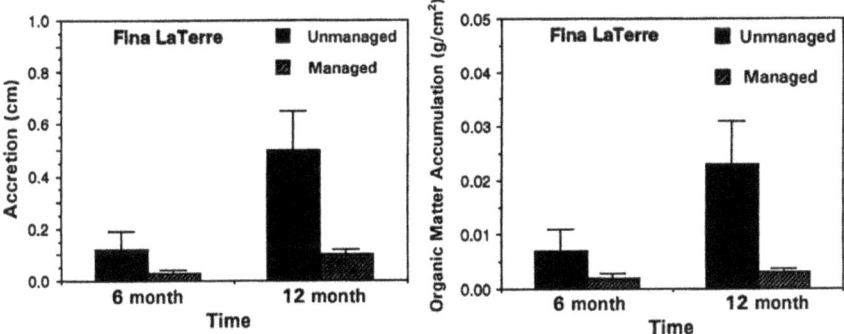

Fig. 5.12 Rates of vertical accretion (left) and organic matter accumulation (right) in managed and unmanaged marshes at Fina LaTerre, Louisiana (means and 1 SE; from Cahoon 1994)

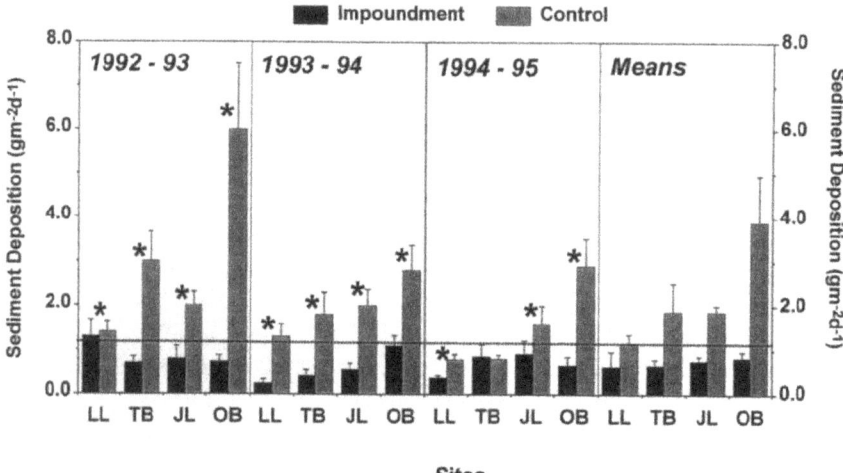

Fig. 5.13 Comparison of marsh surface sediment deposition rates between impounded and reference sites by management year (means ± 1 SE) and means for all years. Stars indicate statistically significant differences at p = 0.05. The horizontal line indicates the 1.1 g m^{-2}day^{-1} of inorganic accumulation required for fresh marsh soils (from Reed et al. 1997)

marshes (0.6 ± 0.8 g m^{-2}d^{-1}) that had natural hydrology. Organic short-term sedimentation was also higher at the unmanaged site (0.9 g m^{-2}d^{-1}) compared to the managed site (0.5 ± 0.6 g m^{-2}d^{-1}).

5.2.4 Saltwater Intrusion

As stated previously, dredged canals tend to preferentially capture water flow from natural channels and, in addition to reducing density of natural channels, promote penetration of salt water into wetlands that were previously isolated from direct exchange with higher salinity water (Sasser et al. 1986; Turner 1987; Wang 1988; Gagliano and Wicker 1989; Turner and Rao 1990; Reed 1995; Bass and Turner 1997; Ko and Day 2004; Ko et al. 2004). Increases in salt concentration in water cause changes in vegetation composition and reduced productivity and/or death of fresh and low salinity marsh species that eventually leads to formation of open water (Mendelssohn and McKee 1988; McKee and Mendelssohn 1989; Mendelssohn and Morris 2000). Introduction of saltwater into previously fresh or very low salinity wetlands also provides sulfate, which can be reduced to sulfides in anaerobic sediments. Hydrogen sulfide is a potent toxin to coastal marsh vegetation. Even in high salinity marshes, saline intrusion can impact marsh health in areas with a high density of canals. Spoil banks reduce sediment input and water exchange and lead to increasing water depths. This interacts with higher salinity and hydrogen sulfide levels to cause multiple stresses on wetland vegetation (Mendelssohn and Morris 2000). Rapid soil decomposition occurs where marsh has died to plant stress due to increased flooding and salinity intrusion via the anaerobic sulfate reduction pathway.

5.2.5 Erosion Related to Oil and Gas Activities

High wave energy on exposed marsh shores can lead to shoreline retreat, enlargement of interior marsh ponds and lakes, and scour of the surface of the marsh (Stevenson et al. 1985; Boumans et al. 1997; Day et al. 1997, 1999). Erosion of the marsh edge around bays, and other open water areas, has been found to be a major mechanism of land loss of estuarine wetlands in the Mississippi River deltaic plain (Day et al. 2000, Barras et al. 2003), as well as other estuaries. For example, in Western Rehoboth Bay, Delaware, marsh erosion rates were as high as 43 cm yr^{-1} (Schwimmer 2001), Pamlico Sound showed a mean erosion rate of up to 91 cm yr^{-1} (Phillips 1986), and Day et al. (1997) reported up to a meter of erosion per year at Venice Lagoon, with high deposition of sediments and high vertical accretion on the marsh surface.

In coastal wetlands, healthy plant roots help hold soil in place to minimize erosion. Erosion is an important factor contributing to wetland loss in the MRD but under natural conditions this process typically occurs mainly around bays and other open water areas and less in large and/or continuous interior marshes (Deegan et al. 1984; Day et al. 2000; Barras et al. 2003; Day et al. 2011). With vegetation death due to saltwater intrusion through canals and/or increased flooding impacts due to impoundments, soils that are less resistant to erosion and can be removed by low energy waves and currents, including export of liquefied soil material, especially from interior marshes (Day et al. 2011).

Oil and gas activity also leads to erosion in interior marshes because construction of canals, which are typically straight channels, increases tidal flow into these marshes. The increase in tidal exchange, coupled with wakes from boat traffic and wind-induced waves, causes rapid morphological changes in marsh edges and spoil banks as wakes and currents removal material, erode the base, undercut the bank and cause sloughing and collapse which leads to canal widening over time (Fig. 5.14;

Fig. 5.14 Shrimp boat traveling down a pipeline canal causing a wake that results in marsh edge erosion (photos by senior author)

Davis 1973; Gagliano and Wicker 1989; Reed 1995; Ko and Day 2004; Ko et al. 2004). As a canal becomes wider and deeper, a greater volume of water can flow through it, which increases erosion rates over time (Doiron and Whitehurst 1974). As the canals widen through erosion or maintenance dredging, marsh loss increases.

Marsh edge erosion in coastal Louisiana has become more apparent over the last half century due the enlargement of bays and interior lakes, as well as navigable waterways due to increased boat traffic and associated wakes as well as wind-induced wave erosion in open water areas with sufficient fetch (Day et al. 2000; Barras et al. 2003). In interior marshes, submergence and water logging cause significant marsh loss due to lack of sufficient sediment input, weakening of soils, and low sediment capture efficiency (Day et al. 2011). An oil and gas field, with a higher concentration of canals and impounded marsh and work boat traffic, exacerbates these processes by reducing sediment input, enhancing saltwater intrusion, increasing subsidence and length of inundation, and diminishing marsh health (Cahoon 1990).

The rate of naturally occurring marsh edge erosion is dependent upon a number of factors, including wind waves (wave height, power, which is related to fetch), currents, sea level rise (SLR), soil characteristics, vegetation, bioturbation, storms, and tidal elevation (Watzke 2004; Wilson and Allison 2008; Feagin et al. 2009; Mariotti and Fagherazzi 2010; Mariotti et al. 2010; Fagherazzi et al. 2012; Trosclair 2013; Bilkovic et al. 2017; Raposa et al. 2018). Anthropogenic activities such as canal dredging and boat wakes can increase erosion rates (Bilkovic et al. 2017). Fagherazzi et al. (2012) used a numerical model to simulate shoreline erosion rates and demonstrated that a scarped marsh had a higher erosion rate per unit energy compared to a non-scarped edge.

Marsh edge erosion can occur gradually as waves break against the shoreline and cause the marsh edge to retreat or it can occur more abruptly when waves undercut the marsh, excavate the fine muds from below the root zone and cause intact portions of marsh to detach and collapse into the water (Fig. 5.15; Valentine and Mariotti

Fig. 5.15 Eroding marsh edge showing the cutting of the marsh edge below the root zone (left); well vegetated eroding marsh edge in Barataria Basin, Louisiana (center; Sapkota and White 2019); and intact marsh that has detached (right; photo by Chris Neill, Woods Hole Research Center 2016)

2019; Sapkota and White 2019; Schwimmer 2001; Marani et al. 2011; Bendoni et al. 2016; Wang et al. 2017; Haywood et al. 2020).

Marsh edge retreat can also be a function of rising sea level (Finkelstein and Hardaway 1988). In Blackwater estuary, on the south-east coast of England, Pethick (1993) observed an approximate balance between the volume eroded from the intertidal zone and that deposited in the sub-tidal channel, suggesting that the response of this estuary to sea-level rise is to redistribute the sediment from the marsh edge to within the estuarine channel to achieve a wider, shallower cross sectional shape. This agrees with the 'The Bruun Rule' that, for a shoreline in equilibrium, a given rate of sea level rise will result in shoreline erosion sufficient to deposit sediment on the nearby marsh surface to a depth equal to the rate of sea level rise (Bruun 1962). Analysis of shoreline morphology and processes indicate that decreasing tidal ranges result in increasing long-term erosion rates (Rosen 1977), which may have significant implications for microtidal coastal Louisiana, where the mean tidal range is about 30 cm.

Feagin et al. (2009) measured marsh erosion rates in a simulated wave flume experiment. The researchers investigated differences in erosion rates in samples with and without plants, with different plant species, and with different soil types. Erosion rates were significantly different between samples from the edge and interior marsh and between samples from restored and natural marshes (Fig. 5.16). They also found linear relationships between erosion and soil characteristics such as percent organic matter and bulk density (Fig. 5.17). As plants proliferate, detritus

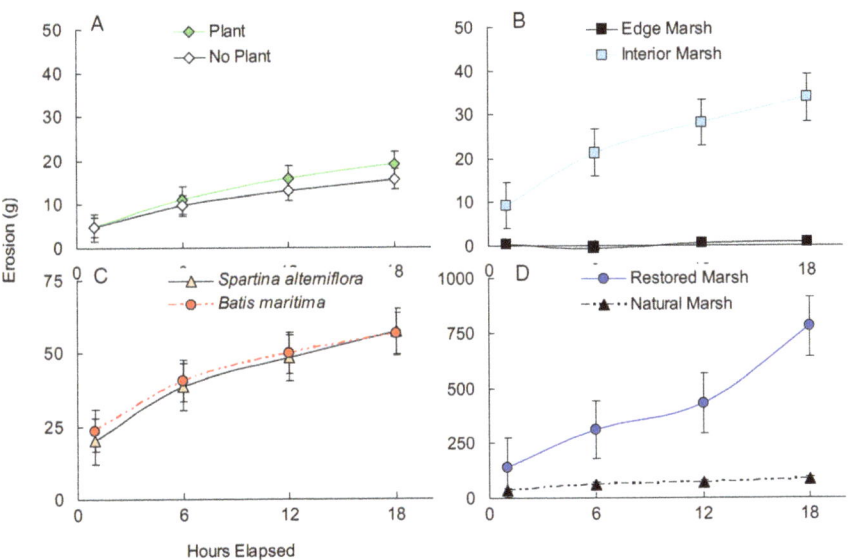

Fig. 5.16 Laboratory flume results. Erosion rates (mean + s.e.) for plant versus no plant samples (**a**), marsh edge versus marsh interior samples (**b**), S. alterniflora versus B. maritime samples (**c**), and restored versus natural marsh samples (**d**) (Feagin et al. 2009)

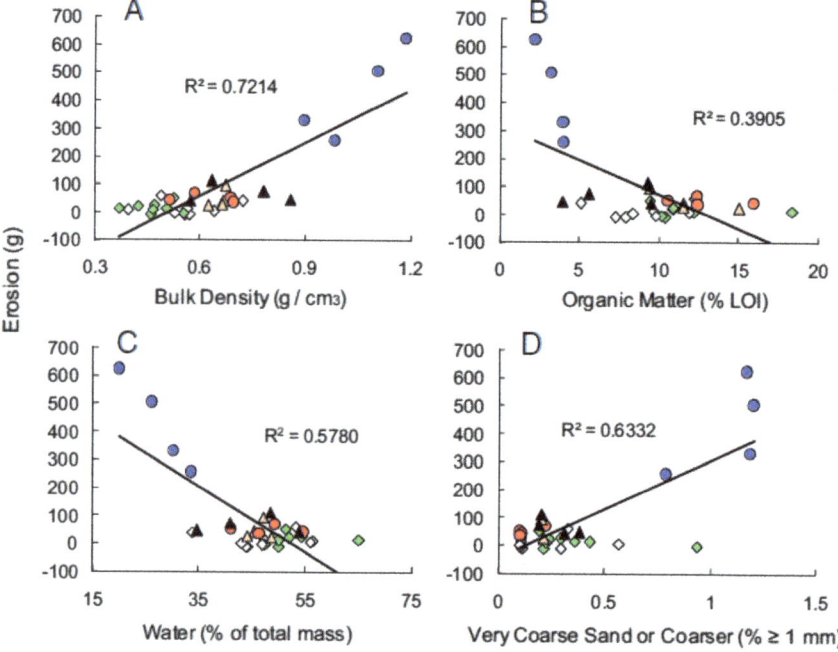

Fig. 5.17 Relationships between erosion and bulk density (**a**), percentage of organic matter (**b**), percentage water (**c**), and percentage of coarse particles. Data points represent sample replicates (Feagin et al. 2009)

and finer-grained sediments become incorporated into the soil matrix through accretion, and soil becomes less dense, less coarse, and more cohesive. These sedimentary properties are associated with resistance to lateral wave-induced erosion.

Boat wakes cause many different impacts to aquatic environments, including significant marsh edge erosion (Fig. 5.18). The amount of edge erosion caused by boat wakes is dependent upon size of the wake, water depth, current direction and velocity, channel and bank morphology, soil characteristics, presence of wind waves, and distance of the vessel from the marsh edge (Macfarlane and Renilson 1999; Price 2005). The size of a wake generated by a boat passage is dependent upon vessel length and speed, hull shape, displacement (loading), draft, and trim. Generally, fast moving vessels displacing large volumes of water produce the largest wakes while vessels displacing less water and moving slowly or at planing speed produce the smallest wakes (Bilkovic et al. 2017).

Hull type affects the size and shape of a boat wake. Planing hulls are designed to ride on top of the water while displacement hulls ride in the water, pushing it to the side as they move forward. Generally, the amount of water displaced is equivalent to the weight of the vessel (Bilkovic et al. 2017). All other factors being equal, a positive correlation exists between the size of a vessel and the size of its wake (Hill et al. 2002; Fonseca and Malhotra 2012). The single best predictor of the size of the

5 Impacts of Oil and Gas Activity in the Mississippi River Delta

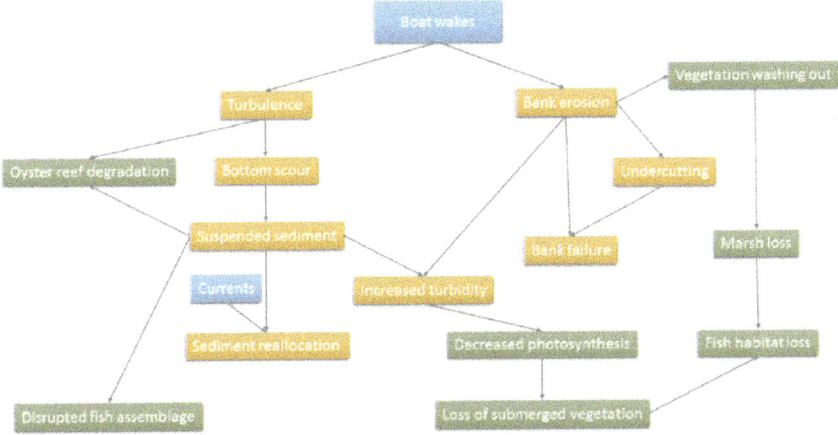

Fig. 5.18 Diagram showing potential impacts from boat wakes to aquatic resources. Blue boxes are drivers of change. Yellow boxes are changes in ecosystem structures and functions. Green boxes are impacts on living resources (Bilkovic et al. 2017)

wake that any given boat will produce is the speed at which the vessel is traveling (Sorenson 1973; Zabawa and Ostrom 1980; Fonseca and Malhotra 2012), although this relationship is not linear for planing hulls (Bilkovic et al. 2017).

For navigation channels, one of the most effective strategies to reduce erosion along channel edges is to construct armaments such as rock wall or concrete bulkheads (Fig. 5.19; Thatcher et al. 2011). However, construction of armaments may

Fig. 5.19 Sediment erosion along the unarmored (left) bank of the Gulf Intracoastal Waterway, coastal Louisiana compared to the armored (right) bank (Used by permission from Clyde Lockwood)

Table 5.2 Mean widening rates of canals on the Rockefeller Refuge (Nichols 1961)

Canal	Mean widening rate (ft month^{-1})	
	Exp and Dev period	Production period
Humble system	1.16	0.42
Superior	1.43	0.35
Deep lake	1.22	0.43
Constance Bayou	1.64	0.28

affect interior marshes by reducing movement of water, nutrients and sediments into the marsh, although the presence of spoil banks along these channels has most likely already reduced exchange of resources.

Nichols (1961) measured oil company canal widths on the Rockefeller Refuge (Humble and Superior systems) and compared those widths to initial widths when the canals were dredged. He found that the canals of the Humble system were, on average, 212% wider than their original widths and those of the Superior system were 242% wider than their original widths. Mean widening rates ranged from 0.28 to 1.64 ft month^{-1} (50 cm month^{-1}) (Table 5.2) and were greater in the exploration and development period of an oil field when canal traffic is the heaviest than in the production period when boat traffic is not frequent. Nichols (1961) concluded that the canals on the Rockefeller Refuge widened at a rate directly related to usage and that the variables that affected this rate, in order of importance, included boat traffic, boat speed, levee construction and maintenance techniques, marsh type, and canal width. Nichols also concluded that the use of construction techniques such as bulkheading and berms could be used to reduce canal erosion.

Browne (2017) conducted an extensive study on factors affecting erosion rates of natural and channelized marsh edges at Hempstead Bay, New York. He found that, on average, the edges of marshes that were channelized retreated horizontally by 27.52 m between 1966 and 2007 while the edges of natural marshes retreated, on average, by 8.38 m, for the same period.

Browne (2017) also further categorized marsh edge erosion by subdividing channels by vessel use category. He found significant differences in edge change based on boat type and use frequency for both channelized and natural marsh edges. Loss of marsh edges was much higher in both channelized and natural marshes with more frequent boat use (Table 5.3).

5.2.6 Changes in Vegetation Health and Deterioration

As described above, spoil banks cause increased inundation duration, vegetation waterlogging stress, reduced biomass production and accretion, and enhanced subsidence (Day et al. 2000, 2011; Mendelssohn and Morris 2000; Snedden et al. 2015). Accretion of both mineral and organic matter helps offset RSLR while healthy plant roots hold soil in place to minimize soil erosion and lead to organic soil formation.

5 Impacts of Oil and Gas Activity in the Mississippi River Delta

Table 5.3 Change in marsh edge from 1966 to 2007 for channelized and natural marshes along navigational channels of different boat use categories. The categories are described below. Negative numbers indicate marsh loss (Browne 2017)

Use	Channelized median (3)	Mean	Number	SD	SE	Natural median (m)	Mean	Number	SD	SE
A	−30.33	−43.68	6	33.27	13.58	−15.42	−15.89	16	13.89	3.47
B	−31.42	−27.51	11	14.6	4.4	−10.07	−13.73	19	11.72	2.69
C	−13.94	−23.02	3	17.42	10.06	−11.49	−14.02	26	15.8	3.10
D	−22.93	−22.93	2	9.63	6.81	−12.38	−13.9	23	7.12	1.48
E	−21.45	−27.62	3	20.3	11.72	−7.82	−10.00	16	9.57	2.39
X	−2.26	−2.26	1	NA	NA	−5.85	−6.77	77	18.88	2.16
All	−26.305	−29.41	26	21.36	4.19	−8.98	−10.62	177	15.72	1.18

A = Main channels. Many large boats, including commercial traffic (coastal tankers, coastal fishing boats, party boats, ocean-going charter boats, casino boats). Peak use could exceed 60 boats hr^{-1}

B = Secondary channels that feed into A. Many boats, some large commercial traffic as defined for A. Peak use could approach 60 boats hr^{-1}

C = Tertiary channels that feeds into B. Frequent small and medium boat traffic. Peak use was about 20 boats hr^{-1} with boats up to about 8 m at water line (AWL)

D = Local use channels, mostly side channels connecting B and C channels. Small boats to about 8 m AWL. Peak use about 10 boats hr^{-1}

E = Low local use channels. Small side channels with up to about 2–3 boats hr^{-1}

X = Not official navigational water. Boat use was rare to non-existent, usually less than 1–2 per day, confined to local fishermen in small shallow draft boats less than 6 m AWL

Increased waterlogging decreases plant productivity and production of organic matter that further reduces accretion and, when combined with enhanced subsidence due to petroleum extraction, leads to vegetation death and the creation of small ponds which then combine to form larger open water bodies (DeLaune et al. 1994; Nyman et al. 1995; Ko et al. 2004). Other factors resulting from increased inundation that reduce vegetation productivity and/or cause death are sulfide toxicity, anoxia, and salt stress (Mendelssohn and McKee 1988; McKee and Mendelssohn 1989; Mendelssohn and Morris 2000).

Based on research conducted at two Louisiana *Spartina alterniflora* marshes (Old Oyster Bayou, a sediment-rich area near the mouth of the Atchafalaya river and Bayou Chitigue, a sediment-poor area about 70 km to the east), Day et al. (2011) developed a conceptual model for wetland loss showing how accretion, subsidence, and waterlogging stress interact to cause marsh loss (Fig. 5.20). As wetland elevation decreases, the duration of inundation becomes longer and further stresses vegetation. Soil strength decreases as inundation inhibits consolidation and compaction and sediment that is trapped is fluid and is easily exported rather than being incorporated into the marsh soil. When stress becomes excessive, vegetation death occurs and much of the dead root mat is rapidly mineralized via anaerobic sulfate decomposition causing elevation loss via peat collapse. Although wave erosion is not a major factor in interior marshes, wakes from boat travel on oil and gas canals does cause erosion in marshes where they are dredged. As wetlands disappear due to oil and gas activities, interior marshes become more exposed and waves cause erosion in previously isolated areas (Day et al. 2011).

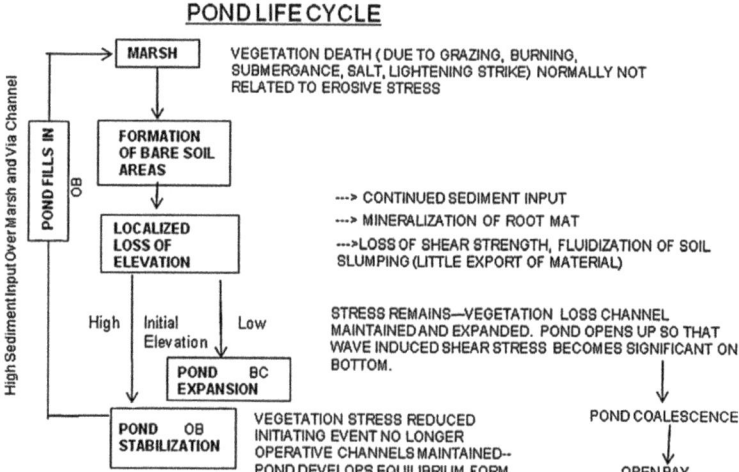

Fig. 5.20 Conceptual diagram of the process of pond formation and expansion in a Louisiana coastal marsh (Old Oyster Bayou (OB) and Bayou Chitigue (BC)). The initial elevation of the marsh surface is critical to whether the pond expands or is stable (from Day et al. 2011)

5.3 Oil and Gas Induced Subsidence

Land subsidence due to subsurface fluid removal has been widely reported worldwide. Some of the most striking examples are due to groundwater withdrawal. In the Houston area, for example, groundwater withdrawal and oil production from shallow deposits led to subsidence of up to 3 m (Ward 1993; White and Calnan 1990; Stumpt and Haines 1998; Kasmarek et al. 2009). In Bangkok, Thailand, subsidence rates due to groundwater pumping have been as high as 120 mm yr^{-1} with subsidence of 2 m since the mid twentieth century (Phien-wej and Nutalaya 2005).

It has been observed that subsidence can be increased dramatically both during and after oil and gas production and that wetland loss is higher in the vicinity of oil and gas fields (Morton et al. 2006; Mallman and Zoback 2007; Dokka 2011; Kolker et al. 2011; Yu et al. 2012; Chang et al. 2014). Thousands of ha of the Po deltaic plain are 1–4 m below sea level due to subsidence caused primarily by extraction of shallow deposits of natural gas with a high water content (Sestini 1992). The background rate of geological subsidence in the Po Delta is 1–3 mm yr^{-1} (Sestini 1992; Bondesan et al. 1995; Tosi et al. 2016) but human induced subsidence has been as high as 5–20 mm yr^{-1} and in a small area in the western Po Delta, subsidence was 100 cm between 1958 and 1962 (Bondesan et al. 1995). In the northeastern Netherlands subsidence due to natural gas extraction over an area of about 400 km^2 from 1992 to 2010 was greater than 2 mm yr^{-1} with some areas being greater than 6 ^{mm} yr^{-1} with total subsidence greater than 1 m (Fig. 5.21; Caro Guenca et al. 2011).

Morton et al. (2006) and Mallman and Zoback (2007) reported higher subsidence rates in the vicinity of oil and gas fields with rates as high as 23 mm yr^{-1} with total surface subsidence in excess of 1 m over the life of a field. Subsidence often continued for decades after a reduction in production (Mallman and Zoback 2007; Morton and Bernier 2010). Processes contributing to increased subsidence in the vicinity of oil and gas fields include slow dissipation of pore pressure (Baú et al. 1999) and creep compaction (Hettema et al. 2002). Yu et al. (2012) suggested that enhanced subsidence was locally related to hydrocarbon production in a field in the Chenier Plain because subsidence increased between 0.5 and 1 m over the period of production for the field. When large volumes of oil and gas are removed, this causes reservoir compaction and/or activating slippage along fault lines (Ko et al. 2004; Morton et al. 2005, 2006; Chan and Zoback 2007; Fig. 5.22). Subsidence and land loss from due to oil and gas production may occur directly above the producing formation or several kilometers away from the producing wells (Morton et al. 2006). Based on analyses of oil and gas production and wetland loss over time, Morton et al. (2006) concluded "wetland loss and fault reactivation typically are attributed to induced subsidence when the area and timing of wetland losses and fault movement coincide with advanced stages of hydrocarbon production" (Fig. 5.23).

Local rates of subsidence in oil and gas fields in south central Louisiana (as much as 23 mm yr^{-1}) were much higher than regional rates in the Mississippi River Delta (about 10 mm yr^{-1}; Morton et al. 2002). Morton et al. (2005) analyzed remote images, elevation surveys, stratigraphic cross sections, and hydrocarbon production data in

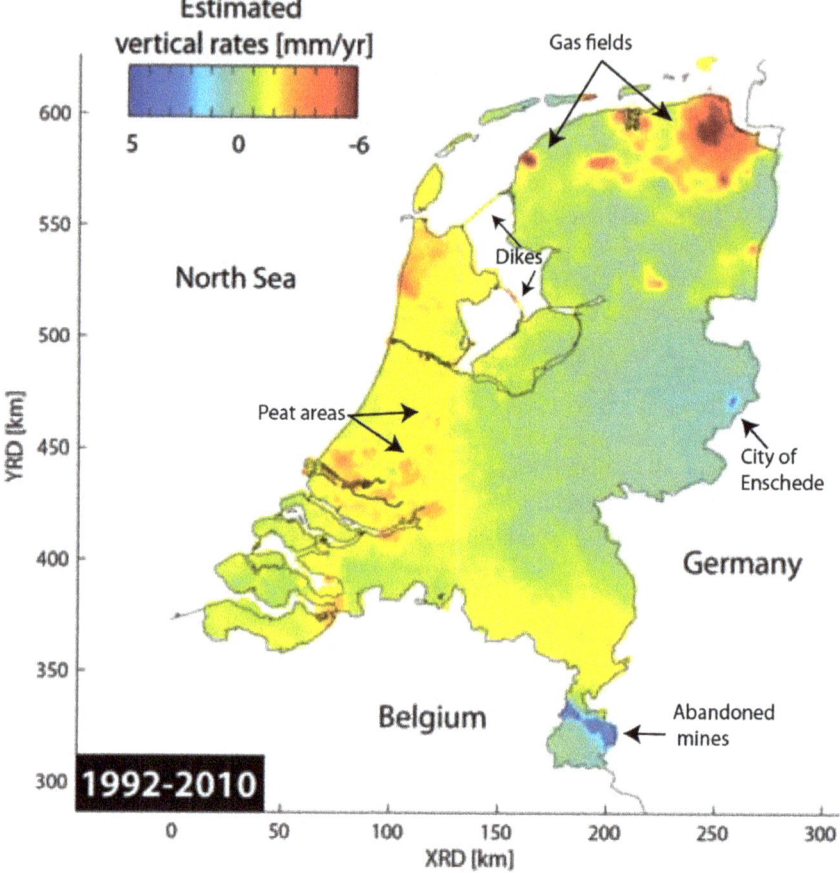

Fig. 5.21 Land subsidence from 1992 to 2010 in the northeastern Netherlands due to natural gas extraction (Caro Guenca et al. 2011)

relation to wetland loss for the Terrebonne-LaFourche basins and found that the highest rates of subsidence coincided with the location of oil and gas fields (Fig. 5.24). Analysis of survey data along Louisiana Highway 1 between Raceland and Leeville, LA showed much greater subsidence rates near producing oil and gas fields and faults than between the fields during two periods of measurement (Fig. 5.25; Morton et al. 2005). This study also showed that subsidence rates accelerated over time and were greater between 1982 and 1993 than between 1965 and 1982 (Morton et al. 2005). Morton et al. (2006) conclude that these rapid changes were likely caused by induced subsidence and fault reactivation resulting from oil and gas production. This resulted in increased inundation that interacted with canal related impacts to cause wetland loss. Chang et al. (2014) reviewed data from the same area as Morton et al. (2006) and concluded that subsidence was caused, in part, by time-dependent deformation of reservoir-bounding shales driven by the slow drainage of pore fluid from the shale

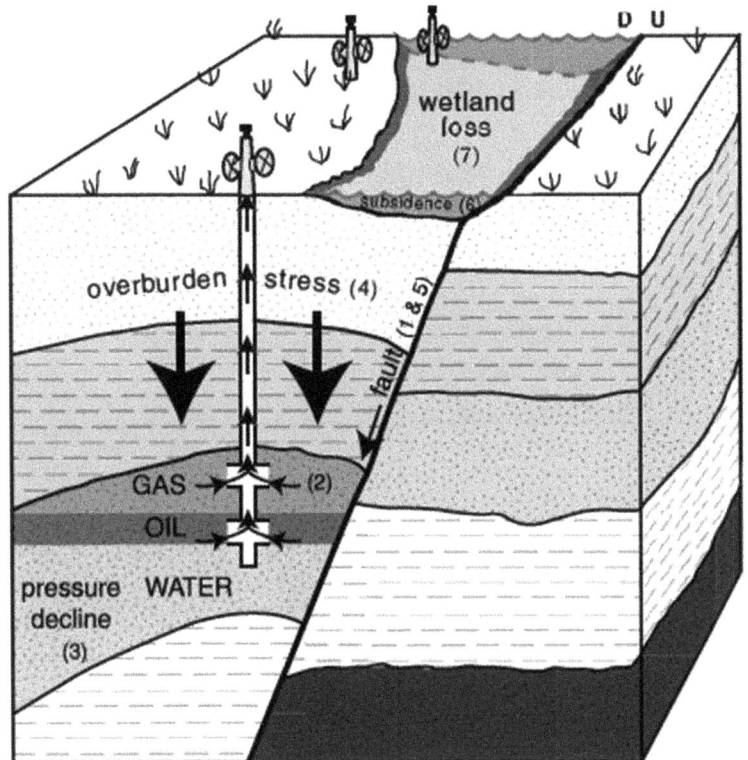

Fig. 5.22 Sequence of production-related subsurface events that may induce land subsidence and reactivate faults. Prolonged or rapid production of oil, gas, and formation water (2) causes formation pressures to decline (3). This increases the effective vertical stress of the overburden (4), which causes compaction of the reservoir rocks and may cause formerly active faults (1) to be reactivated (5). Either compaction of the reservoir and surrounding strata or slip along fault planes can cause land-surface subsidence (6). Where compaction or fault-related subsidence occurs in wetland areas, the wetlands typically are submerged and converted to open water (7). Figure is not to scale (from Morton et al. 2006)

into a severely depleted sand reservoir, a mechanism that was not considered by previous investigators (Fig. 5.26). Modeling by these researchers showed that the calculated subsidence rate due to shale compaction was higher than the subsidence induced by reservoir depletion, suggesting that post-depletion compaction in the reservoir-surrounding shale may explain the observed acceleration of subsidence after depletion (Fig. 5.27).

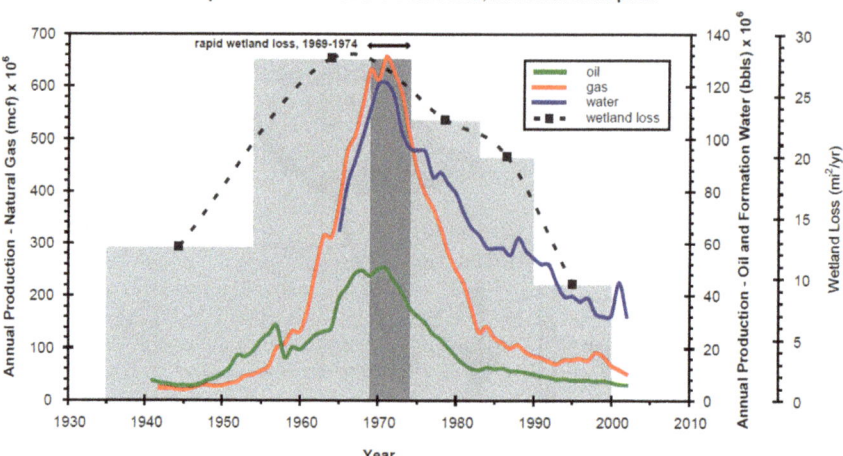

Fig. 5.23 Composite histories of fluid production from oil and gas fields and wetland loss in south Louisiana. The width of the gray bars indicates the time over which land loss rates (dashed line) were calculated (from Morton et al. 2005)

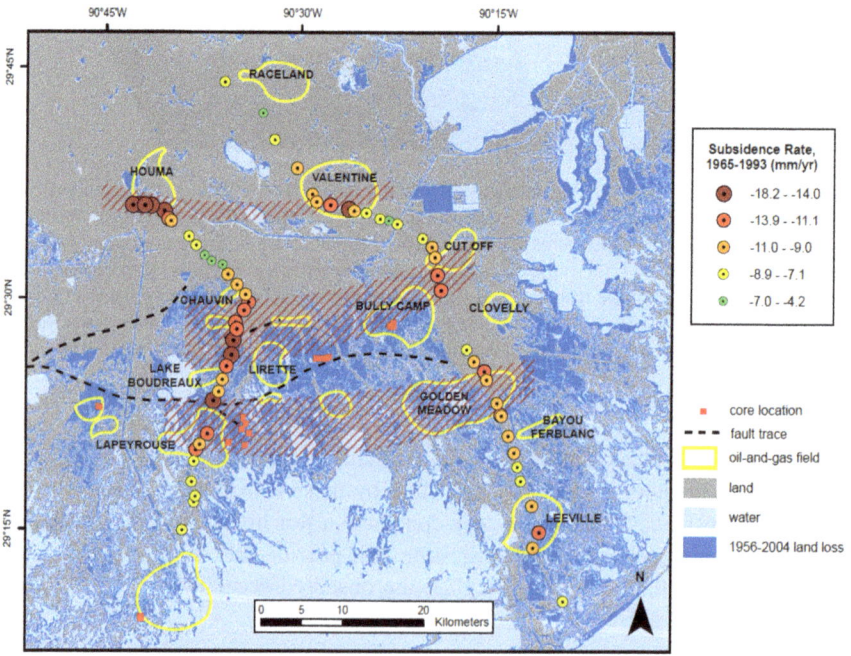

Fig. 5.24 Map showing average subsidence rates between 1965 and 1993 in south Louisiana. Areas of highest subsidence rates (>12 mm yr^{-1}; hatched pattern) correlate closely with locations of oil and gas fields. Lowest average subsidence rates are located between major producing fields (from Morton et al. 2005)

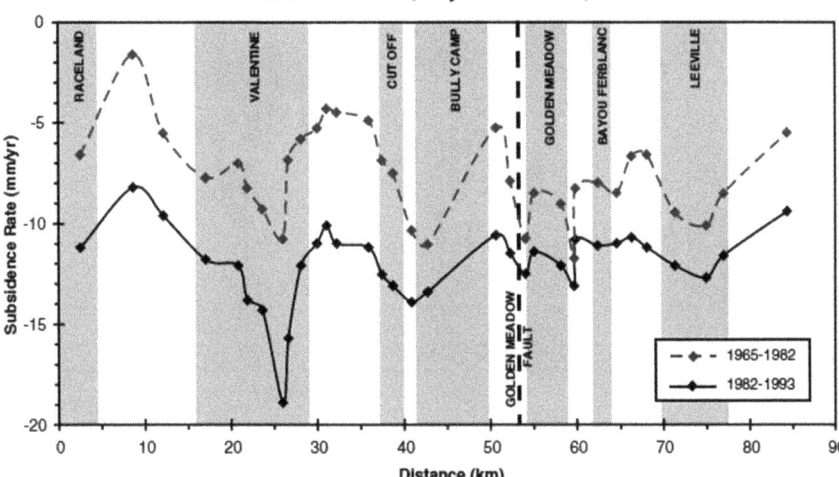

Fig. 5.25 Plots of historical subsidence rates along LA Highway 1 between Raceland and Leeville, Louisiana showing a close spatial correlation between highest subsidence rates, hydrocarbon-producing fields (delineated in grey), and the projected intersection of the Golden Meadow fault zone (from Morton et al. 2005)

5.4 Interactive, Cumulative, and Indirect Impacts of Oil and Gas Impacts

Each of the three types of oil and gas activity (alteration of surface hydrology, induced subsidence, and production of toxic contaminants) causes significant impacts. However, when the effects of all three activities interact in cumulative and indirect ways, the impacts are much greater. In Fig. 5.28, the impacts of each type of activity are shown separately. The direct effect of induced subsidence is a loss of surface elevation due fault activation and/or compaction of subsurface sediments due to fluid removal. Loss of surface elevation leads in turn to stressed wetland vegetation, lower productivity, and ultimately vegetation death and wetland loss. Surface hydrologic alteration due to canals and spoil banks also leads to a loss of surface elevation due to a cascading series of impacts as shown by the relationships in Fig. 5.28. Deep straight canals cause the deterioration of natural tidal channels and changes in regional hydrology and saltwater intrusion. Spoil banks block tidal flooding of marshes and lead to increased inundation of marshes as well as reduced sediment and nutrient input to marshes. These impacts stress marshes so that productivity is lower and organic soil formation is reduced. The combination of lower sediment input and lower organic soil formation causes higher shallow subsidence and ultimately to vegetation death. Saltwater intrusion and vegetation death lead to collapse of the soil column due to rapid decomposition of soil organic matter. Toxic impacts are due to oil spills and produced water discharge which contains very high salinity

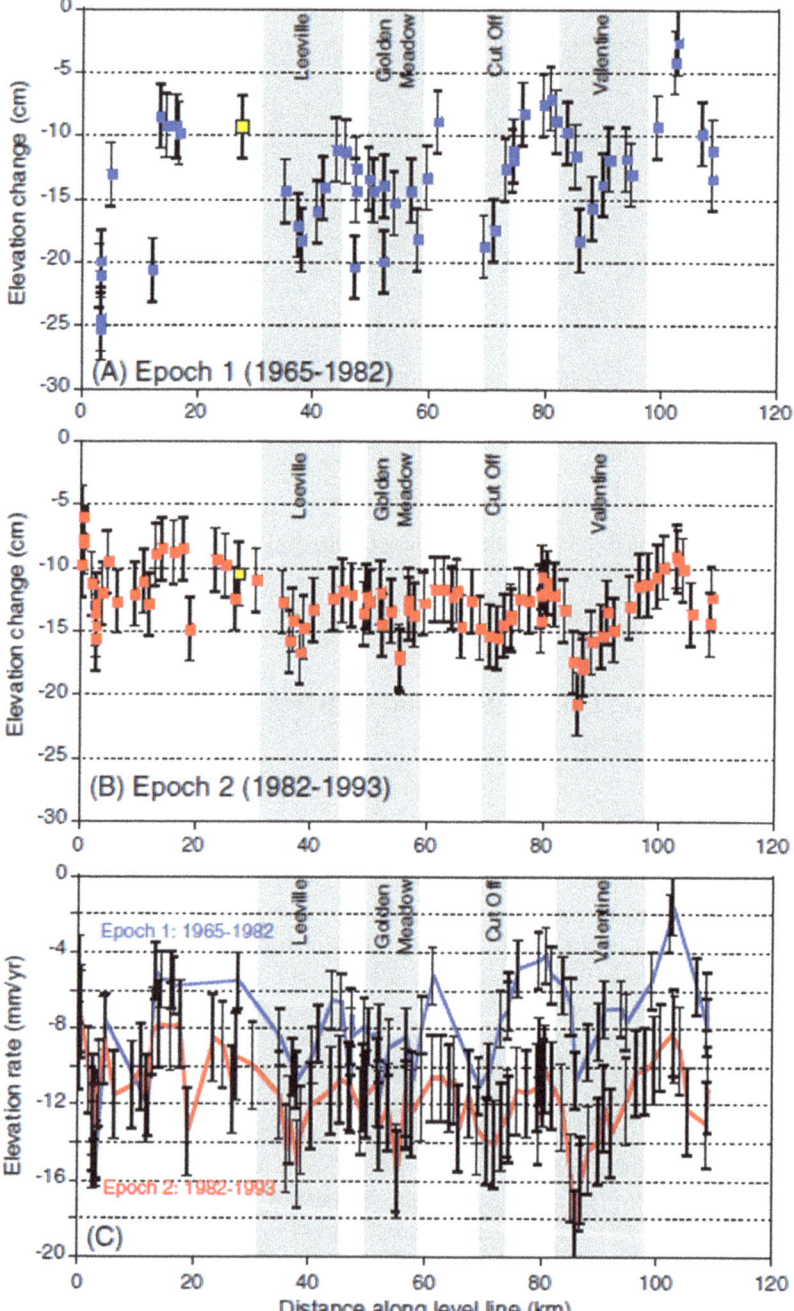

Fig. 5.26 Elevation changes during epoch 1 (1965–1982; **a**) and epoch 2 (1983–1993; **b**), and the rates of subsidence in the two epochs (**c**). Over the entire transect, subsidence rates were greater in epoch 2 than in epoch 1. The yellow squares in **a**, **b** indicate an arbitrarily selected reference station approximately 8 km south and outside of the projected Leeville field, to estimate the magnitudes of local production-related subsidence signals (Chang et al. 2014)

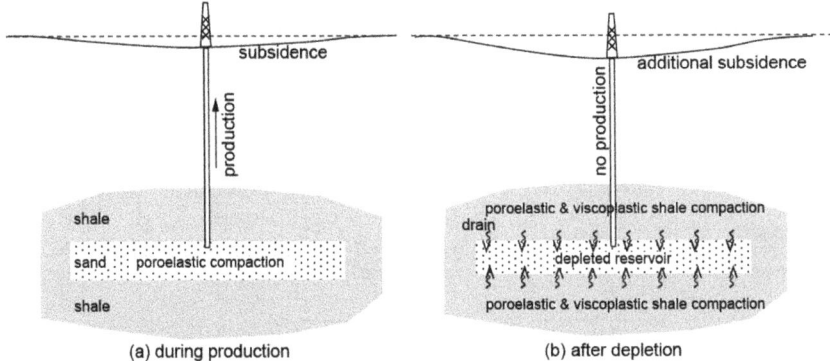

Fig. 5.27 A simple compaction model of a sand reservoir embedded with surrounding shale. It is assumed that during production stage, pore pressure depletion in a sand reservoir induces poroelastic compaction, and after depletion, a slow decrease in pore pressure in the bounding shale induces shale compaction in two independent rheologica modes (poroelastic and viscoplastic) (modified from Chang et al. 2014)

brines and a variety of toxic materials. These stress vegetation and cause wetland mortality and collapse of marsh soils leading to loss of elevation.

Thus, all three types of impacts lead to stressed wetland vegetation, loss of elevation, and vegetation death. The cumulative and indirect impacts are much greater than for individual impacts acting independently. These interact with a range of environmental forcings such as lack of riverine input, sea-level rise, and hurricanes to exacerbate wetland loss even further. For example, wetland loss in the region of the central coast impacted by Atchafalaya River discharge is much lower than most of the rest of the coast even in areas impacted by O&G activity (McGenity et al. 2012; St. Pe 1990). The net effect is that O&G activities make wetlands more susceptible to other forcings that negatively affect wetlands.

Such cumulative and indirect impacts due to O&G production are evident when images of oil fields are viewed over time. The Venice Salt Dome in Plaquemines Parish and the Bully Camp oil field in Terrebonne Parish are just two of hundreds of examples of wetland disappearance over time in coastal Louisiana (Figs. 5.29 and 5.30). In these two fields, pre-oil and gas activity maps show the presence of small natural levee ridges and extensive unbroken marsh. In both fields, the 1950s maps show canals and spoil banks but without significant marsh loss. Over time, marsh loss expands to cover almost the entire area of the field and spoil banks begin to disappear as they subside below water level. The widespread loss of wetlands is due to such cumulative and interacting effects of surface alterations in hydrology, induced subsidence, and toxic effects.

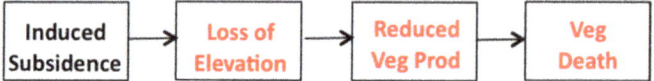
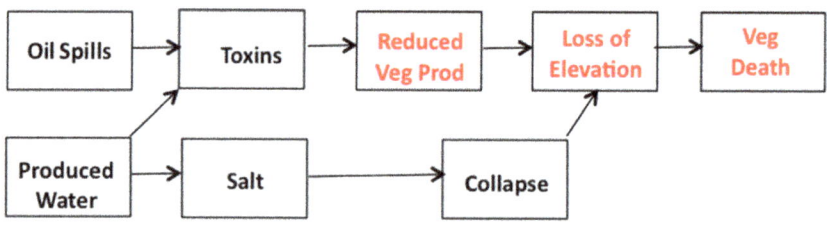
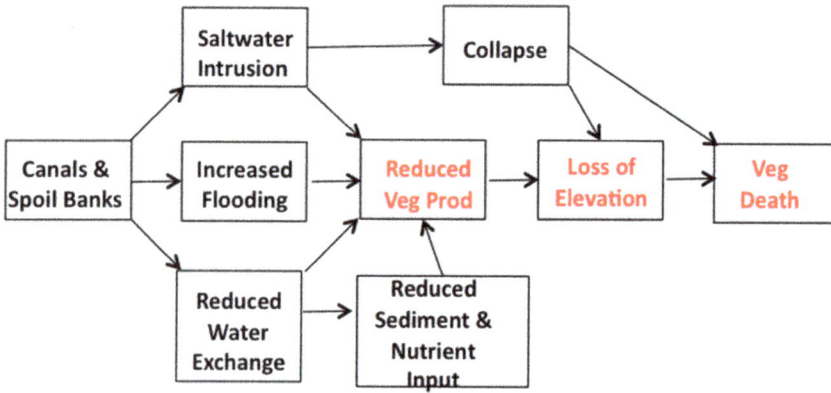

Fig. 5.28 Diagram showing the three general types of impacts from oil and gas activity-induced subsidence, toxicity effects, and altered surface hydrology due to canals and spoil banks. The impacts of these activities are shown separately, and impacts shown in red are common to all three activities. As reduced vegetation productivity, loss of elevation, and vegetation death can result from all three activities, there are pervasive cumulative and indirect impacts that result in overall greater impacts on coastal ecosystems, especially, that would result from the impacts acting separately

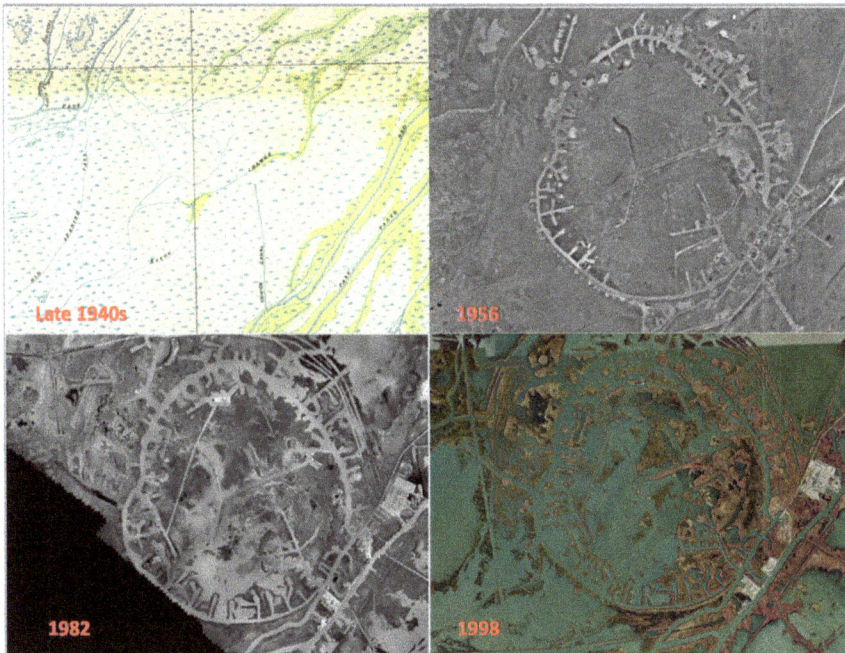

Fig. 5.29 Images of hydrologic changes and wetland loss in the Venice field, Plaquemines Parish, Louisiana over time. This oil and gas field has been termed the "wagon wheel" because of the circular canals and spoil banks that outline the salt dome around which oil and gas containing formations were located. Note that wetland loss occurs both in the vicinity of the canals as well as throughout the area of the field. In addition to wetland loss, many of the spoil banks have subsided below water level and are no longer visible. The green highlighted channels in the 1940s are minor distributary ridges that both conveyed water and were a barrier to horizontal flow perpendicular to these ridges. (Images from USGS topographic maps (late 1940s) and Google Earth)

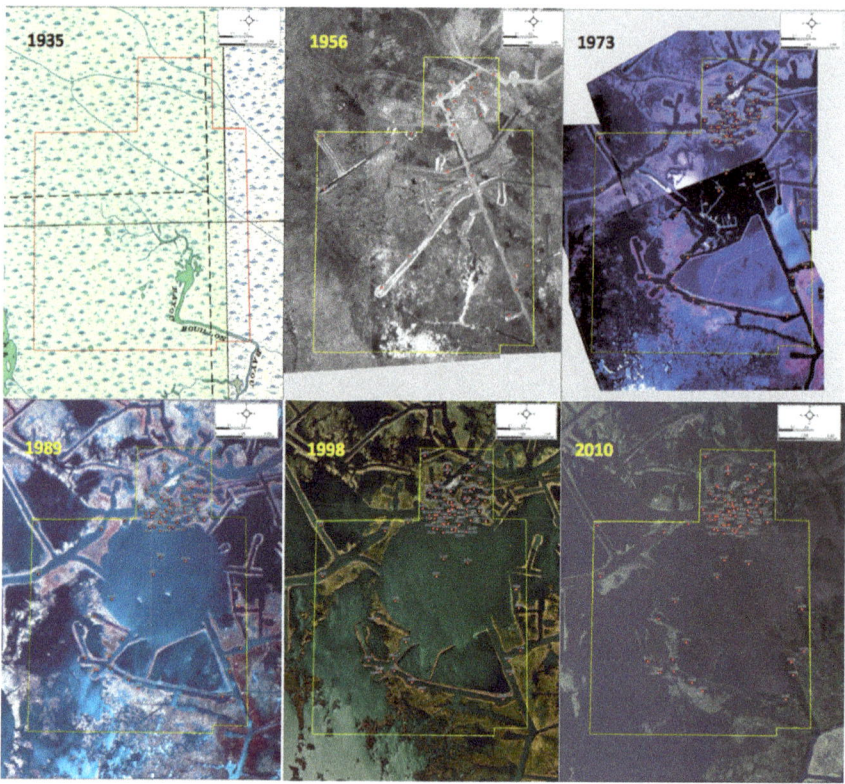

Fig. 5.30 Images of the Bully Camp oil and gas field over time. The yellow line denotes an ownership tract in the field and red dots show locations of oil and gas wells. Note that land loss has been pervasive over the general area of the field. The central water body in the 1989 image is over an area of sulfur production on the top of the salt dome underlying this field. The narrow blue lines in the 1935 map are abandoned minor distributary ridges that were breached by canals (images from McLindon 2016)

References

Asano T (1995) Sediment transport under sheet-flow conditions. J Waterw Port Coast Ocean Eng ASCE 121:239–246

Barras J, Beville S, Britsch D et al (2003) Historical and projected coastal Louisiana land changes: 1978-2050: USGS Open File Report 03-334, 39 p (Revised January 2004)

Bass AS, Turner RE (1997) Relationships between salt marsh loss and dredged canals in three Louisiana estuaries. J Coast Res 13:895–903

Baú D, Gambolati G, Teatini P (1999) Residual land subsidence over depleted gas fields in the Northern Adriatic Basin. Environ Eng Geosci 5:389–405

Baumann RH, Turner RE (1990) Direct impacts of outer continental shelf activities on wetland loss in the central Gulf of Mexico. Environ Geol Wat Sci 15:189–198

Bendoni M, Mel R, Solari L et al (2016) Insights into lateral marsh retreat mechanism through localized field measurements. Water Resour Res 52:1446–1464

Bilkovic D, Mitchell M, Davis J (2017) Review of boat wake wave impacts on shoreline erosion and potential solutions for the Chesapeake Bay. STAC Publication Number 17-002, Edgewater, MD. 68 pp

Boesch DF, Josselyn MN, Mehta AJ, Morris JT, Nuttle WK, Simenstad CA, Swift DJP (1994) Scientific assessment of coastal wetland loss, restoration and management. J Coast Res Spec Issue 20:1–103

Bondesan M, Castiglioni G, Elmi C, Gabbianelli G, Marocco R, Pirazzoli P, Tomasin A (1995) Coastal areas at risk from storm surges and sea-level rise in northeastern Italy. J Coast Res 11:1354–1379

Boumans RM, Day JW (1994) Effects of two Louisiana marsh management plans on water and materials flux and short-term sedimentation. Wetlands 14:247–261

Browne JP (2017) Long-term erosional trends along channelized salt marsh edges. Estuaries Coasts 40:1566–1575

Bruun P (1962) Sea level rise as a cause of shore erosion. American Society of Civil Engineering Proceedings. J Waterways Harbor Div 88:117–130

Bryant JC, Chabreck RH (1998) Effects of impoundment on vertical accretion of coastal marsh. Coast Est Res Fed 21:416–422

Cahoon DR (1990) Field monitoring of structural marsh management. In: Cahoon DR, Groat CG (eds) A study of marsh management practice in coastal Louisiana, vol III, Ecological evaluation. Final report submitted to Minerals Management Service (MMS), New Orleans, Louisiana, Contract Number 14-12-0001-30410. Outer Continental Shelf Study/MMS 90-0077. pp 357–368

Cahoon DR (1994) Recent accretion in two managed marsh impoundments in coastal Louisiana. Ecol Appl 4:166–176

Cahoon D, Turner RE (1989) Accretion and canal impacts in a rapidly subsiding wetland II. Feldspar marker horizon technique. Estuaries 12:260–268

Cahoon DR, Groat CG (1990) A study of marsh management practice in Coastal Louisiana, vol 1. Executive summary. U.S. Dept. of the Interior, Minerals Management Service, Gulf of Mexico OCS Region, New Orleans, Louisiana, 36 pp

Caro Guenca M, Hanssen R, Hooper A, Arikan M (2011) Surface deformation of the whole Netherlands after PSI analysis. In: Proceedings of the Fringe 2011 Workshop, Frascati Italy. ESA SP-697

Chan AW, Zoback MD (2007) The role of hydrocarbon production on land subsidence and fault reactivation in the Louisiana coastal zone. J Coast Res 23:771–786

Chang C, Mallman E, Zoback M (2014) Time-dependent subsidence associated with drainage-induced compaction in Gulf of Mexico shales bounding a severely depleted gas reservoir. AAPG Bull 98(6):1145–1159

Craig NJ, Turner RE, Day JW, Jr (1979) Land loss in coastal Louisiana (U.S.A.). Environ Manag 3:133–144

Davis DW (1973) Louisiana canals and their influence on wetland development. Ph.D. Dissertation, Louisiana State University, 234 pp

Day R, Holz R, Day J (1990) An inventory of wetland impoundments in the coastal zone of Louisiana, USA: historical trends. Environ Manag 14(2):229–240

Day JW Jr, Martin JF, Cardoch L, Templet PH (1997) System functioning as a basis for sustainable management of deltaic ecosystems. Coast Mngt 25:115–153

Day J, Rybczyk J, Scarton F, Rismondo A, Are D, Cecconi G (1999) Soil accretionary dynamics, sea-level rise and the survival of wetlands in Venice Lagoon: a field and modeling approach. Estuar Coast Shelf Sci 49:607–628

Day JW Jr, Britsch LD, Hawes SR, Shaffer GP, Reed DJ, Cahoon D (2000) Pattern and process of land loss in the Mississippi Delta: a spatial and temporal analysis of wetland habitat change. Estuaries 23:425–438

Day JW, Kemp GP, Reed DJ, Cahoon DR, Boumans RM, Suhayda JM, Gambrell R (2011) Vegetation death and rapid loss of surface elevation in two contrasting Mississippi delta salt marshes: the role of sedimentation, autocompaction and sea-level rise. Ecol Eng 37:229–240

Day J, Shaffer G, Cahoon D, DeLaune R (2019) Canals, backfilling and wetland loss in the Mississippi Delta. Estuar Coast Shelf Sci. https://doi.org/10.1016/j.ecss.2019.106325

Day J, Clark H, Chang C et al (2020) Life cycle of oil and gas fields in the Mississippi River Delta: a review. Water 12:1492. https://doi.org/10.3390/w12051492

Deegan LA, Kennedy HM, Neill C (1984) Natural factors and human modifications contributing to marsh loss in Louisiana's Mississippi River deltaic plain. Environ Manag 8:519–528

DeLaune RD, Nyman JA, Patrick WH Jr (1994) Peat collapse, ponding and wetland loss in a rapidly submerging coastal marsh. J of Coast Res 10:1021–1030

DeLaune RD, Pezeshki SR (1994) The influence of subsidence and salt water intrusion on coastal marsh stability: Louisiana Gulf coast, USA. J Coast Res Special Issue 12:77–89

Doiron LN, Whitehurst CA (1974) Geomorphic processes active in the Southeast Louisiana Canal, LaFourche Parish, Louisiana. Research Monographs, Division of Engineering Research—RM 5, 48 pp

Dokka RK (2011) The role of deep processes in late 20th century subsidence of New Orleans and coastal areas of southern Louisiana and Mississippi. J Geophys Res 116:B06403. https://doi.org/10.1029/2010JB008008

Fagherazzi S, Kirwin ML, Mudd SM et al (2012) Numerical models of salt marsh evolution: Ecological, geomorphic, and climatic factors. Rev Geophy 50:1–25

Feagin RA, Lozada-Bernard SM, Ravens TM (2009) Does vegetation prevent wave erosion of salt marsh edges? PNAS 106:10109–10113

Finkelstein K, Hardaway CS (1988) Late holocene sedimentation and erosion of estuarine fringing marshes, York river, Virginia. J of Coast Res 4:447–456

Fonseca MS, Malhotra A (2012) Boat wakes and their influence on erosion in the Atlantic Intracoastal Waterway, North Carolina. Beaufort, NC, NOAA/National Centers for Coastal Ocean Science. 24pp

Gagliano SM (1973) Canals, dredging and land reclamation in the Louisiana coastal zone. Hydrologic and Geologic Studies of Coastal Louisiana. Report no. 14. Center for Wetland Resources, Louisiana State University, Baton Rouge, 104 pp

Gagliano SM (2017) Assessment of environmental impacts associated with oil and gas exploration and development in the Bully Camp oil and gas field Lafourche Parish, Louisiana. Export report prepared for Jones, Swanson, Huddell & Garrison, 601 Poydras St., Suite 2655, New Orleans LA 70130

Gagliano SM, van Beek JL (1970) Geologic and geomorphic aspects of deltaic processes, Mississippi delta system. In: Gagliano SM, Muller R, Light P, Al-Awady M (eds) Hydrologic and geologic studies of coastal Louisiana, vol I. La State Univ, Coastal Studies Institute and Dept of Marine Sciences, 140 pp

Gagliano SM, Wicker KM (1989) Processes of wetland erosion in the Mississippi River Deltaic Plain, pp 28–48. In: Duffy WG, Clark D (eds) Marsh management in Coastal Louisiana: effects and issues—Proceedings of a symposium. United States Fish and Wildlife Service and Louisiana Department of Natural Resources. United States Fish and Wildlife Service Biological Report 89(22).

Gascuel-Odux C, Cros-Cayot S, Durand P (1996) Spatial variations of sheet flow and sediment transport on an agricultural field. Earth Surf Proc Land 21:843–851

Gosselink JG, Hatton R, Hopkinson CS (1984) Relationship of organic carbon and mineral content to bulk density in Louisiana marsh soils. Soil Sci 137:177–180

Haywood BJ, Hayes MP, White JR, Cook RL (2020) Potential fate of wetland soil carbon in a deltaic coastal wetland subjected to high relative sea level rise. Sci Total Environ 711:135–185. https://doi.org/10.1016/j.scitotenv.2019.135185

Hettema M, Papamichos E, Schutjens PMTM (2002) Subsidence delay: field observations and analysis. Oil Gas Sci Technol 57:443–458

Hill DF, Beachler MM, Johnson PA (2002) Hydrodynamic impacts of commercial jet-boating on the Chilkat River, Alaska. Department of Civil and Environmental Engineering, Pennsylvania State University, University Park, PA. 120 pp

Jarvis JC (2010) Vertical accretion rates in coastal Louisiana: a review of the scientific literature. ERDC/EL TN-10-5. U.S. Army Engineer Research and Development Center, Vicksburg, MS, 15 pp

Johnson WB, Gosselink JG (1982) Wetland loss directly associated with canal dredging in the Louisiana coastal zone. In: Boesch DF (ed) Proceedings of the conference on coastal erosion and wetland modification in Louisiana: causes, consequences, and options. U.S. Fish and Wildlife Service, Biological Services Program, Washington, D.C. FWS/OBS-82/59. pp. 60–72

Johnston JB, Cahoon DR, La Peyre MK (2009) Outer continental shelf (OCS)-related pipelines and navigation canals in the Western and Central Gulf of Mexico: relative impacts on wetland habitats and effectiveness of mitigation. U.S. Dept. of the Interior, Minerals Management Service, Gulf of Mexico OCS Region, New Orleans, LA. OCS Study MMS 2009-048, 200 pp

Kasmarek MC (2012) Hydrogeology and simulation of groundwater flow and land-surface subsidence in the northern part of the Gulf Coast aquifer system, Texas, 1891–2009. Scientific Investigations Report 2012-5154. Harris–Galveston Subsidence District, Fort Bend Subsidence District, Lone Star Groundwater Conservation District, TX. https://doi.org/10.3133/sir20125154

Ko JY, Day JW (2004) A review of ecological impacts of oil and gas development on coastal ecosystems in the Mississippi Delta. Ocean Coast Manag 47:597–623

Ko JY, Day J, Barras J, Morton R, Johnston J, Steyer G, Kemp GP, Clairain E, Theriot R (2004) Impacts of oil and gas activities on coastal wetland loss in the Mississippi Delta. In: Cleveland CJ (ed) Encyclopedia of energy, vol 6. Elsevier, pp 397–408

Kolker AS, Allison MA, Hameed S (2011) An evaluation of subsidence rates and sea level variability in the northern Gulf of Mexico. Geophys Res Lett 38:L21404. https://doi.org/10.1029/2011GL049458

Macfarlane GJ, Renilson MR (1999) Wave wake – a rational method for assessment. Proc. RINA Intl. Conf. on Coastal Ships and Inland Waterways, London, UK

Mallman EP, Zoback MD (2007) Subsidence in the Louisiana Coastal Zone due to hydrocarbon production. J Coast Res SI 50:443–449

Mariotti G, Fagherazzi S (2010) A numerical model for the coupled long-term evolution of salt marshes and tidal flats. J Geophy Res 115. https://doi.org/10A640954

Mariotti G, Fagherazzi S, Wiberg PL et al (2010) Influence of storm surges and sea level on shallow tidal basin erosive processes. J Geophys Res 115:C11012

Marani M, d'Alpaos A, Lanzoni S, Santalucia M (2011) Understanding and predicting wave erosion of marsh edges. Geophys Res Lett 38. https://doi.org/10.1029/2011GL048995

McGenity TJ, Folwell BD, McKew BA, Sanni GO (2012) Marine crude-oil biodegradation: a central role for interspecies interactions. Aquat Biosyst 8:10

McGinnis JT, Ewing RA, Willingham CA et al (1972) Environmental aspects of gas pipeline operations in the Louisiana coastal marshes. Battelle Memorial Institute, Columbus Laboratories, Columbus, OH. 125 pp

McKee KL, Mendelssohn IA (1989) Response of a freshwater marsh plant community to increased salinity and increased water level. Aquat Bot 34:301–316

McLindon C (2016) Using oil and gas industry data to achieve coastal sustainability. SOC Block VI. https://sites.law.lsu.edu/coast/2016/06/state-of-the-coast-2016/

Mendelssohn IA, McKee KL (1988) *Spartina alterniflora* die-back in Louisiana: time-course investigation of soil waterlogging effects. J Ecol 76:509–521

Mendelssohn IA, Morris JT (2000) Eco-physiological controls on the productivity of Spartina alterniflora loisel. In: Weinstein MP, Kreeger DA (eds) Concepts and controversies in tidal marsh ecology. Kuwer Academic Publishers, Boston, MA, USA, pp 59–80

Mitsch WJ, Gosselink JG (2015) Wetlands. Wiley, Hoboken, N.J., 582 pp

Monte JA (1978) The impact of petroleum dredging on Louisiana's coastal landscape—A plant biogeographical analysis and resource assessment of spoil bank habitats in the Bayou Lafourche delta. Ph. D. dissertation, Louisiana State University, Baton Rouge, LA. 321 pp

Morris JT, Sundberg K, Hopkinson CS (2013) Salt marsh primary production and its responses to relative sea level and nutrients in estuaries at Plum Island, Massachusetts, and North Inlet, South Carolina, USA. Oceanography 26:78–84

Morton RA, Bernier JC (2010) Recent subsidence-rate reductions in the Mississippi delta and their geological implications. J Coast Res 29:555–561

Morton RA, Buster NA, Krohn MD (2002) Subsurface controls on historical subsidence rates and associated wetland loss in southcentral Louisiana. Trans Gulf Coast Assoc Geol Soc 52:767–778

Morton RA, Bernier JC, Barras JA (2006) Evidence of regional subsidence and associated interior wetland loss induced by hydrocarbon production, Gulf Coast region, USA. Environ Geol 50:261–274

Morton RA, Bernier JC, Barras JA, Ferina NF (2005) Rapid subsidence and historical wetland loss in the south-central Mississippi delta plain: likely causes and future implications. U.S. Geological Survey Open-file Report 2005-1216. http://www.pubs.usgs.gov/of/2005/1216

Neill C, Turner RE (1987) Backfilling canals to mitigate wetland dredging in Louisiana coastal marshes. Environ Manag 11:823–836

Nichols LG (1959) Technical report of the Louisiana Wildlife and Fisheries Commission, Rockefeller Refuge levee study. Louisiana Wildlife and Fisheries Commission, Refuge Division, New Orleans, Louisiana

Nichols LG (1961) Erosion of canal banks on the Rockefeller Wildlife Refuge. Louisiana Wildlife and Fisheries Commission, Refuge Division, New Orleans, LA. 9 pp

Nyman J, DeLaune R, Pezeshki R, Patrick W (1995) Organic matter fluxes and marsh stability in a rapidly submerging estuarine marsh. Estuaries 18:207–217

Olea RA, Coleman JL Jr (2014) A synoptic examination of causes of land loss in Southern Louisiana as related to the exploitation of subsurface geologic resources. J Coast Res 30:1025–1044

Penland S, Wayne L, Britsch LD, Williams SJ, Beall AD, Butterworth VC (2000a) Geomorphic classification of coastal land loss between 1932 and 1990 in the Mississippi River delta plain, southeastern Louisiana: U.S. Geological Survey Open-File Report 00-417, 1 sheet

Penland S, Wayne L, Britsch LD, Williams SJ, Beall AD, Butterworth VC (2000b) Process classification of coastal land loss between 1932 and 1990 in the Mississippi River delta plain, southeastern Louisiana: U.S. Geological Survey Open-File Report 00-418, 1 sheet

Pethick J (1993) Shoreline adjustments and coastal management: physical and biological processes under accelerated sea-level rise. Geograph J 159:162–168

Phien-wej N, Giao PG, Nutalya P (2006). Land subsidence in Bangkok, Thailand. Eng Geol 82:87–201

Phillips JD (1986) Coastal submergence and marsh fringe erosion. J Coast Res 2:427–436

Poland JF, Lofgren BE, Ireland RL, Pugh RG (1972) Land subsidence in the San Joaquin valley, California, as of 1972. USGS Professional Paper 437-H, 78 p

Pratt WE, Johnson DW (1926) Local subsidence of the Goose Creek oil field. J Geol 34:577–590

Price FD (2005) Quantification, analysis, and management of Intracoastal Waterway channel margin erosion in the Guana Tolomato Matanzas National Estuarine Research Reserve, Florida. M.S. Thesis, Florida State University, Tallahassee, FL. 80 pp

Rabalais NN, McKee BA, Reed DJ, Means JC (1992) Fate and effects of produced water discharges in coastal Louisiana, Gulf of Mexico, USA. Produced water. Springer, Boston, MA, pp 355–369

Rabalais NN, McKee BA, Reed DJ, Means JC (1991) Fate and effects of nearshore discharges of OCS produced waters, vol 1. Executive summary. OCS Study/MMS 91-0004. U.S. Dept. of the Interior, Minerals Management Service, Gulf of Mexico OCS Regional Office, New Orleans, LA, 48 pp

Raposa KB, McKinney RA, Wigand C (2018) Top-down and bottom-up controls on southern New England salt marsh crab populations. PeerJ 6:e4876. https://doi.org/10.7717/peerj.4876

Ray GL (2007) Thin layer placement of dredged material on coastal wetlands: a review of the technical and scientific literature. ERDC/EL Technical Notes Collection (ERDC/EL TN-07-1). U.S. Army Engineer Research and Development Center, Vicksburg, MS

Reed DJ (1992) Effect of weirs on sediment deposition in Louisiana coastal marshes. Environ Manag 16:55–65

Reed DJ (1995) Status and historical trends of hydrologic modification, reduction in sediment availability, and habitat loss/modification in the Barataria and Terrebonne estuarine system. BTNEP Publ. No. 20, Barataria-Terrebonne National Estuary Program, Thibodaux, Louisiana, 338 pp, plus appendices

Reed DJ, Cahoon DR (1992) The relationship between marsh surface topography and vegetation parameters in a deteriorating Louisiana *Spartina alterniflora* salt marsh. J Coast Res 8:77–87

Reed DJ, Wilson L (2004) Coast 2050: a new approach to restoration of Louisiana coastal wetlands. Phys Geogr 25:4–21

Reed DJ, de Luca N, Foote AL (1997) Effect of hydrologic management on marsh surface sediment deposition in coastal Louisiana. Estuaries 20:301–311

Rogers DR, Rogers BD, Herke WH (1992) Effects of a marsh management plan on fishery communities in coastal Louisiana. Wetlands 12:53–62

Rosen PS (1977) Increasing shoreline erosion rates with decreasing tidal range in the Virginia Chesapeake Bay. Chesapeake Sci 18:383–386

Sapkota Y, White JR (2019) Long-term fate of rapidly eroding carbon stock soil profiles in coastal wetlands. Sci Total Environ 753. https://doi.org/10.1016/j.scitotenv.2020.141913

Sasser CE, Dozier MD, Gosselink JG, Hill JM (1986) Spatial and temporal changes in Louisiana's Barataria basin marshes, 1945–1980. Environ Manag 10(5):671–680

Scaife WW, Turner RE, Costanza R (1983) Coastal Louisiana recent land loss and canal impacts. Environ Manag 7:433–442

Schwimmer RE (2001) Rates and processes of marsh shoreline erosion in Rehoboth Bay, Delaware, USA. J Coast Res 17:672–683

Sestini G (1992) Implications of climatic changes for the Po delta and Venice lagoon. In: Jeftic L. Milliman J, Sestini G (eds.) Climatic Change and the the Mediterranean. Edward Arnold, London, England. pp 428-494

Snedden GA, Cretini K, Patton B (2015) Inundation and salinity impacts to above- and belowground productivity in Spartina patens and Spartina alterniflora in the Mississippi River deltaic plain: implications for using river diversions as restoration tools. Ecol Eng 81:133–139

St. Pe KM (1990) An assessment of produced water impacts to low-energy brackish water systems in southeast Louisiana. Louisiana Department of Environmentally Quality, Water Pollution Control Division, Baton Rouge, Louisiana, 204 pp

Stevenson JC, Kearney MS, Pendleton EC (1985) Sedimentation and erosion in a Chesapeake Bay brackish marsh system. Mar Geol 67(3–4):213–235

Stone JH, Bahr LM, Day Jr JW (1978) Effects of canals on freshwater marshes in coastal Louisiana and implications for management. In: Good RE, Whigham DF, Simpson RL (eds) Freshwater wetlands—Ecological processes and management potential, Academic Press, New York, NY, pp 291–321

Stumpt RP, Haines JW (1998) Variations in tidal level in the Gulf of Mexico and implications for wetlands. Est Coast Shelf Sci 46:165–173

Swenson EM, Turner RE (1987) Spoil banks: effects on a coastal marsh water-level regime. Estuar Coast Shelf Sci 24:599–609

Swenson EM (1983) Marsh hydrological studies, 1982–1983 data report. Coastal Ecology Institute, Center for Wetland Resources, Louisiana State University, Baton Rouge. Prepared for the National Marine Fisheries Service, Southeast Region, St. Petersburg, Florida. Contract no. NA81-BA-P00006. Publication no. LSU-CEFI-83-18

Taylor NC, Day JW, Neusaenger GE (1989) Ecological characterization of Jean Lafitte National Historical Park, Louisiana: basis for a management plan. In: Marsh Management in Coastal Louisiana: effects and issues—proceedings of a symposium

Taylor NC (1988) Ecological characterization of Jean Lafitte National Historical Park. Louisiana: Basis for a management plan. M.S. Thesis, Louisiana State University, Baton Rouge, LA

Tosi L, Lio CD, Strozzi T, Teatini P (2016) Combining L- and X-band SAR interferometry to assess ground displacements in heterogeneous coastal environments: The Po River Delta and Venice Lagoon, Italy. Remote Sens 8(4):308. https://doi.org/10.3390/rs8040308

Trosclair K (2013) Wave transformation at a saltmarsh edge and resulting marsh edge erosion: Observations and modeling. M.S. Thesis, University of New Orleans, New Orleans, LA. 86 pp

Turner RE (1987) Relationship between canal and levee density and coastal land loss in Louisiana. US Fish Wildlife Service Biological Report 85(14). U.S. Department of Interior, Washington, D.C., 58 pp

Turner RE (1990) Landscape development and coastal wetland losses in the Northern Gulf of Mexico. Am Zool 30:89–105

Turner RE (1997) Wetland loss in the Northern Gulf of Mexico: multiple working hypotheses. Coast Estuar Res Fed 20:1–13

Turner RE, Cahoon DR (eds) (1987) Causes of wetland loss in the coastal Central Gulf of Mexico, vol II: technical narrative. Prepared for Minerals Management Service, New Orleans, LA. Contract No. 14-12-0001-30252. OCS Study/MMS 87-0120, 400 pp

Turner RE, Swenson E (2020) The life and death and consequences of canals and spoil banks in salt marshes. Wetlands. https://doi.org/10.1007/s13157-020-01354-w

Turner RE, Swenson E, Lee J (1994) A rationale for coastal wetland restoration through spoil bank management in Louisiana, USA. Env Mngt 18:271–282

Turner RE, McClenachan G (2018) Reversing wetland death from 35,000 cuts: opportunities to restore Louisiana's dredged canals. PLoS One 13(12):e0207717. https://doi.org/10.1371/journal.pone.0207717

Turner RE, Rao YS (1990) Relationships between wetland fragmentation and recent hydrologic changes in a deltaic coast. Estuaries 13:72–281

Turner RE, Costanza R, Scaife W (1982) Canals and land loss in coastal Louisiana. In: Boesch DF (ed) Proceedings of the conference on coastal erosion and wetland modification in Louisiana: causes, consequences, and options, pp 73–84. U.S. Fish and Wildlife Service, Biological Services Program, 256 pp

USACE (2006) Hydraulic design of deep-draft navigation projects. Engineer Manual EM 1110-2-1613. Washington, DC. 212 pp

Valentine K, Mariotti G (2019) Wind-driven water level fluctuations drive marsh edge erosion variability in microtidal coastal bays. Cont Shelf Res 176:76–89

Wang FC (1988) Dynamics of saltwater intrusion in coastal channels. J Geophys Res 93(C6):6937–6946

Wang H, van der Wal D, Li X, et al (2017) Zooming in and out: scale dependence of extrinsic and intrinsic factors affecting salt marsh erosion. Earth Surf 122:1455–1470

Ward AW, Dixon GL, Jachens RC (1993) Geologic setting of the East Antelope Basin, with emphasis on fissuring on Rogers Lake, Edwards AFB, Mojave Desert, California. U.S. Geological Survey Open-File Report 93-263. 9p

Watzke DA (2004) Short-term evolution of a marsh island system and the importance of cold front forcing, Terrebonne Bay, Louisiana. M.S. Thesis, Louisiana State University, Baton Rouge, LA

White WA, Calnan TR (1990) Sedimentation and Historical Changes in Fluvial-Deltaic Wetlands along the Texas Gulf Coast with Emphasis on the Colorado and Trinity River Deltas. The University of Texas at Austin, Bureau of Economic Geology, Final report prepared for the Texas Parks and Wildlife De-partment, Austin, Texas. 124 pp

Wilson C, Allison MA (2008) An equilibrium profile model for retreating marsh shorelines in southeast Louisiana. Est Coast Shelf Sci 80:483–494

Yu S-Y, Tornqvist TE, Hu P (2012) Quantifying Holocene lithospheric subsidence rates underneath the Mississippi Delta. Earth Planet Sci Lett 331–332:21–30. https://doi.org/10.1016/j.epsl.2012.02.021

Zoback MD, Zinke JC (2002) Production-induced normal faulting in the Valhall and Ekofisk oil fields. In: Trifu CI (ed) The mechanism of induced seismicity. Pageoph topical volumes. Birkhauser, Basel, pp 403–420

Chapter 6
Chemical and Toxin Impacts of Oil and Gas Activities on Coastal Systems

John H. Pardue and Vijaikrishnah Elango

6.1 Introduction

Coastal wetlands lie at the interface of the terrestrial and marine environment and are susceptible to chemical spills from anthropogenic activities on the coastal plain. The northern Gulf of Mexico is one location where coastal ecosystems are coincident with high crude oil exploration and production activity (Day et al. 2020; Delaune et al. 1990; Ko and Day 2004). As a result, spill frequency is high and these systems are regularly exposed to crude oil (Grubesic et al. 2017). Natural seeps also exist in the region which provide another point of exposure to coastal wetlands in the region (Johansen et al. 2020; Krajewski et al. 2018). Spilled crude oil and the highly saline "produced waters" that are liberated during production can cause a range of impacts to these ecosystems, both direct and indirect (Fig. 6.1). Crude oil and brine can be directly toxic to marsh species through coating and smothering mechanisms, osmotic stress due to high salinity, and direct toxicity from components such as polycyclic aromatic hydrocarbons (PAHs). Impacts can also be indirect, through alteration of physical, chemical and biological gradients which define the ecological organization of these unique coastal systems. This chapter will review the impacts of toxins released during oil and gas production and exploration on coastal wetland ecosystems. We will review the current state of knowledge of these impacts with an emphasis on what has been learned from the Macondo oil spill in 2010.

The original version of this chapter was revised: The author "Vijaikrishnah Elango's" affiliation has been updated. The correction to this chapter is available at https://doi.org/10.1007/978-3-030-94526-8_10

J. H. Pardue (✉) · V. Elango
Department of Civil and Environmental Engineering, Louisiana State University, Baton Rouge, LA, USA
e-mail: jpardue@lsu.edu

© The Author(s), under exclusive license to Springer Nature Switzerland AG 2022, corrected publication 2022
J. W. Day et al. (eds.), *Energy Production in the Mississippi River Delta*, Lecture Notes in Energy 43, https://doi.org/10.1007/978-3-030-94526-8_6

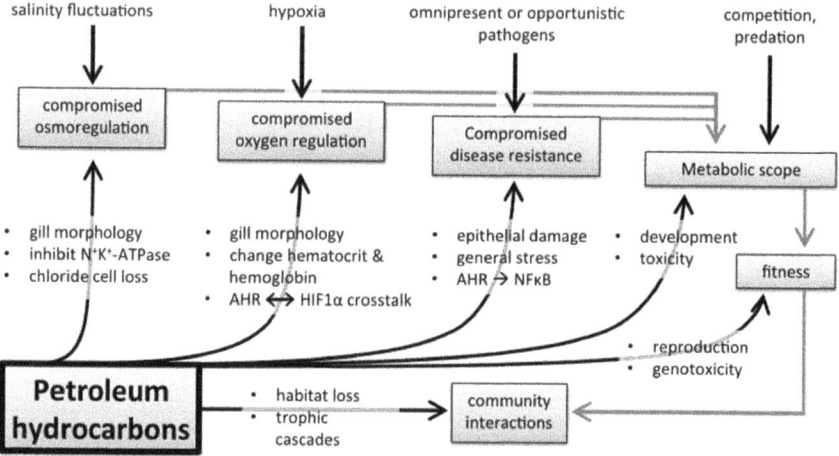

Fig. 6.1 Conceptual model of interactions between petroleum hydrocarbon pollutants and environmental stressors that are naturally encountered in estuaries. Natural ecological stressors (top row) are connected by black down-facing arrows to the physiological functions that are impaired by interactions with petroleum hydrocarbons. Petroleum hydrocarbons are connected to physiological effects by remaining black arrows. The mechanisms that underlie interactions between hydrocarbons and natural stressors are listed and superimposed above black arrows. Gray arrows indicate the flow of downstream effects (Whitehead 2013)

6.2 Crude Oil Composition

Hydrocarbon exploration and production activities generate crude oil, a complex mixture of hydrocarbons, and produced waters, which consist of saline waters that are brought up from the formation during production. The produced waters have chemical characteristics that differ according to basins (Collins 1970) but constituents common to all locations include salts, residual oils, organic and inorganic compounds (Neff 2002) and naturally occurring radioactive materials (NORM), including radium-226 and -228 (Pardue and Guo 1998). Crude oils also retain basin-specific chemical composition that can be separated into saturate, aromatic, resin and asphaltene classes (Raki et al. 2000). Detailed compositional characteristics of these oils have improved with modern analytical tools including multidimensional chromatography (Reddy et al. 2012) and high-resolution mass spectrometry (Huba and Gardinali 2016). Composition of the Macondo oil released during the Deepwater Horizon oil spill is presented in Table 6.1.

As these materials are released into coastal wetland environments, physical, chemical and biological transformation processes occur that steadily change oil and produced water composition (Fig. 6.2). These processes can serve to mitigate or accelerate exposure to crude oil and produced waters. Physical processes include advection and dispersion that move and mix oil and brine, coupled with mass transfer processes that partition contaminants between water, air, biota and sediments. These mixing events are driven by local and regional coastal processes (e.g., the beach

Table 6.1 Composition of Macondo Oil (MW-1[a]) (Reddy et al. 2012)

Analyte	Class	Content (%)
n-alkanes	Saturates	15
Branched alkanes		26
Cycloalkanes		16
Alkylbenzenes and indenes	Aromatics	9
Polycyclic aromatic hydrocarbons		3.9
Biomarkers		1.6
Others		18
Polars	Resins and asphaltenes	10

[a]Sample collected from wellhead on June 21, 2010

aggradation/degradation cycle; natural marsh accretion and subsidence; alongshore sediment transport) that shape these ecosystems (Curtis et al. 2018). Biologically, the organic fraction in crude oils and produced waters can be transformed by microbial and fungal species that enzymatically degrade these compounds to lower molecular weight end products including CO_2 (Elango et al. 2014; Pardue et al. 2014). The sun can also modify composition significantly, leading to compositional changes that are just beginning to be understood (Ray et al. 2014; Ray and Tarr 2014, 2015). Chemical changes also occur such as the coprecipitation of radium with barite near production pits (Pardue and Guo 1998).

As these processes proceed, the intensity and extent of impacts on these coastal systems changes. For the Macondo spill, one of the largest oil spills to date (Fig. 6.3), the compositional changes that occurred as oil and gas moved from the wellhead to the ocean surface and from the ocean surface to the shoreline were significant in understanding the nature of the impact. Oil reached the shoreline as a heavily modified water-in-oil emulsion, missing concentrations of key toxic components (i.e., alkylated naphthalenes) that volatilized at sea (Elango et al. 2014; Lemelle et al. 2014; Urbano et al. 2013). The timing of arrival of this oil to the shoreline was coincident with two tropical weather events, Hurricane Alex and Tropical Storm Bonnie, which served to move emulsified oil further into the shoreline (Curtis et al. 2018). These events and the associated compositional changes resulted in a different exposure profile altogether from a nearshore black oil spill.

Changes in the relative composition of four classes of polycyclic aromatic hydrocarbons (PAHs) during the Macondo spill are presented in Fig. 6.4. As the oil left the wellhead, alkylated naphthalenes were the dominant fraction, yet as the oil reached the surface of the ocean and was carried to the adjacent coastal headland beaches, higher molecular weight, alkylated phenanthrenes became the dominant PAH family. As oil remained on the beach surface or was buried beneath the beach surface, slower biodegradation processes removed 3-ring alkylated phenanthrenes and dibenzothiophenes, leaving the 4-ring chrysenes as the dominant PAH family. Finally, Macondo oil also underwent extreme fractionation during uptake into organisms. In

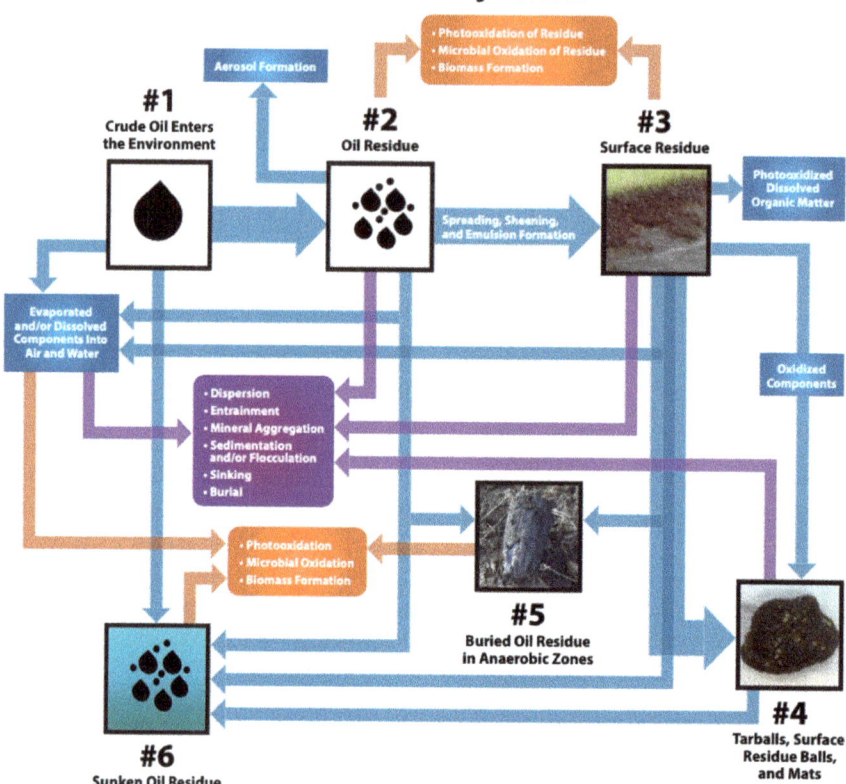

Fig. 6.2 Diagram of the various stages of environmental weathering of crude oil. The initial oil that enters the environment is depicted as #1. After loss of some components by dissolution and evaporation, it becomes an oil residue (#2). Upon further weathering and movement, the residue becomes more viscous and may mix with water to form an emulsion (#3). Further weathering through biodegradation and photooxidation produces a semisolid residue (#4). Some of the surface residue (#3) may be buried in anoxic zones, becoming residue #5, and its further weathering is very slow. Some heavy crude oils can sink after evaporative/dissolution losses, and other oil residues can sink through sedimentation, flocculation, and mineral aggregate formation. Components lost from the oil and oil residues undergo microbial and photooxidations with differing time scales. Microbial and photooxidations also cause changes in the oil residues themselves, including the formation of high molecular weight compounds. Some of the oxidized high and low molecular material can be assimilated into the water column and become dissolved organic matter (Tarr et al. 2016)

mangrove leaf tissue, for example, only the C1- and C2-naphthalenes crossed into the cuticle from contaminated air, even though the percentage of alkylated naphthalenes remaining in the oiled marsh was less than 5% of the total PAHs (Fig. 6.4).

The evolution of crude oil as it is modified though physical, chemical, and biological processes is one of the constant themes that run through all spills and planned releases of exploration and production waste. Others themes also are important in understanding impacts on coastal systems. For example, crude oil spills result in

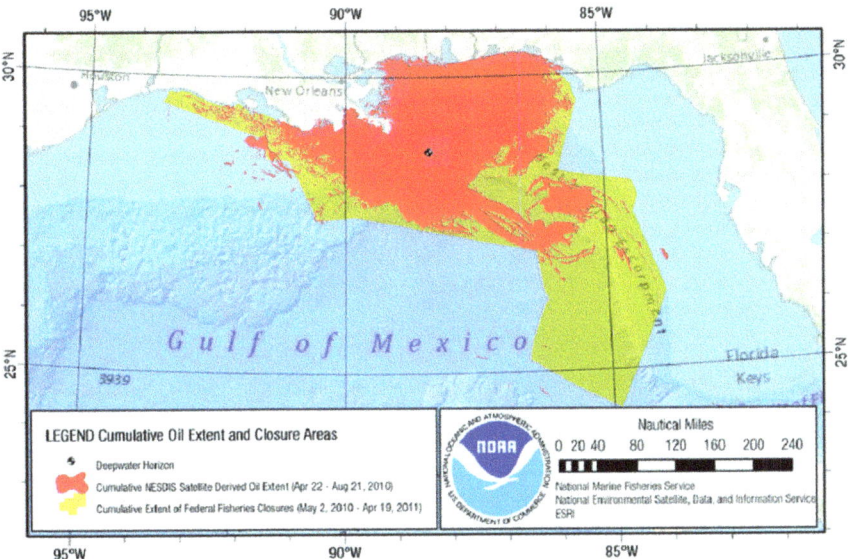

Fig. 6.3 Gulf of Mexico map showing the extent of oiling that occurred from April 22 through August 21 after the Deepwater Horizon platform explosion occurred on April 20, 2010. The black and white circle indicates the wellhead location. The orange shaded areas show the cumulative National Environmental Satellite, Data and Information Service (NESDIS) satellite footprints of the oil. The yellow polygon overlay shows how the federal fisheries closure areas aligned with the oil distribution (Ylitalo et al. 2012)

an increased demand for oxygen in the ecosystem (Shin and Pardue 2001; Shin et al. 2000, 2001a). In coastal systems, the oxygen gradients are defining drivers of ecological function (Levine et al. 2017; Mueller et al. 2020; Reddy and Delaune 2008). As available oxygen changes, functions change. These oxygen impacts are described below. Another theme is the importance of understanding coastal physical processes when studying impacts. Spilled oil in the form of oil: sand aggregates, water-in-oil emulsions, and condensed oil such as "tarballs" are mobilized or attenuated by coastal processes such as erosion, alongshore sediment transport, and beach aggradation/erosion (Curtis et al. 2018; Georgiou et al. 2005) (see Fig. 6.2). As a result, oil cycles in these systems for years after the original spill or permitted release. Finally, in any system, the impact of oil depends on the bioavailability of the crude oil components to the higher organisms potentially impacted by these components or the microorganisms that can degrade them. Bioavailability refers to the complex interactions of physical and chemical processes that allow crude oil components, including those that are sparingly soluble such as PAHs, to enter the organisms where they can potentially cause impacts. Emulsified oil, which characterized the Macondo spill, has a higher surface contact with water because of the droplets of seawater mixed into the oil. Yet when the emulsions reached the marsh, they generally did not disperse into the marshes but were deposited as discrete layers, likely minimizing their bioavailability when compared to a more dispersed oil.

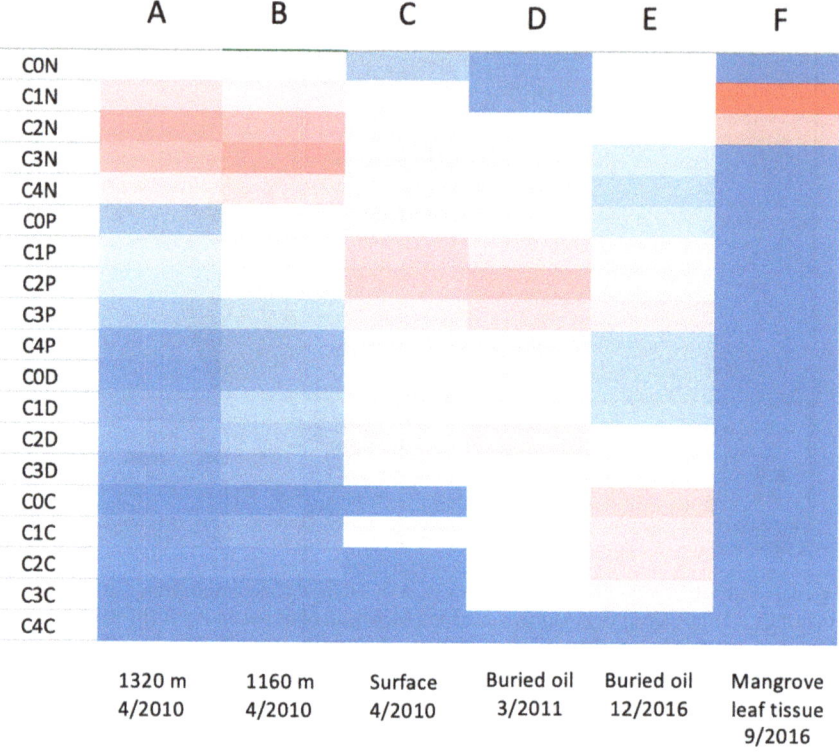

Fig. 6.4 Heat map of distribution of polycyclic aromatic hydrocarbons (PAHs) following the Macondo spill as the oil evolved through **a** site 33, 1320 m depth near Macondo wellhead site (Diercks et al. 2010), **b** 1160 m depth near Macondo wellhead site (Diercks et al. 2010), **c** site 30, ocean surface near Macondo wellhead site (Diercks et al. 2010), **d** buried oil on Fourchon Beach beneath water table, March 1 2011, D-D' (Pardue et al. 2014), E. Buried oil on Fourchon Beach above the water table (Romaine et al. 2021), **f** Mature mangrove leaf, Bay Batiste Barataria Basin, September 30, 2016. Abbreviations: naphthalene (C_0N), C1-naphthalenes (C_1N), C2-naphthalenes (C_2N), C3-naphthalenes (C_3N), C4-naphthalenes (C_4N), phenanthrene (C_0P), C1-phenanthrenes (C_1P), C2-phenanthrenes (C_2P), C3-phenanthrenes (C_3P), C4-phenanthrenes (C_4P), dibenzothiophene (C_0D), C1-dibenzothiophenes (C_1D), C2-dibenzothiophenes (C_2D), C3-dibenzothiophenes (C_3D), chrysene (C_0C), C1-chrysenes (C_1C), C2-chrysenes (C_2C), and C3-chrysenes (C_3C). Dark red represents highest proportion of PAH fractions. Dark blue represents lowest proportion of the PAH fraction

6.3 Impacts of Produced Water Discharges on Coastal Systems

"Produced" waters, which are liberated from the formation during oil exploration and production, are highly saline brines with specific organic (e.g., crude oil components and production chemicals used in the recovery process) and inorganic pollutants (e.g., boron and radioactive salts) that create impacts when released in coastal environments

Table 6.2 Common components of produced water resulting from oil and gas production (Neff et al. 2011)

Category	Constituent
Metals	Aluminum, antimony, arsenic, barium, boron, cadmium, gold, iron, mercury, chromium, copper, lead, zinc, nickel, platinum, silver, strontium, tin, magnesium, molybdenum, titanium, thallium, vanadium
Naturally occurring radioactive materials	Radium-226, radium-228
Monocyclic aromatic hydrocarbons	Benzene, benzoic acid, chlorobenzene, di-n-butyl phthalate, toluene, ethylbenzene, isopropylbenzene, n-propylbenzene, p-chloro-m-cresol, xylenes
Polycyclic aromatic hydrocarbons	Acenaphthene, anthracene, benzo(a)anthracene, chrysenes, dibenzothiophenes, fluoranthene, fluorene, naphthalenes, phenanthrenes, benzo(a)pyrene
Miscellaneous organic chemicals	Phenols, bis(2-ethylhexyl)phthalate, oil and grease
Miscellaneous other components	Calcium, potassium, radon, sodium chloride, total dissolved solids, ammonia, bromides, sulfates, sulfides

(Neff 2002) (Table 6.2). For every barrel of oil produced, 7–10 barrels (280–400 gallons) of water are generated, mostly in the form of a brine (Guerra et al. 2011). Historically, brine discharges occurred directly into coastal waters (Boesch and Rabalais 1989; St. Pe' 1990). Brine composition differs with basin type and location, although the typical produced water has higher salinity than full-strength seawater (Blondes et al. 2018; Collins 1970). The relative proportions contributing to produced water ionic strength can differ significantly from seawater. For example, sulfate can be completely depleted in certain produced waters (Collins 1970). Because the salinity difference between the produced water brines and natural waters along typical estuarine gradients is so large, stratified multiphase flow can occur (Boesch and Rabalais 1989). In this situation, multiphase flow is defined by separation of the brine into a distinct layer that sinks beneath natural waters and moves separately. These multiphase flows damage these ecosystem because the brines lie on the sediment surface, directly impacting benthic populations. These flows also encourage the rapid development of hypoxic conditions because of the limited oxygen solubility of saline waters, high sediment oxygen demand and the lack of mixing (Buzzelli et al. 2002; Ritter and Montagna 1999).

Fate and effects of produced waters in coastal environments

Produced waters directly discharged into coastal water cause a cascade of problems including direct toxicity from salts, specific toxicity from phytotoxins like boron and hypoxic impacts due to multiphase flow. Improper disposal of brines in coastal environments was a common practice along the US Gulf of Mexico until the mid-1990s when more restrictive practices were adopted by the states. At present, discharges are regulated under the federal oil and gas extraction effluent limitation guidelines and pretreatment standards (40 CFR Part 435). These standards apply to offshore and onshore oil and gas producers and centralized waste facilities. Currently, the majority of onshore produced water discharges in the coastal zone are reinjected through Type 1 or Type 2 injection wells which returns these highly saline waters to deep formations. Historically, discharges directly to surface water (including wetlands) occurred across the coastal zone from the exploration activities, with little or no provisions for mixing or dilution using diffusers (Jiang and Law 2013) designed to minimize salinity changes.

Macrophytes are key components of coastal environments and salinity and inundation gradients dictate the location of different species across estuaries (Hester et al. 2001). Discharges of brines can impact vegetation of these systems through salinity stress or through the action of specific phytotoxins (Gacia et al. 2007). In coastal wetlands also stressed from sea level rise, subsidence, and wave action, additional salinity stress results from brine discharges that reduce plant fitness and productivity. Coastal halophytes such as *Spartina* spp. possess various mechanisms to overcome highly saline conditions but they still prefer conditions that are well below the ionic strength of full-strength seawater (Yuan et al. 2019).

In addition to impacts to keystone species like marsh macrophytes, highly saline brines can affect other fauna. Exposure to produced waters and the associated pollutants (salinity, crude oil components, metals, and radioisotopes) represent an additional stressor on these systems (Adams 2005). Alkylphenols and PAHs impacted cod and blue mussels near offshore discharge points in Norway (Bakke et al. 2013). Brine discharges in the Gulf of Mexico produced measurable impacts even in the offshore environment as measured in fish tissues (Schifter et al. 2015). Projecting impacts in the coastal zone when direct discharge occurred to poorly flushed receiving waters may benefit from examining literature on discharge of brine from desalination facilities in coastal regions (Belatoui et al. 2017; Clark et al. 2018).

Current brine disposal guidelines try to minimize salinity changes in receiving water, whether from produced waters from conventional and unconventional oil exploration, discharges from salt mining or from desalination-reject water from membrane facilities. These discharges impact benthos and fish, even in situations when the salinity changes were relatively low. The greatest impacts from desalination brines have been observed when the oceanic rates of flushing were low (Jenkins et al. 2012). Discharge of brines into coastal waters, particularly poorly flushed bays, inlets and tidal streams that drain marshes, have caused significant impacts. This is why direct discharge into coastal waters has been banned since the mid-1990's.

6.4 Impacts to Oxygen Availability

A significant impact of oil and gas exploration waste discharges and spills is the disruption in the availability of oxygen in coastal waters and sediments. Crude oil, drilling fluids and other production chemicals exert an oxygen demand on coastal waters and sediments as bacteria consume these organic materials (Shin and Pardue 2001; Shin et al. 2000, 2001a, b, 2003). The low solubility of oxygen in coastal waters, exacerbated by salinity and high temperatures, makes the system very sensitive to an influx of organic materials and nutrients (Justic et al. 1996). Oxygen serves as a key electron acceptor for metabolism of these compounds by bacteria and is also required for respiration by macrophytes that underpin primary productivity of these ecosystems (Collins et al. 2020). These oxygen gradients provide a key forcing function for both marsh and coastal headland beach systems (Fig. 6.5). In the marsh, inundation and salinity-tolerant vegetation can only survive if oxygen can be transported to maintain root respiration (Hester et al. 2001). In the marsh and the beach, these oxygen gradients also dictate the rate of respiration of organic matter which proceeds under aerobic conditions or via alternate electron acceptors once oxygen disappears (Delaune et al. 1990). Changes in these redox gradients can alter ecosystem functions (Levine et al. 2017).

Spilled oil increases oxygen demand through direct metabolism of crude oil components by aerobic microorganisms and through oxidation of reduced end products from anaerobic metabolism (Collins et al. 2020; Shin and Pardue 2001; Shin et al. 2000, 2001a). Oil increased oxygen demand threefold in an oiled marsh sediment primarily due to stimulation of sulfate reduction and the subsequent oxygen demand from reduced sulfides (Shin et al. 2000). These processes lead to indirect formation of phytotoxins like hydrogen sulfide that cause additional stress to vegetation (Mendelssohn and Morris 2012). Oiled sands from a coastal headland beach had an order of magnitude increased oxygen demand after oiling from the Macondo spill (Collins et al. 2020). In addition to metabolism, crude oil in the form of a slick can alter oxygen transfer as layers of oil serve as barriers for oxygen diffusion from the atmosphere (Anikiev et al. 1988). The presence of a non-aqueous phase liquid (NAPL) can actually increase the possibility of oxygen transfer since solubility of oxygen in NAPL is higher than in water (Ho et al. 1990; Jia et al. 1996). For example, oxygen is eight times more soluble in n-hexadecane (a C16 alkane and a model NAPL) than in pure water (Ho et al. 1990; Jia et al. 1996). Therefore, depending on the form of oil on the surface (free oil; oiled soil; oiled vegetation), this barrier can exert a positive or negative influence on oxygen transfer on the surface of the marsh or the surface of marsh vegetation. These positive or negative changes to the supply of oxygen by crude oil discharges are countered by increased oxygen demand in coastal environments where oxygen can already be limiting (Delaune et al. 1990; Rodrigue et al. 2020).

Oxygen impacts are not limited to crude oil spills as produced water discharges can also impact oxygen dynamics. The large differences in salinity between produced water and coastal receiving waters often leads to stratified multiphase flow: an intact layer of brine lying at the bottom of the water column (St. Pe' 1990). These stratified flows occur naturally in estuaries (Buzzelli et al. 2002; Lin et al. 2006) particularly when mixing processes slow and temperatures are warm. While this type of flow may be advantageous in minimizing exposure to fish and other organisms that can migrate to other points in the water column (i.e., (Houghton and Dabiri 2019), these multiphase flows create some special problems, especially for benthic organisms.

Fig. 6.5 Oil reaching coastal headland beach and marsh shorelines. **a** Heavily oiled Caminida Beach Headlands, LA; **b** Emulsified oil reaching shoreline on Caminada Headlands in July 2010; **c** oil buried beneath the water table where no oxygen is available on Caminada Headlands in 2014; **d** Aerial view of oiled Barataria Basin marsh shorelines (2010); **e** Emulsified oil reaching Barataria Basin marsh shoreline during Tropical Storm Bonnie in 2010; **f** soft asphalt on marsh surface in Bay Batiste, LA in 2018. All images from J. Pardue and F. Travirca

Highly saline waters have very low oxygen solubilities. For example, distilled water at 27 °C has an oxygen solubility of 7.96 mg/L while a brine with a salinity of $215°/_{oo}$ has an oxygen solubility of only 2.14 mg/L (Sherwood et al. 1991). This small amount of oxygen is readily depleted in the presence of oil and other oxygen demanding substances, leading to a stable, hypoxic layer of water that lays on the sediment. Even under natural stratification, salinity differences of five psu were sufficient to generate a hypoxic lower layer (Buzzelli et al. 2002). Direct discharge of produced waters to brackish or saltwater systems create salinity differences that exceed 100 psu (Collins 1970) leading to stratified flow. Benthic organisms depend on oxygen transfer to the sediments that these discharges would prevent.

The Macondo spill in 2010 and the subsequent impacts on the shoreline shed further light on impacts to oxygen concentrations and redox gradients in these systems. Hydrocarbons, including methane, released during the spill were rapidly metabolized at depth in the ocean by methanotrophs using oxygen as an electron acceptor (Kessler et al. 2011). A persistent oxygen anomaly was observed in 2010 that indirectly showed this degradation process occurring even though the relative demand was not high enough to drive anoxic conditions (Kessler et al. 2011). As oil reached the shoreline, the relative balance between oxygen supply and demand was disrupted. On the beach, oil persisted when buried beneath the water table. At these depths on the beach, weathering through biodegradation slowed substantially, likely as the supply of oxygen through tidal exchange deep in the beach profile could not keep up with demand (Collins et al. 2020; Pardue et al. 2014). Specific laboratory evidence for the role of oxygen in oil persistence was demonstrated in laboratory experiments where oxygenated water moving vertically through an oiled lens led to rapid degradation of crude oil components. In experiments where slow diffusion of oxygen was the only source of supply to the oiled lens, biodegradation completely ceased (Collins et al. 2020). These results were confirmed by microprofiles showing consumption of oxygen just prior to reaching the oiled lens (Fig. 6.6). Buried oil on the western Gulf of Mexico was virtually indistinguishable from oil that reached the shoreline in 2010 (Pardue et al. 2014). In the impacted brackish and salt marshes, oxygen also exerted control on the biodegradation of PAHs years after oil reached the shoreline (Rodrigue et al. 2020).

Overall, oxygen concentrations can be impacted by both spilled crude oil and produced waters entering coastal environments. The balance of supply and demand for oxygen is tenuous due to its low solubility in water and hypoxic areas can develop and disappear during mixing events. Brines discharged to marshes without sufficient

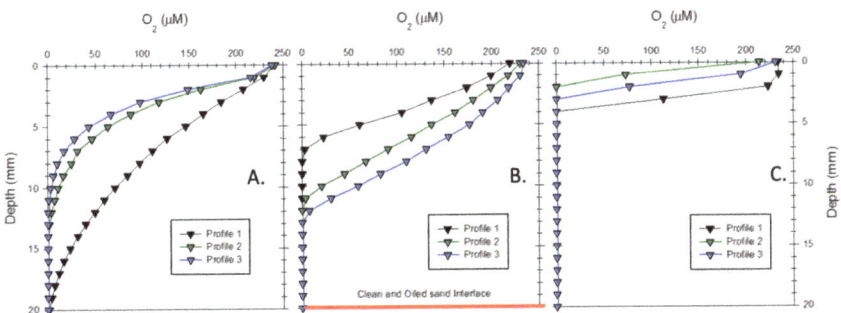

Fig. 6.6 O_2 Microprofiles in **a** unoiled Sands, showing the background oxygen consumption; **b** layered oiled and clean sands, showing the complete consumption of oxygen above the oiled layer due to diffusion of reduced end products and **c** oiled sands showing the impacts of elevated oxygen demand (Collins et al. 2020)

Table 6.3 First-order rate constants for crude oil degradation at Fourchon and Bay Jimmy, LA aerated reactors (Rodrigue et al. 2020)

PAH	Fourchon rate (1/year)	Bay Jimmy rate (1/year)	Significant difference between rates from 2 sites?
C_1P	16 ± 1.5^a	15 ± 3.4	No ($p = 0.772$)
C_2P	13 ± 1.1	11 ± 2.5	No ($p = 0.505$)
C_3P	9.8 ± 0.76	7.5 ± 1.6	No ($p = 0.264$)
C_4P	3.3 ± 0.59	3.5 ± 0.59	No ($p = 0.822$)
C_1D	18 ± 0.88	12 ± 2.4	No ($p = 0.079$)
C_2D	13 ± 1.4	10 ± 2.6	No ($p = 0.367$)
C_3D	8.7 ± 1.6	5.2 ± 0.89	No ($p = 0.129$)
C_1C	3.2 ± 0.40	2.2 ± 0.66	No ($p = 0.625$)
C_2C	2.7 ± 0.10	1.9 ± 0.57	No ($p = 0.239$)
C_3C	0.88 ± 0.71	2.6 ± 0.79	No ($p = 0.093$)

[a] ± 1 standard error

energy to mix them can result in persistent anoxic layers of water and sediments until storms or other high-energy events serve to mix and dilute the brine. The energy required to mix waters of different salinities is proportional to the relative water volumes and salinity differences (Gogate and Pandit 1999). As a result of the large differences in salinities during historical discharges of brines to coastal waters, persistence of stratified flow may occur in the absence of significant storm and wave action in affected areas. A second long-term oxygen impact results from the burial and mixing of oil into marsh and beach sediments after spills. Significant impacts on oxygen and redox gradients were observed in the years after the Macondo spill (Collins et al. 2020; Levine et al. 2017; Pardue et al. 2014; Rodrigue et al. 2020) with little trajectory towards normal conditions under natural recovery processes. Supplying oxygen, however, was sufficient to initiate rapid degradation of weathered oil in marsh soils at Port Fourchon and Bay Batiste, LA (Table 6.3).

6.5 Crude Oil Toxicity

Crude oil components can also exert direct toxicity to organisms. There are four classes of PAHs from crude oil exposure, including *biogenic* PAHS that are produced by organisms, *diagenic* PAHs formed by transformations in soils and sediments, *petrogenic* PAHs that originate from crude oils and *pyrogenic* PAHs that originate from combustion (Neff 1979). An example of biogenic PAHs are the phenanthrenes and 9,10-dihydrophenanthrenes produced by root tissue of *Juncus spp.* that have antimicrobial properties (Toth et al. 2016). In coastal ecosystems, however, petrogenic and pyrogenic sources via water and air, respectively, are the primary toxicity concern (Hylland 2006). Structurally, the petrogenic PAHs are distinguished from pyrogenic PAHs by methyl and ethyl groups on the aromatic rings that allow separate consideration by source. These compounds impact many components of coastal ecosystem food webs. Most organisms, however, possess the ability to metabolize PAHs, forming more reactive metabolites that exhibit toxicity. For example, PAHs in fish undergo Phase 1 metabolism by cytochrome-450 in the liver, forming hydroxylated metabolites that can exhibit carcinogenicity and other toxic responses (Hylland 2006). Subsequently, Phase 2 metabolism also occurs and the water-soluble PAH conjugates are excreted (Hylland 2006).

PAHs originating in crude oil initiate exposure to coastal ecosystem food webs through direct uptake (through water or gills) or through diet (Hylland 2006). The exposure to the biota from PAHs originating in crude oil and produced waters results from complex interactions of organic and inorganic solids and the oil consistency itself. During the Macondo spill, oil reached the shoreline as a water-in-oil emulsion or mousse, increasing its volume but also changing the nature of exposure to organisms (Lemelle et al. 2014; Urbano et al. 2013). Petrogenic PAHs are generally more bioavailable than pyrogenic sources that typically enter these system as soot. Yet depending on the form of oil and the geometry of oil deposition, the bioavailability of crude oil components to coastal ecosystem biota depends on how readily mass transfer can occur to water. As a result, toxicological studies on crude oil typically focus on a water-accommodated fraction (WAF). PAHs readily bioconcentrate in fish, bivalves (Noh et al. 2018), zooplankton (Almeda et al. 2013), amphipods (Lotufo et al. 2016), and macrophytes such as *Spartina alterniflora* (Mohammed et al. 2014; Watts et al. 2006). Since PAH metabolism occurs in many trophic levels in the system (through Phase 1 and 2 metabolism), PAHs do not typically bioaccumulate so trophic exchange is less of a concern than direct chronic or acute toxicity to segments of the food web. Because of this metabolism, crude oil components have different considerations than chlorinated persistent organic compounds like PCBs that are not readily metabolized by the organisms themselves.

The Macondo spill drove significant inquiry into the impacts of spilled oil on coastal ecosystems. Gulf killifish (*Fundulus grandis*), a dominant marsh fish along the northern Gulf of Mexico, was studied extensively (Dubansky et al. 2012, 2017; Whitehead et al. 2012). Killifish in impacted areas had higher cytochrome C4501A mRNA and protein abundances in their liver as well as other tissues including intestinal tissues, indicative of ingestion as a route of exposure (Dubansky et al. 2012, 2017). As described above, these biomarkers are indicative of Phase 1 metabolism in these fish species and indicate that these species are being exposed to hydroxylated PAHs and other reactive metabolites that can damage DNA and cause other developmental abnormalities (Fig. 6.7). Several years later, population studies were unable to detect changes in fish community composition in impacted and reference areas, but these studies did not control for additional stressors (Able et al. 2015). These processes are not confined to killifish species, PAHs are similarly metabolized to reactive compounds in most fish species and hydroxylated metabolites are observed in fish even after exposure to lower molecular weight PAHs like naphthalenes and phenanthrenes (Pulster et al. 2017), the most common PAHs found after the Macondo spill.

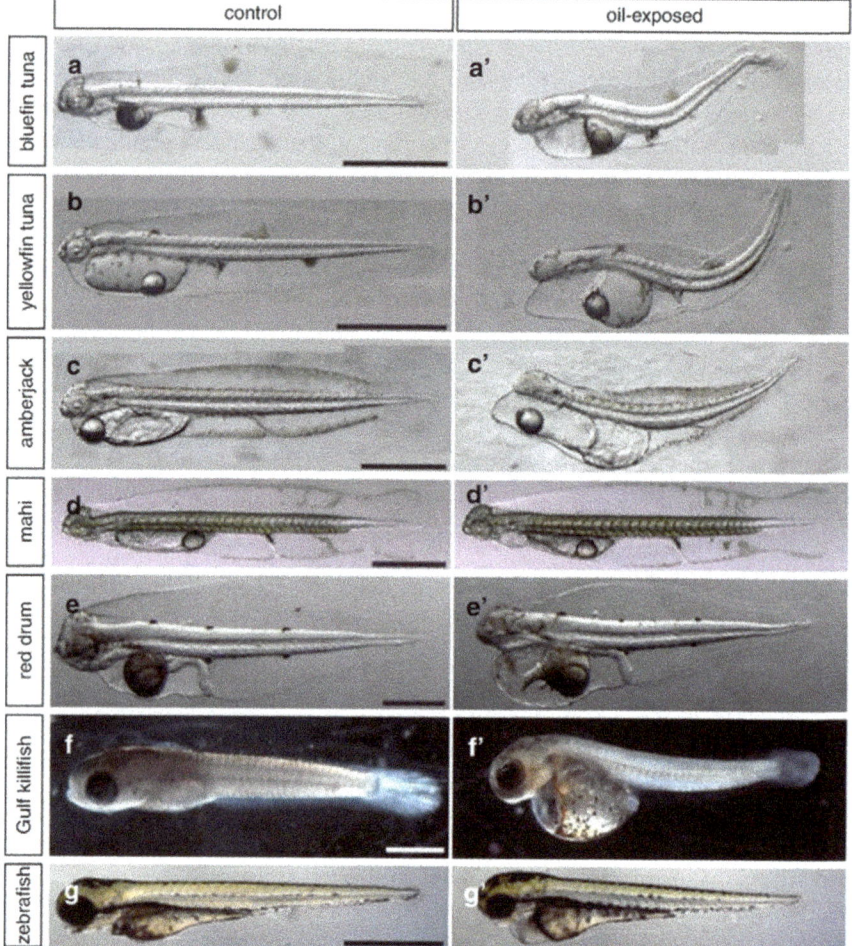

Fig. 6.7 Larval phenotypes of different species exposed to Deepwater Horizon-MC252 crude oil during embryogenesis. Five marine pelagic and two demersal species are represented, with control larvae on the left and exposed larvae on the right: **a** Bluefin tuna, **b** yellowfin tuna, **c** yellowtail amberjack, **d** mahi mahi, **e** red drum, **f** Gulf killifish, and **g** zebrafish. All were exposed to high-energy water accommodated fractions, except for killifish **f** which was exposed to oil-contaminated marsh sediment collected from the Louisiana shoreline (Grand Terre Island). Scale bars are 1 mm (Incardona and Scholz 2018)

A sentinel marsh species, the periwinkle snail (*Littoraria irrorata*), had major population declines in oiled areas (Deis et al. 2020; Gamer et al. 2017; Zengel et al. 2015, 2016) that continue to persist nearly a decade after the spill. Chamber studies demonstrated that snails could not navigate through oiled areas to unoiled vegetation (Gamer et al. 2017) (Fig. 6.8) and subsequent exposures resulted in significant snail mortality. At the field scale, reductions in snail densities as high as 73% were observed

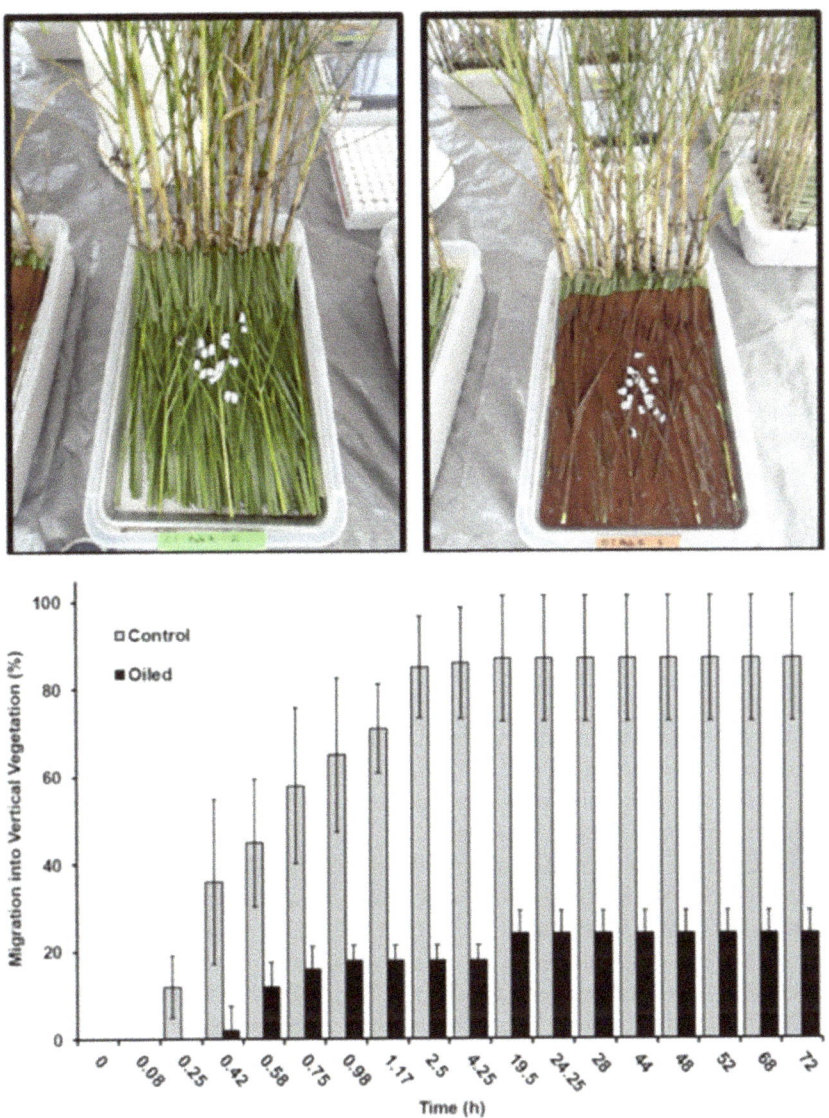

Fig. 6.8 Dimensions of experimental chambers for Movement Assay 1 (**a**) and Movement Assay II (**b**). Black dot indicates location where snails were placed at the beginning of each assay (Gamer et al. 2017)

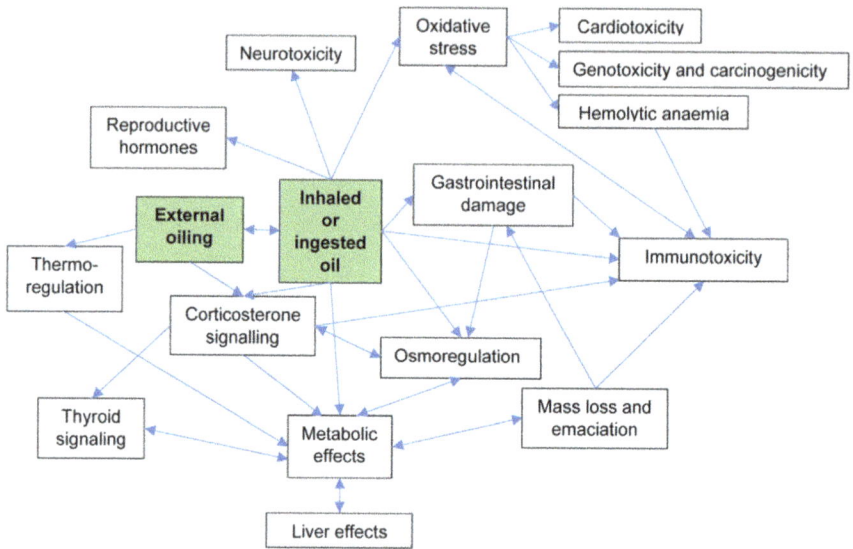

Fig. 6.9 Conceptual diagram of petroleum exposure's (green shaded boxes) effects (white boxes) on avian biology (King et al. 2021)

at the marsh edge and interior (Zengel et al. 2017). Remarkably, significant recovery still lags with lower densities and different population size structures than control sites (Deis et al. 2020) nine years later. Full recovery may not occur for one to two decades in more heavily oiled shorelines.

Impacts of the Deepwater Horizon spill were not limited to aquatic species. Terrestrial species like the seaside sparrow, which relies on salt marshes for habitat, were also impacted (Bonisoli-Alquati et al. 2016, 2020; Olin et al. 2017; Perez-Umphrey et al. 2018). Transcriptomic analysis of sparrow liver tissues identified 295 genes differentially expressed between oil-exposed and control birds (Bonisoli-Alquati et al. 2020), including genes responsible for liver metabolism and energy production in these terrestrial birds (Fig. 6.9).

References

Able KW, Lopez-Duarte PC, Fodrie FJ, Jensen OP, Martin CW, Roberts BJ, Valenti J, O'Connor K, Halbert SC (2015) Fish Assemblages in: Louisiana Salt Marshes effects of the Macondo Oil Spill. Estuaries Coasts 38:1385–1398

Adams SM (2005) Assessing cause and effect of multiple stressors on marine systems. Mar Pollut Bull 51:649–657

Almeda R, Wambaugh Z, Chai C, Wang ZC, Liu ZF, Buskey EJ (2013) Effects of crude oil exposure on bioaccumulation of polycyclic aromatic hydrocarbons and survival of adult and larval stages of gelatinous zooplankton. Plos One 8

Anikiev VV, Mishukof VF, Moiseevsky GN, Tkalin AV (1988) The effect of oil films on water evaporation and oxygen content in sea water. GeoJournal 16:19–24

Bakke T, Klungsoyr J, Sanni S (2013) Environmental impacts of produced water and drilling waste discharges from the Norwegian offshore petroleum industry. Mar Environ Res 92:154–169

Belatoui A, Bouabessalam H, Hacene OR, de-la-Ossa-Carretero JA, Martinez-Garcia E, Sanchez-Lizaso JL (2017) Environmental effects of brine discharge from two desalination plants in Algeria (South Western Mediterranean). Desalination Water Treat 76:311–318

Blondes M, Gans K, Engle M, Kharaka Y, Reidy M, Saraswathula V, Thordsen J, Rowan E, Morissey E (2018) U.S. Geological Survey National Produced Waters Geochemical Database in: Survey, U.G (ed), ver. 2.3 ed

Boesch D, Rabalais N (1989) Environmental impacts of produced water discharges in Louisiana. Mid Continent Oil and Gas, p 312

Bonisoli-Alquati A, Stouffer PC, Turner RE, Woltmann S, Taylor SS (2016) Incorporation of Deepwater Horizon oil in a terrestrial bird. Environ Res Lett 11:114023

Bonisoli-Alquati A, Xu W, Stouffer PC, Taylor SS (2020) Transcriptome analysis indicates a broad range of toxic effects of Deepwater Horizon oil on seaside sparrows. Sci Total Environ 720:12

Buzzelli CP, Luettich RA, Powers SP, Peterson CH, McNinch JE, Pinckney JL, Paerl HW (2002) Estimating the spatial extent of bottom-water hypoxia and habitat degradation in a shallow estuary. Mar Ecol Prog Ser 230:103–112

Clark GF, Knott NA, Miller BM, Kelaher BP, Coleman MA, Ushiama S, Johnston EL (2018) First large-scale ecological impact study of desalination outfall reveals trade-offs in effects of hypersalinity and hydrodynamics. Water Res 145:757–768

Collins A (1970) Geochemistry of some petroleum associated waters from Louisiana. Bureau of Mines

Collins AW, Elango V, Curtis D, Rodrigue M, Pardue JH (2020) Biogeochemical controls on biodegradation of buried oil along a coastal headland beach. Mar Pollut Bull 154:111051

Curtis D, Elango V, Collins AW, Rodrigue M, Pardue JH (2018) Transport of crude oil and associated microbial populations by washover events on coastal headland beaches. Mar Pollut Bull 130:229–239

Day JW, Clark HC, Chang CD, Hunter R, Norman CR (2020) Life cycle of oil and gas fields in the Mississippi river delta: a review. Water 12:29

Deis DR, Fleeger JW, Johnson DS, Mendelssohn IA, Lin QX, Graham SA, Zengel S, Hou AX (2020) Recovery of the salt marsh periwinkle (Littoraria irrorata) 9 years after the Deepwater Horizon oil spill: size matters. Mar Pollut Bull 160:10

Delaune RD, Gambrell RP, Pardue JH, Patrick WH (1990) Fate of petroleum hydrocarbons and toxic organics in Louisiana coastal environments. Estuaries 13:72–80

Diercks AR, Highsmith RC, Asper VL, Joung DJ, Zhou ZZ, Guo LD, Shiller AM, Joye SB, Teske AP, Guinasso N, Wade TL, Lohrenz SE (2010) Characterization of subsurface polycyclic aromatic hydrocarbons at the Deepwater Horizon site. Geophys Res Lett 37:L20602

Dubansky B, Bodinier C, Rice CD, Whitehead A, Galvez F (2012) Effects of exposure to crude oil from the Deepwater Horizon oil spill on populations of gulf killifish (Fundulus grandis) in Barataria Bay, Louisiana. Integr Comp Biol 52:E49–E49

Dubansky B, Rice CD, Barrois LF, Galvez F (2017) Biomarkers of aryl-hydrocarbon receptor activity in gulf killifish (Fundulus grandis) from northern Gulf of Mexico marshes following the Deepwater Horizon oil spill. Arch Environ Contam Toxicol 73:63–75

Elango V, Urbano M, Lemelle KR, Pardue JH (2014) Biodegradation of MC252 oil in oil: sand aggregates in a coastal headland beach environment. Front Microbiol 5

Gacia E, Invers O, Manzanera M, Ballesteros E, Romero J (2007) Impact of the brine from a desalination plant on a shallow seagrass (Posidonia oceanica) meadow. Estuar Coast Shelf Sci 72:579–590

Gamer TR, Hart MA, Sweet LE, Bagheri HTJ, Morris J, Stoeckel JA, Roberts AP (2017) Effects of Deepwater Horizon oil on the movement and survival of marsh periwinkle snails (Littoraria irrorata). Environ Sci Technol 51:8757–8762

Georgiou IY, FitzGerald DM, Stone GW (2005) The impact of physical processes along the Louisiana coast. J Coast Res 72–89

Gogate PR, Pandit AB (1999) Mixing of miscible liquids with density differences: effect of volume and density of the tracer fluid. Can J Chem Eng 77:988–996

Grubesic TH, Wei R, Nelson J (2017) Optimizing oil spill cleanup efforts: a tactical approach and evaluation framework. Mar Pollut Bull 125:318–329

Guerra K, Dahm K, Dundorf S (2011) Oil and gas produced water management and beneficial use in the western United States. US Deparment of Interior, Bureau of Reclamation, Denver, CO, p 113

Hester MW, Mendelssohn IA, McKee KL (2001) Species and population variation to salinity stress in Panicum hemitomon, Spartina patens, and Spartina alterniflora: morphological and physiological constraints. Environ Exp Bot 46:277–297

Ho CS, Ju LK, Baddour RF (1990) Enhancing penicillin fermentations by increased oxygen solubility through the addition of n-hexadecane. Biotechnol Bioeng 36:1110–1118

Houghton IA, Dabiri JO (2019) Alleviation of hypoxia by biologically generated mixing in a stratified water column. Limnol Oceanogr 64:2161–2171

Huba AK, Gardinali PR (2016) Characterization of a crude oil weathering series by ultrahigh-resolution mass spectrometry using multiple ionization modes. Sci Total Environ 563:600–610

Hylland K (2006) Polycyclic aromatic hydrocarbon (PAH) ecotoxicology in marine ecosystems. J Toxicol Environ Health Part A Curr Issues 69:109–123

Incardona J, Scholz N (2018) Case study: the 2010 Deepwater Horizon oil spill and its environmental developmental impacts. In: Burggren W, Dubansky B (eds) Development and environment. Springer

Jenkins S, Paduan J, Roberts P, Schlenk D, Weis J (2012) Management of brine discharges to coastal waters, recommendations of a science advisory panel, Southern California coastal waters research project. State Water Resources Control Board, p 113

Jia SR, Park YS, Okabe M (1996) Enhanced oxygen transfer in tower bioreactor on addition of liquid hydrocarbons. J Ferment Bioeng 82:191–193

Jiang B, Law AW-K (2013) Non-interfering multiport brine diffusers in shallow coastal waters. J Appl Water Eng Res 1:148–157

Johansen C, Macelloni L, Natter M, Silva M, Woosley M, Woolsey A, Diercks AR, Hill J, Viso R, Marty E, Lobodin VV, Shedd W, Joye SB, MacDonald IR (2020) Hydrocarbon migration pathway and methane budget for a Gulf of Mexico natural seep site: Green Canyon 600. Earth Planet Sci Lett 545:13

Justic D, Rabalais NN, Turner RE (1996) Effects of climate change on hypoxia in coastal waters: a doubled CO2 scenario for the northern Gulf of Mexico. Limnol Oceanogr 41:992–1003

Kessler JD, Valentine DL, Redmond MC, Du MR, Chan EW, Mendes SD, Quiroz EW, Villanueva CJ, Shusta SS, Werra LM, Yvon-Lewis SA, Weber TC (2011) A persistent oxygen anomaly reveals the fate of spilled methane in the deep Gulf of Mexico. Science 331:312–315

King MD, Elliott JE, Williams TD (2021) Effects of petroleum exposure on birds: a review. Sci Total Environ 755

Ko JY, Day JW (2004) A review of ecological impacts of oil and gas development on coastal ecosystems in the Mississippi Delta. Ocean Coast Manag 47:597–623

Krajewski LC, Lobodin VV, Johansen C, Bartges TE, Maksimova EV, MacDonald IR, Marshall AG (2018) Linking natural oil seeps from the Gulf of Mexico to their origin by use of Fourier transform ion cyclotron resonance mass spectrometry. Environ Sci Technol 52:1365–1374

Lemelle KR, Elango V, Pardue JH (2014) Distribution, characterization and exposure of MC252 oil in the supratidal beach environment. Environ Toxicol Chem 33:1544–1551

Levine BM, White JR, DeLaune RD, Maiti K (2017) Crude oil effects on redox status of salt marsh soil in Louisiana. Soil Sci Soc Am J 81:647–653

Lin J, Xie L, Pietrafesa LJ, Shen J, Mallin MA, Durako MJ (2006) Dissolved oxygen stratification in two micro-tidal partially-mixed estuaries. Estuar Coast Shelf Sci 70:423–437

Lotufo GR, Farrar JD, Biedenbach JM, Laird JG, Krasnec MO, Lay C, Morris JM, Gielazyn ML (2016) Effects of sediment amended with Deepwater Horizon incident slick oil on the infaunal amphipod Leptocheirus plumulosus. Mar Pollut Bull 109:253–258

Mendelssohn IA, Morris JT (2012) Eco-physiological controls on the productivity of Spartina alterniflora loisel. In: Weinstein MP, Kreeger DA (eds) Concepts and controversies in tidal marsh ecology. Springer, Dordrecht

Mohammed Y, Elango V, Pardue JH (2014) Uptake and deposition of pyrogenic and petrogenic PAHs on Spartina leaves at two field sites. In: 2014 Gulf of Mexico oil spill and ecosystem science conference, Mobile, AL

Mueller P, Granse D, Nolte S, Weingartner M, Hoth S, Jensen K (2020) Unrecognized controls on microbial functioning in Blue Carbon ecosystems: the role of mineral enzyme stabilization and allochthonous substrate supply. Ecol Evol 10:998–1011

Neff J (2002) Bioaccumulation in marine organisms: effect of contaminants from oil well produced water. Elsevier

Neff JM (1979) Polycyclic aromatic hydrocarbons in the aquatic environment. Sources, fates and biological effects. Applied Sciences, Barking, Essex, England

Neff JM, Lee K, DeBlois EM (2011) Produced water: overview of composition, fates and effects. Produced water. Springer, New York, NY, pp 3–54

Noh J, Kim H, Lee C, Yoon SJ, Chu S, Kwon BO, Ryu J, Kim JJ, Lee H, Yim UH, Giesy JP, Khim JS (2018) Bioaccumulation of polycyclic aromatic hydrocarbons (PAHs) by the Marine Clam, Mactra veneriformis, chronically exposed to oil-suspended particulate matter aggregates. Environ Sci Technol 52:7910–7920

Olin JA, Burns CMB, Woltmann S, Taylor SS, Stouffer PC, Bam W, Hooper-Bui L, Turner RE (2017) Seaside Sparrows reveal contrasting food web responses to large-scale stressors in coastal Louisiana saltmarshes. Ecosphere 8:18

Pardue JH, Guo TZ (1998) Biogeochemistry of Ra-226 in contaminated bottom sediments and oilfield waste pits. J Environ Radioact 39:239–253

Pardue JH, Lemelle KR, Urbano M, Elango V (2014) Distribution and biodegradation potential of buried oil on a coastal headland beach. In: International oil spill conference, Savannah, GA, pp 1073–1086

Perez-Umphrey AA, Burns CMB, Stouffer PC, Woltmann S, Taylor SS (2018) Polycyclic aromatic hydrocarbon exposure in seaside sparrows (Ammodramus maritimus) following the 2010 Deepwater Horizon oil spill. Sci Total Environ 630:1086–1094

Pulster EL, Main K, Wetzel D, Murawski S (2017) Species-specific metabolism of naphthalene and phenanthrene in 3 species of marine teleosts exposed to Deepwater Horizon crude oil. Environ Toxicol Chem 36:3168–3176

Raki L, Masson JF, Collins P (2000) Rapid bulk fractionation of maltenes into saturates, aromatics, and resins by flash chromatography. Energy Fuels 14:160–163

Ray PZ, Chen H, Podgorski DC, McKenna AM, Tarr MA (2014) Sunlight creates oxygenated species in water-soluble fractions of Deepwater horizon oil. J Hazard Mater 280:636–643

Ray PZ, Tarr MA (2014) Petroleum films exposed to sunlight produce hydroxyl radical. Chemosphere 103:220–227

Ray PZ, Tarr MA (2015) Formation of organic triplets from solar irradiation of petroleum. Mar Chem 168:135–139

Reddy CM, Arey JS, Seewald JS, Sylva SP, Lemkau KL, Nelson RK, Carmichael CA, McIntyre CP, Fenwick J, Ventura GT, Van Mooy BAS, Camilli R (2012) Composition and fate of gas and oil released to the water column during the Deepwater Horizon oil spill. Proc Natl Acad Sci USA 109:20229–20234

Reddy K, Delaune RD (2008) Biogeochemistry of wetlands: science and applications. CRC Press, Boca Raton FL

Ritter C, Montagna PA (1999) Seasonal hypoxia and models of benthic response in a Texas bay. Estuaries 22:7–20

Rodrigue M, Elango V, Curtis D, Collins AW, Pardue JH (2020) Biodegradation of MC252 polycyclic aromatic hydrocarbons and alkanes in two coastal wetlands. Mar Pollut Bull 157:14

Romaine Z, Elango V, Fitch L, Pardue JH (2021) Weathering of surface and buried oil on a coastal headland beach. In review

Schifter I, Gonzalez-Macias C, Salazar-Coria L, Sanchez-Reyna G, Gonzalez-Lozano C (2015) Long-term effects of discharges of produced water the marine environment from petroleum-related activities at Sonda de Campeche, Gulf of Mexico. Environ Monit Assess 187

Sherwood JE, Stagnitti F, Kokkinn MJ, Williams WD (1991) Dissolved oxygen concentrations in hypersaline waters. Limnol Oceanogr 36:235–250

Shin WS, Pardue JH (2001) Oxygen dynamics in crude oil contaminated salt marshes: I. Aerobic respiration model. Environ Technol 22:845–854

Shin WS, Pardue JH, Choi SJ (2001a) Oxygen dynamics in crude oil contaminated salt marshes: II. Carbonaceous sediment oxygen demand model. Environ Technol 22:855–867

Shin WS, Pardue JH, Jackson WA (2000) Oxygen demand and sulfate reduction in petroleum hydrocarbon contaminated salt marsh soils. Water Res 34:1345–1353

Shin WS, Pardue JH, Jackson WA, Choi SJ (2001b) Nutrient enhanced biodegradation of crude oil in tropical salt marshes. Water Air Soil Pollut 131:135–152

Shin WS, Park JC, Pardue JH (2003) Oxygen dynamics in petroleum hydrocarbon contaminated salt marsh soils: III. A rate model. Environ Technol 24:831–843

St. Pe' K (1990) An assessment of produced water impacts to low-energy, brackish water systems in southeast Louisiana. In: St.Pe KM (ed) Louisiana Department of Environmental Quality, Baton Rouge, LA, p 204

Tarr MA, Zito P, Overton EB, Olson GM, Adhikari PL, Reddy CM (2016) Weathering of oil spilled in the marine environment. Oceanography 29:126–135

Toth B, Liktor-Busa E, Kusz N, Szappanos A, Mandi A, Kurtan T, Urban E, Hohmann J, Chang FR, Vasas A (2016) Phenanthrenes from Juncus inflexus with antimicrobial activity against methicillin-resistant staphylococcus aureus. J Nat Prod 79:2814–2823

Urbano M, Elango V, Pardue JH (2013) Biogeochemical characterization of MC252 oil: sand aggregates on a coastal headland beach. Mar Pollut Bull 77:183–191

Watts AW, Ballestero TP, Gardner KH (2006) Uptake of polycyclic aromatic hydrocarbons (PAHs) in salt marsh plants Spartina alterniflora grown in contaminated sediments. Chemosphere 62:1253–1260

Whitehead A (2013) Interactions between oil-spill pollutants and natural stressors can compound ecotoxicological effects. Integr Comp Biol 53:635–647

Whitehead A, Dubansky B, Bodinier C, Garcia TI, Miles S, Pilley C, Raghunathan V, Roach JL, Walker N, Walter RB, Rice CD, Galvez F (2012) Genomic and physiological footprint of the Deepwater Horizon oil spill on resident marsh fishes. Proc Natl Acad Sci U S A 109:20298–20302

Ylitalo GM, Krahn MM, Dickhoff WW, Stein JE, Walker CC, Lassitter CL, Garrett ES, Desfosse LL, Mitchell KM, Noble BT, Wilson S, Beck NB, Benner RA, Koufopoulos PN, Dickey RW (2012) Federal seafood safety response to the Deepwater Horizon oil spill. Proc Natl Acad Sci USA 109:20274–20279

Yuan F, Guo JR, Shabala S, Wang BS (2019) Reproductive physiology of halophytes: current standing. Front Plant Sci 9:13

Zengel S, Bernik BM, Rutherford N, Nixon Z, Michel J (2015) Heavily oiled salt marsh following the Deepwater Horizon oil spill, ecological comparisons of shoreline cleanup treatments and recovery. PLoS ONE 10:27

Zengel S, Montague CL, Pennings SC, Powers SP, Steinhoff M, Fricano G, Schlemme C, Zhang MN, Oehrig J, Nixon Z, Rouhani S, Michel J (2016) Impacts of the Deepwater Horizon oil spill on salt marsh periwinkles (Littoraria irrorata). Environ Sci Technol 50:643–652

Zengel S, Weaver J, Pennings SC, Silliman B, Deis DR, Montague CL, Rutherford N, Nixon Z, Zimmerman AR (2017) Five years of Deepwater Horizon oil spill effects on marsh periwinkles Littoraria irrorata. Mar Ecol Prog Ser 576:135–144

Chapter 7
The Impact of Oil and Gas Activities on the Value of Ecosystem Goods and Services of the Mississippi River Delta

David Batker and Tania Briceno

7.1 Oil and Gas Industry and Deltas

Environmental goods and services have long been recognized as subject to damage by the oil and gas industry. In the 1930s, the State of Louisiana passed protections for wildlife and limits to brine effluent. A 1972 Battelle study commissioned by the Offshore Pipeline Committee, which included ten pipeline companies, reported that land loss due to pipeline canals construction was a problem in coastal Louisiana and had significant and quantifiable damage to valuable environmental goods and services (McGinnis et al. 1972). The report further discusses cumulative and continuing processes of oil and gas canal-caused land loss through boat wakes, tides, scouring and other hydrological changes. These damages were noted in the report and should have been anticipated with further industry development, considering the scope, scale and dynamics of Coastal Louisiana and the Mississippi River Delta.

The Mississippi River Deltaic Plain wetlands and Chenier Plain are outstandingly productive (Day et al. 2013, 2019; Day and Erdman 2018). This productivity includes valuable ecosystem goods and services directly benefiting people including food, water, oxygen, biodiversity and wildlife, water, carbon sequestration, hurricane storm surge and flood buffering, navigation, recreation, energy resources and more (Batker et al. 2010). Additionally, the sheer scale of the larger Mississippi River Delta grants this area as remarkable and critical within North American ecosystems, containing 40% of all US coastal wetlands (Day et al. 2014). Wetlands of the Mississippi River Delta provide stocks and flows of these ecosystem goods and services. These physical

---

The original version of this chapter was revised: The author "Tania Briceno's" affiliation has been updated. The correction to this chapter is available at
https://doi.org/10.1007/978-3-030-94526-8_10

D. Batker (✉)
Batker Consulting, LLC, 14420 Duryea Lane S, Tacoma, WA 98440, USA

T. Briceno
Conservation Strategy Fund (CSF), Washington, DC, USA

stocks and flows produce economic value. Ecosystems can be seen for both their intrinsic value, and their economic value as natural capital assets that yield income and benefit flows.

As in many deltas globally, subsurface oil and natural gas reserves occur in the Mississippi River Delta of Louisiana across inland, wetland, coastal water bodies and offshore areas. Oil and gas are products of geologic processes that have concentrated and trapped the organic remains of organisms under anoxic conditions across tens of millions of years. Accessing oil and gas reserves powered much of twentieth century economic growth and as the twentieth century advanced, oil and gas replaced coal as the world's premier fossil fuels.

Louisiana's oil provinces span upland areas to offshore oil fields and at the surface are straddled by one of the world's most extensive wetland complexes. For millennia, oil and gas reserves deep beneath the surface lay isolated from the surface ecology as the Mississippi River migrated across the delta as it generally expanded. Yet, in the twentieth century, surface and subsurface infrastructure and activities for the purpose of oil and gas extraction have led to dramatic changes in the hydrology and ecology of coastal ecosystems (Day et al. 2020). Oil and gas infrastructure and effluent changed the landscape with canals, spoil banks, impoundments, contamination, produced water brine effluent, as well as oil and gas recovery induced subsidence and faulting. The impact on coastal ecosystems has been significant, particularly in causing the conversion of wetlands to open water. These significant physical and ecological impacts of the oil and gas industry have profound and long-term economic consequences.

About one quarter of the wetlands of Coastal Louisiana have been lost in the last 80 years due to human activities including the impact of oil and gas industry (Blum and Roberts 2009; Day et al. 2020). Today, over 2.2 million people live in Coastal Louisiana (CPRA 2017). As the wetlands disappear, the lives of people in these coastal communities become more tenuous, both economically and physically (Colten et al. 2018; Day et al. 2019). The sustainability of vast engineered systems, such as the Mississippi River navigation, ports and levees are also threatened (Day et al. 2021). These systems are of national and international importance (CPRA 2017).

Furthermore, the impact of the oil and gas industry on deltaic and coastal wetlands is emblematic of challenges globally. "Our unsustainable engagement with Nature is endangering the prosperity of current and future generations." This headline statement in the 2021 publication The Economics of Biodiversity: The Dasgupta Review (Dasgupta 2021) highlights that the transition to sustainability that must take place and a more prosperous 21st century economy requires the transition from more damaging energy, agricultural and technology systems to more sustainable systems. The proliferation of international agreements such as the International Whaling Commission, Law of the Sea, London Dumping Convention, Climate Convention, Biodiversity Convention, Stockholm Convention on persistent organic pollutants, CITES convention on wildlife trafficking and other multilateral environmental agreements further point to the importance of a process for reducing environmentally damaging impacts and restoring the natural processes, biocapacity and biodiversity of natural systems from atmospheric to geological processes.

Fortunately, deltas are systems adapted to change and wetland restoration works. The Louisiana Coastal Protection and Restoration Authority (CPRA) is implementing wetland and coastal projects from large Mississippi River sediment diversions to local weirs and wetland restoration actions (CPRA 2017). Not all projects have been successful, but over the last 50 years, it has become clear that much of the damage done by the oil and gas industry can be undone. One important question is who should pay for restoration. The overarching pool to pay and help solve land loss involves all of society. However, there is also a responsibility for those responsible for causing land loss to do their share in repairing the damage that they caused. This chapter considers these questions of damages caused by the oil and gas industry with a primary focus on natural capital and the ecosystem goods and services it produces.

7.2 Natural Capital Impacts and Ecosystem Goods and Services

7.2.1 What Are Ecosystem Goods and Services?

Economies need nature. Indeed, past and present human economies alike are ultimately built from and powered by nature's resources. Compounds, elements, organics and energy derived from nature literally compose and energize everything including our food, housing, mobility and phones. Human ingenuity has created local to global economic systems for gathering resources and energy, combining and manufacturing goods and services; distributing and consuming the products, and finally recycling, disposing of, or emitting it all back into nature's atmosphere, waters or landscapes.

For much of the twentieth Century, nature's bounty was assumed to be limitless and far less important as a productive form of capital necessary for all economic and human activity (compared to technology, money, and labor). Marketable resources, namely those that can be privately sold and bought in discrete and exclusive forms, such as barrels of oil, tons of ore, or bushels of crops, were recognized and treated as valuable resources, while the role that natural processes have in creating vital goods and services and making them available for the benefit of people was largely ignored (Costanza et al. 1997; Daly and Farley 2004). For example, the price of fish caught from natural ecosystems is well understood and is determined by supply and demand while the value of wetlands that provide habitat for fish larvae, and food for adult fish have no market value for fisheries production. In particular, natural resources that are public in nature, being non-rivalrous and non-excludable, have not been sufficiently considered in economic decisions and land use planning. These goods and services include benefits such as water production and purification, storm protection, habitat, air quality, and food, to name a few. These are highly valuable and essential to ensure not only prosperous economies but also healthy sustainable communities and economies.

Fundamentally, ecosystem goods and services are the benefits people obtain from nature (Costanza et al. 1997). They include physical goods, such as water and food as well as intangible services, such as storm protection. Ecosystems such as wetlands are composed of living and non-living components including plants, water and soils. These have functions and processes, such as photosynthesis, water cleansing, and organisms that reproduce. Some of the resources produced through ecosystem functions and processes are renewable resources such as wind, fish production and water purification, and nonrenewable resources, such as oil and gas reserves. Importantly, nature provides goods and services free of charge through the material, living and non-living components that make up ecosystems. As these goods and services are public in nature and many times nonexcludable, they do not easily fit the frameworks of economic markets and so for many there tend to be no markets regulating their production or distribution. In contrast when they are tangible and excludable, like oil and gas, the resources produced are amenable to markets.

Ecosystem assets, functions and processes that are valuable to people can be classified as assets or physical stocks, such as the carbon stock in soil, or as physical and non-physical flows such as a flow of goods (water, fish, oxygen) and less tangible services (hurricane buffering, water filtration, recreation) sometimes referred to as a flux (Daly and Farley 2004).

Unlike built capital assets, most ecosystem natural assets are self-regenerative, when managed sustainably, and are ultimately critical for economic production (Dasgupta 2021). The concept of ecosystem goods and services has been advanced to highlight these self-regenerative qualities of natural processes which are crucial to life and economies. Figure 7.1 shows how natural capital assets generate ecosystem functions which produce ecosystem goods and services that are valuable to people. Wetlands, for example, provide elevation and drag on hurricane storm surges originating in the Gulf of Mexico. As the storm approaches wetlands hold water, vegetation slows and dampens wave action. These natural functions reduce the incoming

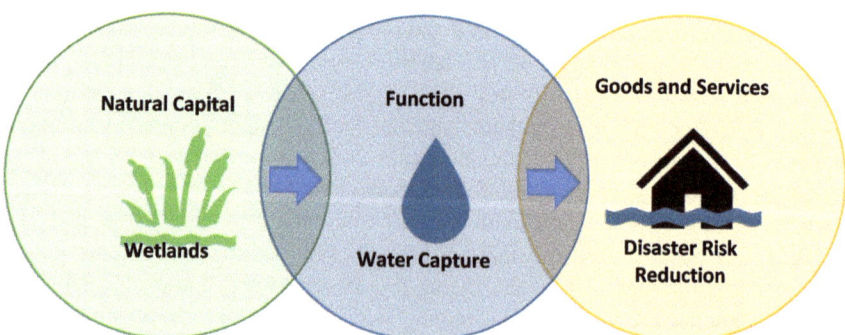

Fig. 7.1 Natural capital assets, functions, goods and services as represented by the role of wetlands in reducing flood risk

storm surge and reduce the damage to built structures and people, providing storm buffering. That storm buffering and protection function is valuable to people.

The Millennium Ecosystem Assessment (MEA) marked a key advancement in the conceptualization of human dependence on nature. This framework classified ecosystem goods and services into broad categories that reflect specific benefits to people (MA 2005). The MEA's four major categories are provisioning goods, regulating services, supporting services, and information services. These categories are defined below with examples from the Mississippi River Delta.

1. ***Provisioning goods and services*** provide physical materials and energy for society. The Mississippi River Delta provides oysters, black drum, and other seafood. The Mississippi River provides fresh water for New Orleans and other coastal communities. Provisioning goods are often traded in markets, and they are easy both to quantify physically and to value monetarily through direct market transactions.

2. ***Regulating services*** are benefits obtained from the natural control of ecosystem processes. Healthy ecosystems regulate water quality, climate, storm impact, and soil formation, and they keep disease organisms in check. For example, in the Mississippi River Delta area, barrier islands, wetlands, cypress trees, and tupelo swamp all provide storm buffering by reducing wave heights and slowing and dispersing storm surges. Wetlands also provide waste treatment value and water quality improvements by absorbing nitrogen, phosphorus, and other drinking water contaminants. Many of these services are non-market services. Hurricane buffering, for example, is quite valuable and naturally provided by wetlands, but it is not sold in markets. Thus, non-market economic analysis is required in order to establish dollar values. For example, after Hurricane Rita, surveys of the debris lines on levees provided a comparison between the storm surge extent in open water areas vs wetlands and showed how much the wetlands between the levee and coast decreased wave action and storm surge. Reduced damages could then be calculated to show the avoided cost provided by a wetland area for hurricane buffering.

3. ***Supporting services*** relate to the availability of refuge and reproductive habitat for wild plants and animals. The Mississippi Delta is one of the world's great bird migration flyways, providing habitat for a multitude of species. In addition, the Delta helps maintain biological and genetic diversity. Habitat for many species has no calculated dollar value (such as that of a spicebush or swallowtail butterfly), however habitat for some wildlife may have a calculable dollar value. Habitat, with the biological diversity of plants, bacteria, animals, fungi provides a supporting service for other goods and services including provisioning goods, such as fisheries, food and timber production. Habitat also supports regulating services including flood and hurricane risk reduction. Habitat also supports cultural services such as recreation, by providing habitat for recreational activities.

4. ***Cultural services*** provide humans with meaningful interaction with nature. This category includes information, science, education, bequest, recreation, cultural

value and other benefits. Louisiana is the "Sportsman's Paradise." There are over 40 recreational activities in Louisiana, including hunting, fishing, hiking, camping, biking, swimming, bird watching, and vehicle-based recreation such as boating or all-terrain vehicle use. Information services also include spiritually significant species and places such as plants used for spiritual ceremonies by the Houma Tribe. Culturally valuable attributes include parks and historic sites such as Fort Jackson. Natural systems also provide scientific and educational value. In the case of recreation, the dollar value can be estimated monetarily by the costs of travel, lodging, and food related to the recreational experience. Other values, such as spiritual value, have no dollar equivalent.

The publication of the MEA with its broad ecosystem service categories was a landmark change in the view of ecosystem services for economies and human well-being. For the purposes of categorizing losses associated with biophysical landscape changes evaluated in this technical report, this framework adopts a more detailed classification system than first initiated by the MEA.

Ecosystem goods and service production is rooted in the land and seascape habitats that produce those goods and services. Each connection to people and the economy lend added value to geographically specific natural systems.

The four broad categories of ecosystem goods and services including supporting, provisioning, regulating and cultural, provide benefits for people and these can be broadly classified as security, basic materials for a good life, health, good social relations and overall enabling freedom of choice and action. Figure 7.2 shows the links connections between ecosystem goods and services and well-being, as well as the connections and the strength.

For the purposes of categorizing losses associated with biophysical landscape changes evaluated in Louisiana's coastal wetlands, this framework adopts a more detailed classification system first initiated by de Groot et al. (2002), further developed in the Millennium Ecosystem Assessment (Ash et al. 2010), and subsequently improved by the authors. Table 7.1 provides a typology with an explanation of the beneficial good or service provided.

The attribution (or lack of) of property rights to ecosystems and ecosystem services has been identified as creating obstacles to implementing sustainable management systems. Given that many ecosystem services are public, dynamic, and fluid—fish swim, birds fly, rivers flow—the assignment of property rights to these in order to ensure incentives for protection is difficult, to say the least. Without property rights regulations to ensure sustainability, community and cultural agreement or other arrangements securing stewardship, management of ecosystems become a legal, economic, and social challenge. The study of ecosystem services has focused on creating knowledge systems and incentives for ensuring the continued existence, restoration and health of natural systems.

Negative impacts that result from environmental mismanagement are referred to as externalities. These often involve the loss of ecosystem services that are public in nature. Externalities are external to markets, but in no other way physically "external" to the inhabited world. Contamination and land loss from oil and gas activities

Fig. 7.2 Linkages between ecosystem services and human well-being. From the millennium ecosystem assessment synthesis, 2005

have caused externalities, in terms of losses of ecosystem goods and services. These physical impacts drive losses in welfare that result because of a lack of market accountability of the goods and services being provided by nature (Daly and Farley 2004). For example, when land erodes, it leads to downstream water quality impacts that are borne by the public at large but that are not easily attributable to the source of the problem. The public and fluid nature of water makes it an easy resource to forget. Many industrial activities use water and other ecosystems without paying for the impacts they impose (e.g. ecosystem degradation) or services they use (e.g. waste assimilation).

In Louisiana's coastal ecosystems, some of the most important ecosystem services being produced include storm protection provided by wetlands' reduction of storm surge and attenuation of waves (Barbier et al. 2013; Costanza et al. 2008), water quality improvements performed by wetlands' absorption of nutrients and other pollutants, habitat to ensure stable fish and bird populations, among others (Day et al. 2004; Hunter et al. 2018).

Louisiana's coast consists of large areas of wetlands. Many of the fish consumed in the U.S. spend at least part of their life cycle in wetlands (Pauly and Yáñez-Arancibia 2013). Wetlands are also important for their ability to retain and control floods (Shaffer et al. 2009), absorb nutrients and chemicals from water that passes through them (Hunter et al. 2018), for storing and sequestering carbon (Lane et al. 2017), as

Table 7.1 Typology for 21 ecosystem services

Good/Service	Economic benefit to people
Provisioning services	
Food	Producing crops, fish, game, and fruits
Medicinal resources	Providing traditional medicines, pharmaceuticals, and assay organisms
Ornamental resources	Providing resources for clothing, jewelry, handicrafts, worship, and decoration
Energy and raw materials	Providing fuel, fiber, fertilizer, minerals, and energy
Water storage	The quantity of water held by a water body (surface or ground water) and its capacity to provide a reliable water supply
Regulating services	
Air quality	Providing clean, breathable air
Biological control	Providing pest and disease control
Climate stability	Supporting a stable climate at global and local levels through carbon sequestration and other processes
Disaster risk reduction	Preventing and mitigating natural hazards such as floods, hurricanes, fires, and droughts
Pollination and seed dispersal	Pollinating wild and domestic plant species
Soil formation and erosion control	Creating soils for agricultural and ecosystem integrity among other uses and maintaining soil fertility. Retaining arable land, slope stability, and coastal integrity. The accretion of soil by wetlands
Soil quality	Improving soil quality by decomposing human and animal waste and removing pollutants
Water quality	Improving water quality by decomposing human and animal waste and removing pollutants
Water capture, conveyance, and supply	Providing natural irrigation, drainage, groundwater recharge, river flows, drinking water supply, and water for industrial use
Navigation	Maintaining water depth that meets draft requirements for recreational and commercial vessels
Supporting services	
Habitat and nursery	Maintaining genetic and biological diversity, the basis for most other ecosystem functions; promoting growth of commercially harvested species
Cultural services	
Aesthetic information	Enjoying and appreciating the presence, scenery, sounds, and smells of nature
Cultural value	Using nature as motifs in art, film, folklore, books, cultural symbols, architecture, media, and for religious and spiritual purposes

(continued)

Table 7.1 (continued)

Good/Service	Economic benefit to people
Future value: bequest	Having resources for future generations
Recreation and tourism	Experiencing the natural world and enjoying outdoor activities
Science and education	Using natural systems for education and scientific research

habitat for many species, and places of recreation. Beneath these wetlands, a significant amount, up to one third, of the U.S.'s oil and gas supply has been and continues to be extracted or transported every day. These natural resources have high market values and are being prioritized heavily in economic and political decisions. This situation, unfortunately, comes to the detriment of many other ecosystem services that are currently provided by Louisiana's wetlands to local communities and beyond. As oil and gas extraction and transport activities compromise the space and environmental quality needed for other ecosystem services' production, these competing uses come as a zero-sum game -or worse-, which is the economic terminology for a situation where the gain of one party entails losses to another party.

Much of the study of ecosystem goods and services has been informed by and developed in Louisiana, partially in response to the impact of oil and gas recovery on coastal wetlands. The journal Ecological Economics was founded at Louisiana State University by Herman Daly, Robert Costanza and Steve Farber while they were working on the economic aspects of wetland loss, including valuation and restoration. Further expanding the concept, Costanza led a team of scientists and economists, (Costanza et al. 1997) estimating the world's annual ecosystem services at $16–54 trillion, substantially larger than global gross domestic product.

Given growing pressures on natural resources and economic development, their value must be integrated into land management decisions including physical measures of damages and restoration, and because land loss and restoration involve valuable assets, goods and services these should also be valued to understand the costs of land loss and the benefits of restoration. To make better informed economic decisions on landscape planning and natural resource management, the trade-off between oil and gas activities and the provision of other ecosystem services needs to be well understood.

7.2.2 Federal Agency Use of Ecosystem Goods and Services Analysis

The value of ecosystem goods and services (EGS) has been recognized worldwide (MA 2005) and nationally in the United States. The U.S. federal government mandated the inclusion of ecosystem services in economic analysis through the Office of Management and Budget in 2016. U.S. agencies, like FEMA (2016, 2013) have further incorporated these values into policy and planning. In the case of FEMA, EGS have been built into disaster risk planning and mitigation efforts.

Duke University's National Ecosystem Services Partnership highlights natural resource agencies that have integrated ecosystem goods and services into their work such as the National Oceanic and Atmospheric Administration (NOAA), the U.S. Bureau of Land Management (BLM), U.S. Environmental Protection Agency (USEPA), U.S. Forest Service (USFS), U.S. Fish and Wildlife Service (USFWS), and U.S. Army Corps of Engineers (National Ecosystem Services Partnership n.d.; Schaefer et al. 2015). Similarly, state agencies such as the Louisiana Coastal Protection and Restoration Authority, recognize ecosystem services and has integrated EGS valuation into project planning (CPRA 2017). Indeed, the extensive and essential value that wetlands, identified, quantified and monetized, provides the economic reasoning behind CPRA's $50 billion plan for restoring wetlands in coastal Louisiana.

7.2.3 Valuing Ecosystem Goods and Services

As environmental resources have become increasingly scarce and mismanaged as a result of a lack of formal market incentives to conserve them, the discipline of economics has developed tools and methods to quantitatively account for the contributions that natural processes provide economies and people. Valuation methods for ecosystem services consist of first identifying the ecosystem services being provided, then quantifying them, and finally estimating their monetary value.

Well-defined goods and services traded within markets can be directly valued through market prices. For example, the price of fish should reflect current demand for fish and hence its current economic value. However, ecosystem services that are provided outside markets require further study to estimate what their value would be within a well-functioning market setting. These estimates of value are often referred to as non-market methods and they use the concept of shadow prices to estimate value. The term shadow price reflects the fact that one must rely on the study of substitute or complementary market goods and services to infer demand. Survey techniques are also used to assess people's preferences and willingness to pay for goods and services outside the market. Some of the most common methods to value ecosystem services are summarized in Table 7.2.

Oil and gas markets are defined, somewhat regulated, and tracked and hence there are market mechanisms in place that help manage the production and consumption of these goods, though without sufficient consideration of the damages (negative externalities) that oil and gas extraction have imposed on natural systems and people tied to those systems. In order to make trade-off decisions about the management of landscapes and ecosystems, valuations of ecosystem goods and services provide more complete information on the foregone value or cost of losing ecosystem services outside market transactions.

Table 7.2 Valuation methods for ecosystem services

Method	Description
Market pricing	The current market value for goods and services produced by an ecosystem are set by prices established in competitive markets
Replacement cost	The cost of replacing the services provided by functional natural systems with built infrastructure
Avoided cost	The cost of damages that would be incurred by communities in the absence of ecosystem services
Production function approach	The value of increased output resulting from ecosystem services
Mitigation and/or restoration cost	The costs of recovering from and preventing further damages due to ecosystem degradation
Contingent valuation	Value elicited by posing hypothetical valuation scenarios, through a controlled survey that seeks to estimate Willingness to Pay or Willingness to Accept for changes in ecosystem services. Choice Modelling is another variant of this method
Travel cost	The costs of travel and opportunity cost of time for a large sample of people is collected and used to build a demand curve to infer the implicit price of an ecosystem service
Hedonic pricing	Extracting characteristics of a marketed good to isolate the value of the environmental good or quality level. For example, measuring the difference in sales value between properties with an ocean view and those without to estimate ocean view aesthetic value
Benefit transfer method	The benefit transfer method is an alternative to the above approaches, where a value that has been estimated for an ecosystem service in one location is "transferred" to a sufficiently similar location where it has yet to be valued. It is simply not feasible to conduct a primary analysis of every ecosystem service and environmental amenity in every corner of the world. This method is often the only practical, cost-effective option for producing reasonable estimates of the wide range of services provided by ecosystems, especially for regions that are large and have high biodiversity. It can be likened to a house appraisal, where values are assigned to individual assets based on condition and market factors, and then the sum of those values represents an estimate of the total value of the home

7.3 How Are EGS Impacted by Oil and Gas Activities?

Trade-offs between oil and gas activities and the provision of other ecosystem services can be studied by first understanding the biophysical changes that oil and gas activities have on the landscape (Day et al. 2020; Ko and Day 2004). These changes translate into alterations to ecosystem structures and functions that are used to assess

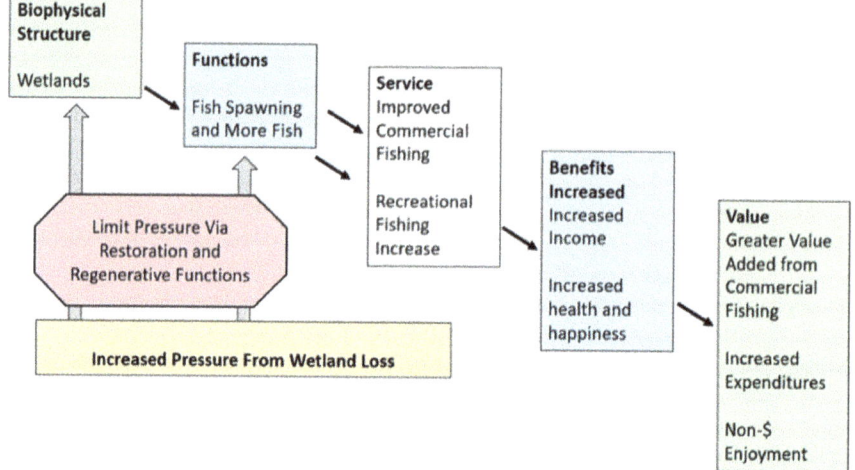

Fig. 7.3 Cascade framework for fishing ecosystem services in Louisiana

the provision of ecosystem services. The quantity and quality of the ecosystem goods and services being provided can then be valued using either market prices or shadow prices for those that are provided completely outside the market. These steps have been presented in a framework called the cascade framework (Haines-Young and Potschin 2010) for assessing ecosystem services this format is adapted for Louisiana and shown in Fig. 7.3.

In the case of oil and gas production activities, the biophysical change begins with impacts on hydrological flows, subsidence, land loss, and contamination of water and land (as discussed earlier in this book). These changes affect critical ecosystem functions such as land retention and erosion control, soil accretion, waste assimilation and vegetation loss, among others. The loss of these functions translates into impacts or losses of goods and services that people consume, use, or value. For example, the loss of fish-catch or birding opportunities are impacts that are closer to people's perceptions and valued activities. These are often used as end points for valuation. Once, these losses are quantified, for example a change in fish population or water quality, then a monetary value can be assigned with one of the standard methods described above.

Land loss and land and water contamination from oil and gas activities have different pathways in the causation of damages, depending on the biophysical change and the type of good or service being damaged. There are also indirect impacts that relate to the transport of oil and gas products, and the contributions to climate change resulting from fossil fuel combustion. Losses are additive where the economic impacts are not duplicative. For example, the value of carbon sequestration is not overlapping with the value of fisheries harvests or hurricane buffering. Thus, these are values that are additive. Where land loss is sustained over time, and cumulative impacts drive further land loss the damages can be compounded.

7.4 Cumulative Impacts

To estimate the full value of a resource, the flow of benefits through time is calculated, which results in an asset value. For example, the value of a house or a firm can be calculated as the opportunity cost of not paying rent for the life span of the house or the expected revenue stream of the factory through its productive life. Similarly, the asset value of an ecosystem can be estimated through the sum of the benefits it provides over a temporal horizon. The temporal horizon can be a historical assessment of the benefits already received (or foregone) by the beneficiary population. It can also be an expected flow of value in the future.

Both historical assessments and future scenarios are generally based on a comparison of the flow of value with a structural ecosystem change (a future with impact) and a baseline scenario of no structural change (a future with no impact). Future valuations in economic analysis are often discounted, under the assumption that there is an opportunity cost of capital and that future benefits are valued less than the same benefits provided today. This assumption requires the use of a discount rate. In addition, built capital assets typically fall apart, depreciate, and lose value over time. Under the same assumptions, the valuation of the ecosystem services provided in the future would be their value today discounted by a given percentage or discount rate.

However, it is important to understand the inherent weaknesses and flaws of this approach when applied to natural capital. First, it assumes that the goal of resource management is to maximize present value. This is reasonable for an individual or corporation particularly with short-lived investments, but is unreasonable for society concerning long-term investments because it has an intergenerational bias for the current generation over future generations, a bias for depletion or liquidation and not sustainability. Fundamentally, the discount rate it treats damages to people in the future as less than damages to people today and can lead to decisions that yield short-term gains with significant long-term costs (Arrow et al. 2004; Weitzman 1998). This sets out a potentially unsustainable decision-making framework. Discounting fundamentally biases decision-making to pull benefits into the non-discounted present, and push damages from the present into the discounted future. Thus, residents of the present benefit while people in the future may suffer greater damages (discounted today, but fully impactful in the future). Taking a sustainability view and treating all generations equally across a time period relieves this bias and can be accomplished by using a zero rate for ecosystem goods and services across an arbitrary period, for example 100 years.

Governments, including the U.S. federal government use a zero-discount rate for budgets and planning. It is insightful to show the future damages with land loss and contamination due to oil and gas activities using both a positive discount rate and a zero-discount rate. This allows decision-makers to see the perspective of the present and future.

In addition, as ecosystems become scarcer and increasingly degraded, their value is expected to increase relative to other goods. Though this could increase expenditures with higher prices for substitute goods, in the case of natural resources it can result in reduced overall benefits. The water cleansing capacity of Coastal Louisiana's wetlands is large. Were it lost, only at great expense could built filtration plants replace some of the wetland's capacity. The Gross Domestic Product would rise with the added expense, however, people in coastal Louisiana would be worse off with less water filtration capacity, more nitrogen and other contaminants, as well as the opportunity cost of money spent on filtration plants that could have been spent elsewhere were wetlands abundant.

Ecosystems are complex. Some impacts to their functionality may result in irreversible losses as critical thresholds are surpassed. When valuing components of an ecosystem separately and only in terms of goods and services that people perceive easily, certain processes and functions may be overlooked (Getzner et al. 2006). When keystone species are lost, they unleash a series of impacts on ecological processes that surpass a valuation of the keystone species by itself. Changes in one species can affect populations and productivity of other species through food chains, competition, or pathogen spread. The collapse of the North Atlantic cod fishery from overfishing, caused cascading impacts and feedback loops supressing the recovery of the cod stock (Sguotti et al. 2019). This may be an irreversible ecological impact. This highlights critical issues. Ecosystems have thresholds that if surpassed can result in dramatic collapse or transformation. Some ecosystem impacts are cascading, resulting in far larger and more comprehensive damages in geography, time, function and biomass. Ecological impacts can damage social systems, culture and institutions. Land loss results in reduced coastal population. This results in a smaller tax base, reduced private enterprise, and shrinking government services, lower income and employment.

7.5 Overview of EGS Valuations (Case Studies) Carried out in Relation to Oil and Gas Activities

7.5.1 Habitat

Every species requires functioning habitat, as it provides biological and physical conditions that allow organisms to survive, reproduce and thrive (Morrison et al. 1999). Wetlands in the U.S. provide essential habitat for about 5,000 plant species, 190 amphibian species, and protect approximately a third of endangered and threatened species within the United States (U.S. Environmental Protection Agency (EPA) 2018). The Mississippi and Gulf of Mexico flyways are also critical for North American bird migrations.

Habitat is both a precondition of the flow of ecosystem services and an ecosystem service in itself (MA 2005). Broadly, habitat serves two major functions. First, habitat is a supporting service. It supports fisheries, hunting, recreational opportunities, agriculture, pest and biological control, biodiversity, and other aspects of life. These habitat-dependent services have high commercial and non-commercial values and are presented in other EGS categories separately. Second, habitat is a valuable ecosystem service in its own right, as people value it directly, holding intrinsic or non-use values for it (García-Llorente et al. 2012; Jin et al. 2016). Additionally, it is almost impossible to account for all the species and services habitat supports, and the lack of scientific knowledge is a further pervasive limit to fully valuing habitat through other ecosystem services. Every year there are many new discoveries of how habitat, biodiversity, and specific species do or can support life and economic activities. For these reasons, economists have concluded that the value of habitat is distinct from that of other ecosystem goods and services (Fromm 2000; Gren et al. 1994).

Within Louisiana, it has long been recognized that the loss of wetlands also results in the loss of habitat (Day et al. 2003; Franklin et al. 2002; O'Connell et al. 2017; USACE 2016). This loss is evidenced through the loss of species diversity and abundance. Studies have shown that populations have decreased as a function of wetland loss by reducing their available habitat (Burdick et al. 1989; Valente et al. 2011). Additionally, studies show that fragmentation of wetlands increases their vulnerability, leading to further land and habitat loss (Couvillion et al. 2016; Hinshaw et al. 2017; Lam et al. 2018).

Often valuation studies only value one aspect of habitat, shorebird habitat, for example. Several valuations have been conducted to value some facet of Louisiana's wetland habitat. Willingness to Pay (WTP) methods are often used as a measure of value. Interis and Petrolia (2016) as well as Petrolia et al. (2014) derived the value of habitat through contingent valuations and estimated how much people were willing to pay to increase wildlife habitat or contribute to a habitat restoration project along the Louisiana or Alabama coast. Other researchers, such as Ko and Johnston (2007), found comparable habitat value results via a "replacement cost" study, which analyzed the amount of money invested in habitat restoration projects for wading birds within the greater Galveston Bay area in Texas (Rutherford et al. 2018).

As oil and gas activities erode and contaminate wetlands, the habitat these wetlands provide also disappears. Drawing from the values placed on habitat restoration projects related to bird populations in the Barataria-Terrebonne Estuary (Interis and Petrolia 2016), a value between $1,873 and $3,122 per acre per year was estimated. This estimate is based on extrapolating the survey values to all households in the state of Louisiana. These values do not fully represent all the habitat benefits provided by wetlands, or the loss in habitat from damaged, but not lost, wetlands, since Interis and Petrolia (2016) only considered some bird habitat.

7.5.2 Commercial Fisheries

Commercial fishing is a significant contributor to the Louisiana coastal economy (Chesney et al. 2000; LDWF 2019; Louisiana Sea Grant n.d.), and land loss reduces the output of this industry (Lynne et al. 1981). Recreational fishing is considered within the recreational category.

The wetlands in Louisiana support most of the Gulf's fisheries (Lassuy 2001). Over 95% of commercially harvested fish in Louisiana rely on coastal wetlands, and 16% of all fisheries harvest in the United States comes from Louisiana's coast (Chesney et al. 2000; Louisiana Sea Grant n.d.). Louisiana is second only to Alaska in fisheries production dwarfing that of neighboring states with longer coastlines including Texas and Florida. This is because wetlands, and not open water are the primary driver of coastal fishery and shellfish production. Many commercially valuable fish and shellfish are critically dependent on wetlands during their lifecycles. Wetland-dependent nekton (aquatic animals such as fish and shellfish larvae that move independent of water currents) disperse across wetlands. If an acre of wetlands is present, they utilize it. If it is lost, they cannot utilize that acre, and the population of fish/shellfish and fish stocks that would have been produced by that acre is lost. The catch lost due to wetland loss can be estimated. A loss in fish catch also trickles through the wholesale and retail markets to create further economic losses. In addition to the loss of catch, there is also a loss in stocks of fish. The uncaught stocks of fish and shellfish are valuable because they produce the next generation of fish and shellfish. Many fisheries in the U.S. have collapsed due to the reduction of the fish stock; the Atlantic cod is perhaps the most catastrophic of these collapses (Sguotti et al. 2019). Redfish too were once a commercial species and are now restricted to recreational catch only because the stock was reduced.

Furthermore, declining fish populations due to loss of wetlands causes a chain of economic losses. If an acre of wetlands is lost, that acre produces no fish. This results in fewer fish being caught, sold in wholesale markets, sold in retail markets, and sold in restaurants. There are three economic impacts to commercial fishing that can be measured: (1) impacts to fish catch each year (a flow of value); (2) impacts to the stock of wild fish (uncaught but important for reproduction and to ensure future catch); and (3) impacts to the full value chain, from dockside to the dinner plate, including all the intermediary actors that make a living in the commercial fisheries sector.

7.5.3 Recreation

Louisiana is the 'Sportsman's Paradise.' The identity of its residents and their quality of life in coastal Louisiana is tied to recreation. Outdoor recreation includes a diverse set of activities—over 200 categories of activities statewide (Cooper et al. 2019). Recreational activities are closely tied to the landscape whether they be baseball (requiring a field), swimming (requiring water), boating, hiking, hunting, or fishing. In coastal Louisiana, many recreational activities are tied to the wetlands, and numerous studies have noted that a decrease in wetland area reduces recreation and the economic value of that recreation (Bell 1997; Bergstrom et al. 1990; Lindstedt 2005). Examining five recreational activities which rely on wetlands for participation within Louisiana and associated literature shows significant values. Table 7.3 describes the effect of fewer wetland acres on the activities and their participants.

Table 7.3 Land loss effect on recreation opportunities

Activity	Environmental response	Human response	Outcome
Recreational fishing	A reduction in essential fish habitat reduces catch for recreational fishers (Williams et al. 2016; Yeager et al. 2016)	Increased catch effort (Bell 1997) decreases the value of the experience (Hunt et al. 2019)	A decrease in the market and non-market value of recreational fishing on the property
Boating	Less wetlands results in less available space and increased boater congestion	Increased congestion reduces the number of boaters over time (Lee et al. 2015)	Fewer boating trips (Bergstrom et al. 1990)
Waterfowl hunting	Bird populations and diversity decrease with wetland loss (Hunter et al. 2017; Iles 2019)	Stricter bag limits for hunters (Iles 2019)	Fewer licenses and less hunting
Wildlife viewing (birding)	Bird populations and diversity decrease with wetland loss (Hunter et al. 2017; Selman et al. 2016)	Less attractive to birders (Steven et al. 2017)	A decrease in market and non-market value of wildlife viewing
Land lease	Bird populations and diversity decrease with wetland loss (Hunter et al. 2017; Iles 2019)	Stricter bag limits for hunters (Iles 2019)	Less demand for land rentals for hunters

7.5.4 Storm Protection

The oil and gas industry is both responsible for wetland loss yet benefits from the value of wetlands for protecting oil and gas infrastructure from storm damage. This presents real incentives for the industry to further embrace a significant scale of restoration. Hurricanes Katrina and Rita resulted in tremendous damage to oil and gas infrastructure at the well head, to pipelines, refineries and to retail gas stations. Storm damage to pipelines continued through 2020. Category 4 "Hurricane Laura Wrecks Havoc on Gulf of Mexico Pipelines" was a headline in the Pipeline Technology Journal (2020) noting that the Garden Banks gas pipeline which moves 1 billion cubic feet of natural gas per day was disrupted.

Wetlands absorb water and put drag on storm surge reducing the height and volume of storm surge as well as reducing and attenuating storm surge waves. This reduces hurricane storm surge impact to levees, buildings, roads, oil and gas infrastructure and other property. Ultimately, by reducing the intensity of storms including storm surge, wetlands help protect people's lives. The valuation of the benefits of hurricane and storm buffering has been a subject of much academic research. A further differentiation can be made between protection against extreme weather events such as coastal storms and hurricanes and riverine floods. Coastal storm protection can be measured through avoided damages using an expected damage function (EDF), where damages can be estimated with respect to wetland size, configurations, density, and vegetation type that would have protected communities from the storm.

While valuation methods have not yet advanced to fully measure all these benefits, there are several models that examine flood- and storm-related risks. Climate and hydrological modelling are overlaid with topographic information, flood depth/damage, wind, social risk, and other vulnerability factors. For example, the Coastal Louisiana Risk Assessment (CLARA) model estimates hurricane flood depths, coastal vulnerability, damages, and the impact of mitigation projects (Fischbach et al. 2012). Several other studies have looked at the storm risk reduction benefits provided by wetlands and estimated partial economic values (Barbier et al. 2013; Boutwell 2016; Costanza et al. 2008).

The study by Boutwell (2016) focused on avoided storm damages in Louisiana relative to wetland size. This study includes coastal storms between 1995 and 2010 with damages greater than $50,000 and uses data recorded in the FEMA's region IV Coastal Flood Loss Atlas (CFLA) data to create the damage function. The results that are calculated in this report find similar results to recent wetland studies that value storm protection (Costanza et al. 2008). The damages cost estimate is based on property and infrastructure damage but not on business disruptions or other social costs. The author further acknowledges that his estimates are low estimates because rising costs from sea-level rise, erosion rates, and future land loss rates based on historical data underestimates wetland buffering benefits for future storm damage reduction (Boutwell 2016). Boutwell calculated costs at the parish level across 13 coastal parishes. Another advancement is to calculate the value of coastal wetlands for reduced hurricane impacts further inland. Degradation of a hurricane at the coast with wetland buffering provides reduced storm damages across the full inland path of the storm. This would likely increase the value of coastal wetlands several fold.

7.5.5 Water Quality

Water quality improvement processes, and the role wetlands employ have been extensively studied in Louisiana (Breaux et al. 1995; Hunter et al. 2018; Lane et al. 2016). Research shows wetlands can improve water quality beyond regulatory standards and beyond the capacity of many waste treatment plants employing sophisticated and expensive technology. However, wetlands require nutrients and sediment deposition to carry out water regulation functions and maintain local water quality. As oil and gas activities disrupt hydrological flows and land stability, water quality functions are impaired or lost. Information about the multiple water quality improvement functions performed by wetlands is outlined in Table 7.4.

Water pollution is a concern nationally. Excess nutrients such as nitrogen and phosphorous cause rapid, substantial algae growth that leads to hypoxia. Reduced oxygen lowers fisheries production. Nutrient pollution has contributed to the Gulf of Mexico Dead Zone, which is estimated to be over 7,800 square miles, roughly the size

Table 7.4 Water quality benefits provided by wetlands in Louisiana

Water quality benefit	Cost/metric ton trad. treatment	Wetland absorption/treatment tons/acre
PPCP Pharmaceuticals and Personal Care Products (These are organic compounds with a wide range physiological effects on people and wildlife)	PPCPs have a wide range of chemical properties, and traditional filtration methods cannot remove many types of PPCPs (Suárez et al. 2008). Those which can be removed have an estimated cost of $13 k–$111 k per metric ton of water (Molinos-Senante et al. 2013)	Removal efficiencies for both planted and unplanted wetlands range from 89–96% for common PPCPs (Özengin and Elmaci 2016)
Total suspended solids (mostly particulate organic but also mineral, matter, not all sediment)	Total suspended solids removal ranges from $31 per ton in rural agricultural settings to $14,432 per ton in urban settings (Whitman et al. 2002)	Wetlands reduce TSS in contaminated waters from 77–94% in combination waters (University of Hawaii Agriculture-Based Remediation Program 2001)
Biological Oxygen Demand (BOD)	BOD removal ranges from $470–$1,160 per metric ton (Damon S. Williams Associates 1999)	Wetlands naturally and reliably remove BOD in ecosystems as part of a healthy process which is not harmful (EPA 2000)
Phosphorus	$1,127–$3,700,000/metric ton (Whitman et al. 2002)	A meta-analysis of 93 studies found an average phosphorus removal rate of 1.2 g per square meter of wetlands per year (Land et al. 2016)
Nitrogen (tertiary treatment) Wetlands divided analysis: nitrate rapidly absorbed, ammonia over a larger area	$282–$1,885,000/metric ton (Whitman et al. 2002; Widney et al. 2017)	A meta-analysis of 93 studies found an average nitrogen removal rate of 93 g per square meter of wetlands per year (Land et al. 2016)
Fecal Coliform reduced/eliminated Other bacteria are not measured or regulated	Trad. Systems reduce to permit limit	Wetlands reduce to standard or permit levels
Wastewater other testing (lengthy list of metals, pesticides, benzene, organics) Lead and other heavy metals	Only tested for occasionally in traditional systems	Heavy metals absorbed/bound in wetland sediment, other contaminants broken down or bound (Kahn et al. 2009)
Biological control (pests)		Mosquito control

Source Adapted from Hunter et al. (2018)

of Massachusetts (NOAA 2019). This dead zone inevitably harms the commercial fishing industry physically and economically (Jordan 2018; Smith et al. 2017). In general, water quality impairments have high economic costs. These costs include impacts on recreation and tourism, property values, human health, drinking water treatment costs, mitigation of algal blooms, and restoration costs (EPA 2015).

Wetlands in Louisiana are often used as waste assimilation (or removal) areas. Industrial and municipal waste is channeled to wetlands to replace secondary and tertiary treatments at waste treatment plants (Breaux 1992; Breaux et al. 1995; Cardoch et al. 2000; Day et al. 2004, 2006; Hunter et al. 2018). The use of green infrastructure (such as wetlands) for water quality improvements is more economical than building and maintaining new treatment plants, thus generating cost savings to local governments and industries (Breaux et al. 1995; Cardoch et al. 2000).

Wetlands are not only responsible for removing nitrogen and phosphorus. They can also filter out harmful pharmaceutical compounds, synthetic industrial compounds (Carina Storrs 2016; García et al. 2020; Gross et al. 2004), and heavy metals (Khan et al. 2009). Importantly, many synthetic, pharmaceutical, personal care, and other industrial and consumer pollutants that enter into water systems are not removed through traditional wastewater treatment processes, but are reduced and/or removed by wetlands, making wetlands uniquely valuable for improving water quality (Carina Storrs 2016). Unfortunately, as wetlands are lost or contaminated, so too are the water quality improvements they provide.

There are different methods for valuing water quality improvements provided by wetlands. Some studies are based on benefits to industries while others look at community benefits. Other studies are done from the production function side, i.e., how much nitrogen is removed by an acre of wetlands, and some considered the value attributed by citizens to general water quality improvements achieved by wetlands. Valuation studies provide a large range of wetland water quality values, some as high as $100,000 per acre (Widney et al. 2017). A value of $1,841 (in 2018 dollars) was provided by Breaux et al. (1995) for Louisiana-based industries, in terms of avoided water quality treatment costs. This is an underestimate as it includes a limited number of pollutants.

7.5.6 Climate Stability

The Gulf of Mexico has been warming and this is increasing the size and likely the frequency of hurricanes and large storm events (Marsooli et al. 2019). It has also resulted in sea level rise, and these are functions of a warming atmosphere and climate (Marsooli et al. 2019). Photosynthesis in wetlands and plant material deposition in anoxic wetland soils sequesters carbon and builds land. Deltaic processes with compaction and dewatering result in natural subsidence which pulls carbon from the troposphere to the lithosphere and more permanent sequestration. At the same time, wetlands in a healthy delta with sediment, nutrient and freshwater inputs can outpace sea level rise and subsidence in many areas.

Wetlands sequester carbon because they support a high density of plant life, with high net primary productivity (Duan et al. 2018; Lane et al. 2016; McLatchey and Reddy 1998; Stagg et al. 2017; Stevenson and Cole 1999). The low oxygen wetland soils help prevent the release of carbon from plant decomposition, which is why wetland soils can range from 93 to 98% organic matter (Nyman et al. 1990). This combination of high plant diversity and low oxygen soils is why wetlands hold 20–25% of the world's soil carbon despite making up only 2–6% of the earth's surface (Amthor et al. 1998; Whiting and Chanton 2001), with temperate and subtropic wetlands, like Louisiana, being the most effective storers of carbon (Whiting and Chanton 2001).

There are important differences between short-term carbon storage and habitats for long-term soil carbon accumulation and burial. Fires and drought can release carbon stores from forests and terrestrial soils. Healthy deltaic systems sequester carbon and through subsidence provide long-term storage. Baustian et al. (2021) recently showed that the vast scale and capacity for carbon storage in the Mississippi River Delta and that this storage rate could be cut in half by land loss over the next 50 years. The qualitative difference between carbon storage in systems that can rapidly release carbon back into the atmosphere and deltaic systems that can provide long-term burial increases the value of wetland carbon sequestration, storage and burial.

Voluntary and regulatory carbon markets have helped establish market prices and value for carbon sequestration, while academic studies have included the social costs related to carbon to provide more robust values for carbon sequestration and climate stability. Coastal Louisiana, has a high rate of sequestration, is not fire prone, and is one of the best areas in North America to sequester carbon permanently.

The U.S. Federal government has adopted a dollar value for sequestered carbon recognizing the climate stability value of carbon sequestration in wetlands and deltas. The damages of increased disasters, such as more frequent and stronger hurricanes, drought, fire, increased spread of disease, greater costs for stormwater systems with greater rainfall events, and more are collectively known as the social cost of carbon (SCC) (IWG 2015). Losing land, and carbon sequestration, results in climate change and negative impacts to society such as larger hurricanes and greater storm damage. There are now many financial incentives for reducing carbon emissions and sequestering carbon. The largest carbon market in the US has been developed in California.

In September of 2006, California passed Assembly Bill 32, also known as California's Global Warming Solutions Act, which created a market-based solution to cut greenhouse gas emissions (AB 32 2006). Under this cap and trade system, a 'cap' is placed on the total amount of emissions which can be produced, and those who either sequester pollutants, or are under their allowances, can 'trade' credits which offset polluters' carbon emissions within the cap (Hsia-Kiung et al. 2014). California has collected over $9.3 billion in the climate cap and trade system charging for carbon emissions (California Air Resources Board 2019). This money has funded energy efficiency; public housing; carbon sequestration in farms, tree planting, land conservation; rebates for zero emission vehicles and economic development. Of the fund, 57% is dedicated to community development.

Louisiana could receive some of this funding in the future. Louisiana-based Tierra Resources produced a report highlighting the opportunity of using Louisiana river diversions, wetland assimilation, mangrove planting, and hydrologic restoration to produce allowances which could be traded on the California carbon market (Mack et al. 2014). Tierra's methodology has been approved by the American Carbon Registry; and is now being considered for approval by the California Air Resources Board. If approved, Louisiana landowners could trade carbon allowances to polluters in California. Louisiana has had financial transactions for carbon sequestration.

The regulation of our climate provides value through the prevention of damages. Carbon in our atmosphere helps to keep the planet warm by trapping energy from the sun, and the addition of more carbon enhances this effect (Bolin and Doos 1989). While this greenhouse effect is necessary for sustaining life on earth, too high a concentration of greenhouse gases in the atmosphere can shift the earth's climate and cause economic, environmental, and social harms (Agarwal et al. 2002; Bretschger and Karydas 2019; Hoegh-Guldberg and Bruno 2010; Mantyka-pringle et al. 2012). These damages are collectively known as the social cost of carbon (SCC) (IWG 2015). Losing land and carbon sequestration marginally contributes to higher temperatures in areas including the Gulf of Mexico, as well as a rise in the size and intensity of hurricanes and other storms. This in turn inflicts greater damages to coastal areas including coastal Louisiana. Climate stability has a sequestration (flow) value and a storage (stock) value. In Louisiana these have been measured on the order of $1,235/acre and $87,130/acre based on sequestration rates and the social cost of carbon.

7.5.7 Cultural Value

"As the people leave, our culture goes with it. We are looking for a place we can be a community. That means a place we can care for each other, celebrate with each other and be together as family and friends on a daily basis." Stated Boyo Billiot, Deputy Chief for the Isle de Jean Charles Band of the Biloxi-Chitimacha-Choctaw Tribe ("Isle De Jean Charles Relocation" n.d.). The IDJC Tribe is currently relocating from their traditional lands on the coast in Terbonne Parish to land acquired further inland to retain IDJC Tribal members and culture.

Land is inextricably linked to culture (Keul 2011). Coastal Louisiana's wetlands have cultural value reflecting their influence in forming value systems, uses, history, religion, traditions, music, language, foods, artistic expressions and freedom embraced by individuals, families and communities. In damaging land, the cultural damages are often significant and additional to other damages, but difficult to quantify and monetize. Land in Southern Louisiana has shaped the traditions, customs, foods, knowledge, recreation, music, and all other aspects of life in the communities that have grown around them. The culture around agriculture, fishing, hunting, and trapping exists thanks to the bayous, wetlands, and farmlands that support these activities (Gramling and Hagelman 2005).

The connection between nature and culture is explicitly recognized in the National Environmental Policy Act (NEPA) in 1970 (Courselle 2010). Two of the six major goals of NEPA are:

1. "Assure for all Americans safe, healthful, productive, and aesthetically and culturally pleasing surroundings"
2. "preserve important historic, cultural, and natural aspects of our national heritage, and maintain, wherever possible, an environment which supports diversity, and variety of individual choice;"

The State of Louisiana Office of Cultural Development has a mission and philosophy valuing Louisiana's cultural heritage and states: "Louisiana's cultural assets enhance communities, the economy, education and quality of life in our state. It is the Office of Cultural Development's duty to serve the people of Louisiana by preserving, supporting, developing and promoting our archaeology, arts, French Language and historic places" (Louisiana Office of Cultural Development n.d.). Native American, Coastal Louisiana African American, Creole and Cajun cultures evolved in the Mississippi River Delta.

Culture provides identity, security, joy, a sense of belonging, is central to wellbeing and sustains familial and community cohesion and builds local economies. This tie between culture, landscape and commitment to community in rural and coastal Louisiana is discussed by Bailey et al. (2014). Bailey et al. note that Louisiana residents have a "…storehouse of ecological knowledge based on generations…" and further that "Strong emotional ties link people to the land and water of coastal Louisiana as well as to their cultural communities."

Cultural values are extremely important but commonly intangible; a reason often cited as an explanation for their poor monetization and appraisal (Milcu et al. 2013). Yet the physical, emotional, and mental benefits that make up cultural ecosystem services are fundamental (Kenter et al. 2011).

Cultural values are crucial aspects of personal wellbeing. Economist Max-Neef (1992), notes that fundamental human needs include subsistence, protection, affection, understanding, participation, recreation, creation, identity and freedom are tied to being, having, doing, and interacting with others and the natural world. The health of culture in communities affects individuals' self-worth, physical health, and overall wellbeing as well as community cohesion, crime rates, political and institutional effectiveness, among many other things. Additionally, nature is a very effective satisfier of human needs, as Max-Neef (1992) points out. It provides a space for the individual to self-realize and for communities to thrive in many ways. In addition, there is often a power imbalance between those damaging cultural value through damage to the landscape and those suffering the damages. Snyder et al. (2003) point out "The links between natural resources and culture loss are further complicated when groups of people with different value systems attempt to value each other's natural resources, or more typically, when powerful interests value the resources of the less powerful." This is certainly the case for the oil and gas industry in Louisiana.

Another marker for valuing cultural damage is recognizing that the cost of cultural loss and damage is larger than damage to economic assets and resources. Snyder et al. state that: "In the case of the Exxon-Valdez there is much evidence to suggest that the destruction of indigenous social fabric was the most serious outcome" (Snyder et al. 2003). Perhaps the most dramatic impact land loss due to sea level rise and climate change, oil and gas canals and subsidence driving retreat from the coast by coastal communities. The Isle de Jean Charles Choktaw Tribe has lost 98 percent of their land to open water and the remaining area is flooded even on high tides. The Tribe has blamed oil and gas drilling for the damage. Facing the loss of tribal lands and people as tribal members moved out of the increasingly vulnerable traditional tribal lands, in a difficult decision, the tribe decided to relocate with assistance from the State of Louisiana and a National Disaster Resiliance Competition grant from the Federal Government ("National Disaster Resilience—HUD Exchange" n.d.). In this case, the tribe experienced natural capital damages with land loss, built capital damages with the impact to roads and houses, human capital loss as people moved away and cultural damage as the tribe was disbursed and traditional landscape upon which culture depends was lost. Land loss presents a serious threat to the cultures of coastal Louisiana as people move away from the coast, tribes lose membership, and communities are reduced in population. The path to protecting culture built on wetlands is to restore lost wetlands.

7.5.8 *Contamination and Health*

Contaminants present from oil and gas operations present a threat to human health. Exposure through direct contact, or exposure via air, water, soil or food presents a public health threat. Often the probability of harms associated with exposure risks are difficult to calculate. There is insufficient space here to cover this topic more fully. However, it is important to note several points. Contamination and the associated health risks from exposure (some acute) to common oil and gas industry contaminants are known to occur in Louisiana oil fields as a result of drilling and recovery. Some common contaminants are listed in Table 7.5 and the health risks are noted from the U.S. Agency for Toxic Substances and Disease Registry.

7.6 Contamination Cases

The severity of the impacts on human health, both physical and mental, depends on the scale and chemical composition of the contaminants as well as exposure parameters and vulnerabilities of the individuals and population. If an oil spill is sudden like the 2010 Deepwater Horizon (DWH) spill event, then effects to human

Table 7.5 Contaminants commonly found in oil and gas operations

Contaminant	Contaminant (Spec)	Health risks
NORM (Naturally occurring radioactive material present at depth and brought up with oil associate fluids and brine water)	Radium 226	Anemia, cataracts, fractured teeth, bone cancer, and death
	Radium 228	Anemia, cataracts, fractured teeth, bone cancer, and death
Produced water		
Drilling fluids	Chromium	Breathing problems (asthma, cough, shortness of breath, wheezing), ulcers, anemia, lowered reproductive function, lung cancer, death
	Barium	Changes in heart rhythm, paralysis, vomiting, abdominal cramps, diarrhea, difficulties breathing, changes in blood pressure, numbness in the face, and muscle weakness
	Arsenic	Arsenic is most known as being a human poison with large doses resulting death. Lower levels result in nausea, stomachache, vomiting, diarrhea, decreased production of red and white blood cells, changes in skin, increased risk of liver, bladder, and lung cancer
	Oil	Human health effects include impacts to the central nervous system, death, fatigue, headache, nausea, drowsiness, paralysis, difficulty breathing, pneumonia, weakening of the immune system, and damage to the liver, spleen, kidneys, developing fetuses, and lungs
Aromatic hydrocarbons		
Hydrocarbons	Benzene	Long-term exposure can cause leukemia, damage to bone marrow, decrease in red blood cells, and anemia
	Toluene	Headaches, dizziness, unconsciousness, incoordination, cognitive impairment, vision loss, hearing loss, fetus development effects, harm to kidneys and liver, harm to the immune system

(continued)

Table 7.5 (continued)

Contaminant	Contaminant (Spec)	Health risks
	Xylene	Irritation of the skin, eyes, nose, and throat, difficulty breathing, impaired lung function, delayed response to visual stimulus, impaired memory, stomach, discomfort, changes in liver and kidneys, headaches, muscle coordination, dizziness, confusion, and death
	Ethyl Benzene	Hearing loss, kidney damage, and cancer. Ingestion has been shown to cause hearing damage
Heavy metals	Lead	Decreased cognitive ability, decreased nervous system functions, weakness in fingers, wrists, or ankles, increase in blood pressure, anemia, kidney damage, death, miscarriage, and is believed to be a carcinogen
	Mercury	Damage to brain and kidneys, personality changes, tremors, changes in vision, deafness, muscle incoordination, loss of sensation, difficulties with memory, trouble breathing, nausea, vomiting, diarrhea, increases in blood pressure, and damage to stomach and intestines
	Zinc	Metal fume fever, nausea, vomiting, anemia, damage to the pancreas, and decrease levels of HDL cholesterol
	Cadmium	Breathing cadmium can severely damage lungs, kidneys, or cause lung cancer or death. Oral exposure can lead to vomiting, diarrhea, kidney damage, bone fragility, and death
	Selenium	While selenium is a dietary requirement for people, long or high exposure can cause it to build up in the liver, kidneys, blood, lungs, heart, and testes. Health complications include lung irritation, dizziness, fatigue, bronchitis, and brittle hair and nails
	Silver	Contact with silver can cause argyria, breathing difficulty, and stomach pain

(continued)

Table 7.5 (continued)

Contaminant	Contaminant (Spec)	Health risks
	Strontium	Contact with strontium by children can lead to problems with bone growth. Radioactive strontium can cause a drop in blood cell counts, leukemia, and other cancers

Source ATSDR (n.d.)

health can be sudden, and its lasting effects are dependent on spill response time. If a contamination event is slower, and more drawn out, the effects can be more pernicious (Baker et al. 1993). There is a large record of petroleum and gas related spills. Two notable cases and the damages related to these spills are summarized here.

7.6.1 Exxon Valdez

The Exxon Valdez tanker ran aground on March 24, 1989, releasing 11 million gallons of crude oil into Prince William Sound. The Exxon Valdez (1989) oil spill natural resources damages assessment was a turning point for ecological economics because of the validation of passive use value (existence or nonuse value) estimation techniques as approved by NOAA and the Federal Government. Carson et al. (2003) found that the original contingent valuation study value of $2.8 billion (1990) dollars of lost passive use values was highly conservative, and that statistically defensible estimates ranged from 4.87 to 7.19 billion dollars.

7.6.2 BP Oil

The British Petroleum (BP) Deepwater Horizon (DWH) oil spill in April 2010 left thousands of miles of Gulf coastline contaminated with crude oil. In order to determine the dollar amount of damages, the trustees involved had to first measure the impact of the DWH oil spill on the value of ecosystem services. Typically, in contamination cases, a Natural Resources Damages Assessment (NRDA) is conducted to determine the type and amount of restoration needed to compensate the public for harm to natural resources as a result of a spill or accident. Following a NRDA approach, trustees investigated how the DWH accident led to change in ecosystems, and how those changes led to changes in the provision and value of ecosystem services. Measuring such changes required an estimation of the difference in the provision and value of ecosystem services present before and after the oil spill. In general, when assessing injuries via the NRDA process, trustees determine whether there is: an exposure pathway between the source and a natural resource, an adverse change in that natural resource as a result of exposure to the discharge, or an injury to a natural resource or impairment of a natural resource service as a result of response actions or a substantial threat of discharge. Damages to natural resources were valued at the cost to restore the damage from the spill and on April 6th, 2016, the court issued a settlement agreement for the damages. It awarded approximately $8.8 billion in damages to Trustees of the states of Texas, Louisiana, Alabama, Mississippi, and Florida. Approximately $4 billion was specifically allocated to the restoration of 'Wetlands, Coastal, and Nearshore Habitats' in Louisiana. To-date, approximately 65 projects within the Gulf have been approved, at an estimated cost of $886 million.

7.7 Systemic Impacts from Oil and Gas Activities

There is increasing evidence in Coastal Louisiana that land loss is a systemic driver of reduced population. Areas of coastal Louisiana have experienced significant population loss, which is in contrast to areas on the U.S. Gulf of Mexico that experience similar storm impacts, have no land loss and see rapid population increases. While large hurricanes will result in evacuations, over time the population of the U.S. Gulf has been growing at a faster pace than other coastal areas in the U.S. and now has over 14 million residents. Between 1960 and 2008, the population along the Gulf of Mexico, soared by 150%, more than double the rate of increase of the nation's population as a whole during the period (Fig. 7.4) (Wilson and Fischetti 2010).

Yet areas in coastal Louisiana have seen dramatic declines in population. This disparity in population change evidences the impact of land loss. Unlike most of the Gulf Coast, all of which experiences hurricanes, the Louisiana Coast, with significant land loss has experienced far lower growth than the U.S. Gulf Coast with dramatic declines in population in some Parishes, such as Cameron Parish, with over 25%

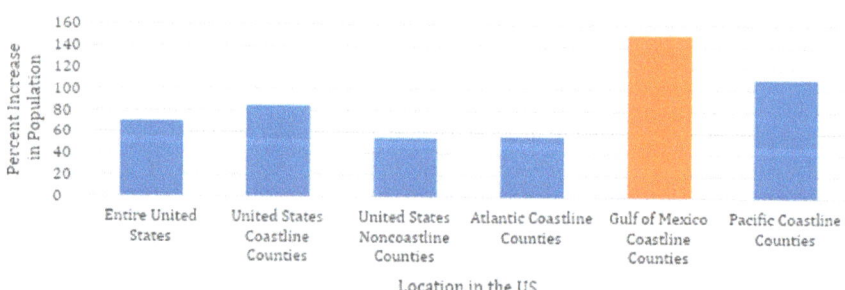

Fig. 7.4 U.S. percent change in population 1960–1980

7 The Impact of Oil and Gas Activities on the Value of Ecosystem Goods...

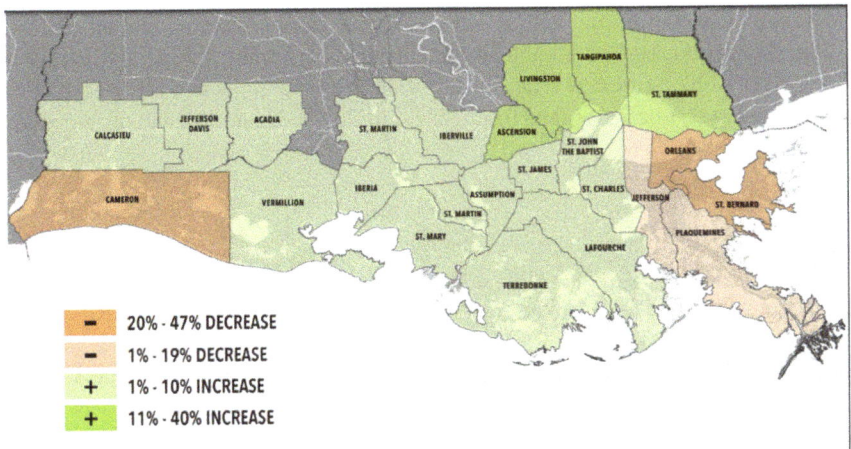

Fig. 7.5 Population change in Louisiana's coastal parishes between 2000 and 2010 (CPRA 2017 Coastal Master Plan)

reduction in population since 2005. Figure 7.5 shows the population trends in Coastal Louisiana between 2000 and 2010.

Land loss has been a driver of this negative population trend (Burley 2010). The loss of population has profound implications for the tax base of cities, parishes and the State of Louisiana. As population leaves an area further impacts include lost expenditures and reduced demand for market goods and services reducing income and expenditures within parishes. Coastal parishes with shrinking populations, such as Cameron, have a smaller "multiplier" for expenditures within the parish as compared to Calcasieu Parish because businesses have moved out of the Parish and people must drive to Calcasieu Parish to shop. Demographic loss impacts everything from schools, health care, roads, stores, utilities and other services. These impacts are beginning to be measured (CPRA 2017).

7.8 Restoration

Restoration represents a path forward for halting the current rate of land loss and increasing land creation and retention across the Louisiana Coast. Restoring the natural hydrology and sediment flows that created and maintained the wetlands is key to rebuilding and retaining these areas. Diversions including the Atchafalaya and Davis Pond have shown a record of restoration and the State of Louisiana has now approved the construction of two large sediment and water diversions on the Mississippi River. The economics of restoration have become more convincing. Calculations in 2010 showed gains of $47 billion (Batker et al. 2010) with restoration on a reduced scale from what is currently underway. Restoration in many oil and gas fields can be accomplished with a wide variety of known restoration methods (See Chap. 8 in this book). Overall, the biocapacity, and economic viability of the Mississippi River Delta ultimately relies on restoring lost land, reducing current land loss, sustaining land and gaining land through restoration and restoring natural processes.

7.9 Conclusions

The asset value of the Louisiana Coastal Zone including the Mississippi River Delta is between $300 billion and $1.3 trillion (Batker et al. 2010). The loss of 1.2 million acres of wetlands in the last 80 years represents a tremendous loss of value. Overall, the oil and gas industry has caused land loss and contamination in the Mississippi Delta and Chenier Plain at a significant scale and over many decades which has damaged natural processes, ecosystem goods and services and resulted in systemic impacts on populations and local city and parish economies. Though currently, the loss of 1.2 million acres of wetlands has resulted in reduced ecosystem goods and services as well as reduced population and economic activity in the Louisiana Coastal Zone, restoration can increase biocapacity and the foundation for the coastal economy. Restoring lost wetlands requires significant effort and investment. Among other sources, restoration should be funded by the oil and gas industry where the industry has been responsible for land loss and contamination.

References

AB 32 (2006) California Global Warming Solutions Act of 2006, AB 32

Agarwal A, Narain S, Sharma A (2002) The global commons and environmental justice—climate change. In: Environmental justice international discourses in political economy. Routledge, New York

Amthor J, Cushman J, Dale V, Edwards N, Farrell M, Garten C, Gunderson C, Hanson P, Hildebrand S, Huston M, Jacobs G, Kanciruk P, King A, Loar J, Luxmoore R, Marland G, McLaughlin S, Mulholland P, Norby R, Wullschleger S (1998) Terrestrial ecosystem responses to global change: a research strategy. ORNL Technical Memorandum

Arrow K, Dasgupta P, Goulder L, Daily G, Ehrlich P, Heal G, Levin S, Mäler K-G, Schneider S, Starrett D, Walker B (2004) Are we consuming too much? J Econ Perspect 18:147–172. https://doi.org/10.1257/0895330042162377

Ash N, Blanco B, Brown C, Garcia K, Henrichs T, Lucas N, Raudsepp-Hearne C, Simpson RD, Scholes R, Tomich T, Vira B, Zurek M (2010) Ecosystems and human wellbeing: a manual for assessment practitioners—UNEP-WCMC (Manual). Island Press

ATSDR (n.d.) Agency for toxic substances and disease registry [WWW Document]. https://www.atsdr.cdc.gov/index.html. Accessed 27 Oct 20

Bailey C, Gramling R, Laska SB (2014) Complexities of resilience: adaptation and change within human communities of Coastal Louisiana. In: Day JW, Kemp GP, Freeman AM, Muth DP (eds) Perspectives on the restoration of the Mississippi Delta: the once and future delta, estuaries of the world. Springer Netherlands, Dordrecht, pp 125–140. https://doi.org/10.1007/978-94-017-8733-8_9

Baker JM, Leonardo GM, Bartlett PD (1993) Long-term fate and effects of untreated thick oil deposits on salt marshes. Presented at the International oil spill conference proceedings, March 1993

Barbier EB, Georgiou IY, Enchelmeyer B, Reed DJ (2013) The value of wetlands in protecting Southeast Louisiana from hurricane storm surges. PLoS ONE 8:e58715. https://doi.org/10.1371/journal.pone.0058715

Batker D, de la Torre I, Costanza R, Swedeen P, Day J, Boumans R, Bagstad K (2010) Gaining ground: wetlands, hurricanes, and the economy: the value of restoring the Mississippi River Delta. Environ Law Rep News Anal 40:11106

Baustian MM, Stagg CL, Perry CL, Moss LC, Carruthers TJB (2021) Long-term carbon sinks in marsh soils of Coastal Louisiana are at risk to wetland loss. J Geophys Res Biogeosci 126. https://doi.org/10.1029/2020JG005832

Bell FW (1997) The economic valuation of saltwater marsh supporting marine recreational fishing in the southeastern United States. Ecol Econ 21:243–254. https://doi.org/10.1016/S0921-8009(96)00105-X

Bergstrom JC, Stoll JR, Titre JP, Wright VL (1990) Economic value of wetlands-based recreation. Ecol Econ 2:129–147. https://doi.org/10.1016/0921-8009(90)90004-E

Blum MD, Roberts HH (2009) Drowning of the Mississippi Delta due to insufficient sediment supply and global sea-level rise. Nat Geosci 2:488–491. https://doi.org/10.1038/ngeo553

Bolin B, Doos BR (1989) Greenhouse effect. Whiley and Sons Inc., New York, NY

Boutwell J (2016) The vulnerability of Louisiana to hurricane damage and the value of wetlands for hurricane risk reduction. LSU Dr. Diss

Breaux A (1992) The use of hydrologically altered wetlands to treat wastewater in Coastal Louisiana. Louisiana State University

Breaux A, Farber S, Day J (1995) Using natural coastal wetlands systems for wastewater treatment: an economic benefit analysis. J Environ Manag 44:285–291. https://doi.org/10.1006/jema.1995.0046

Bretschger L, Karydas C (2019) Economics of climate change: introducing the Basic Climate Economic (BCE) model. Environ Dev Econ 24:560–582. https://doi.org/10.1017/S1355770X19000184

Burdick DM, Cushman D, Hamilton R, Gosselink JG (1989) Faunal changes and bottomland hardwood forest loss in the Tensas Watershed, Louisiana. Conserv Biol 3:282–292. https://doi.org/10.1111/j.1523-1739.1989.tb00088.x

Burley DM (2010) Losing ground: identity and land loss in Coastal Louisiana. Univ. Press of Mississippi

California Air Resources Board (2019) Report: cap-and-trade spending doubles to $1.4 billion in 2018 | California Air Resources Board [WWW Document]. https://ww2.arb.ca.gov/news/report-cap-and-trade-spending-doubles-14-billion-2018. Accessed 27 Oct 2020

Cardoch L, Day JW, Rybczyk JM, Kemp GP (2000) An economic analysis of using wetlands for treatment of shrimp processing wastewater—a case study in Dulac, LA. Ecol Econ 33:93–101. https://doi.org/10.1016/S0921-8009(99)00130-5

Carina Storrs (2016) Designing wetlands to remove drugs and chemical pollutants [WWW Document]. Yale E360. https://e360.yale.edu/features/designing_wetlands_to_remove_drugs_and_chemical_pollutants. Accessed 19 Oct 2020

Carson RT, Mitchell RC, Hanemann M, Kopp RJ, Presser S, Ruud PA (2003) Contingent valuation and lost passive use: damages from the Exxon Valdez oil spill. Environ Resour Econ 25:257–286. https://doi.org/10.1023/A:1024486702104

Chesney EJ, Baltz DM, Thomas RG (2000) Louisiana estuarine and coastal fisheries and habitats: perspectives from a fish's eye view. Ecol Appl 10:350–366. https://doi.org/10.1890/1051-0761(2000)010[0350:LEACFA]2.0.CO;2

Colten CE, Simms JRZ, Grismore AA, Hemmerling SA (2018) Social justice and mobility in Coastal Louisiana, USA. Reg Environ Change 18:371–383. https://doi.org/10.1007/s10113-017-1115-7

Cooper W, LeGrange R, Jack T, Simms S, Albarez R, Baker B, Broyles ST, Domingue MW, Bourgeois D, Credeur D, Windham S, Dixon J, Tuma T (2019) 2014–2019 Louisiana statewide comprehensive outdoor recreation plan. Department of Culture, Recreation, and Tourism

Costanza R, d'Arge R, de Groot R, Farber S, Grasso M, Hannon B, Limburg K, Naeem S, O'Neill RV, Paruelo J, Raskin RG, Sutton P, van den Belt M (1997) The value of the world's ecosystem services and natural capital. Nature 387:253–260. https://doi.org/10.1038/387253a0

Costanza R, Pérez-Maqueo O, Martinez ML, Sutton P, Anderson SJ, Mulder K (2008) The value of coastal wetlands for hurricane protection. AMBIO J Hum Environ 37:241–248. https://doi.org/10.1579/0044-7447(2008)37[241:TVOCWF]2.0.CO;2

Courselle D (2010) We (used to?) make a good gumbo—The BP deepwater horizon disaster and the heightened threats to the unique cultural communities of the Louisiana Gulf Coast. Tulane Environ Law J 24:19–39

Couvillion BR, Fischer MR, Beck HJ, Sleavin WJ (2016) Spatial configuration trends in Coastal Louisiana from 1985 to 2010. Wetlands 36:347–359. https://doi.org/10.1007/s13157-016-0744-9

CPRA (2017) Louisiana's comprehensive master plan for a sustainable coast. Office of Coastal Protection and Restoration, Baton Rouge, LA

Daly H, Farley J (2004) Ecological economics: principles and applications. Island Press

Damon S. Williams Associates (1999) Operations and maintenance report multi-agency benchmarking study. King County California

Dasgupta P (2021) Final report—the economics of biodiversity: the Dasgupta review. United Kingdom

Day JW, Erdman JA (2018) Mississippi delta restoration: pathways to a sustainable future. Springer

Day JW, Yañéz Arancibia A, Mitsch WJ, Lara-Dominguez AL, Day JN, Ko J-Y, Lane R, Lindsey J, Lomeli DZ (2003) Using Ecotechnology to address water quality and wetland habitat loss problems in the Mississippi basin: a hierarchical approach. Biotechnol. Adv. VI International Symposium on Environmental Biotechnology 22:135–159. https://doi.org/10.1016/j.biotechadv.2003.08.012

Day J, Ko J-Y, Rybczyk J, Sabins D, Bean R, Berthelot G, Brantley C, Cardoch L, Conner W, Day JN, Englande A Jr, Feagley S, Keenan E, Lane R, Lindsey J, Mistich J, Reyes E, Twilley R (2004) The use of wetlands in the Mississippi Delta for wastewater assimilation: a review. Ocean Coast Manag 47:671–691. https://doi.org/10.1016/j.ocecoaman.2004.12.007

Day JW, Westphal A, Pratt R, Hyfield E, Rybczyk J, Paul Kemp G, Day JN, Marx B (2006) Effects of long-term municipal effluent discharge on the nutrient dynamics, productivity, and benthic community structure of a tidal freshwater forested wetland in Louisiana. Ecol Eng 27:242–257. https://doi.org/10.1016/j.ecoleng.2006.03.004

Day J, Lane R, Moerschbaecher M, DeLaune R, Mendelssohn I, Baustian J, Twilley R (2013) Vegetation and soil dynamics of a Louisiana estuary receiving pulsed Mississippi River water following hurricane Katrina. Estuaries Coasts 36:665–682. https://doi.org/10.1007/s12237-012-9581-0

Day JW, Kemp GP, Freeman A, Muth DP (eds) (2014) Perspectives on the restoration of the Mississippi delta: the once and future delta, estuaries of the world. Springer Netherlands. https://doi.org/10.1007/978-94-017-8733-8

Day JW, Colten C, Kemp GP (2019a) Mississippi delta restoration and protection: shifting baselines, diminishing resilience, and growing nonsustainability. In: Coasts and estuaries: the future. Elsevier, pp 167–186

Day JW, Hunter RG, Lane RR, Shaffer GP, Day JN (2019b) Long-term assimilation wetlands in coastal Louisiana: review of monitoring data and management. https://doi.org/10.1016/j.ecoleng.2018.09.019

Day JW, Clark HC, Chang C, Hunter R, Norman CR (2020) Life cycle of oil and gas fields in the Mississippi River Delta: a review. Water 12:30

Day JW, Hunter R, Kemp GP, Moerschbaecher M, Brantley CG (2021) The "problem" of new Orleans and diminishing sustainability of Mississippi River management—future options. Water 13:26. https://doi.org/10.3390/w13060813

De Groot R, Wilson M, Roelof M (2002) A typology for the classification, description and valuation of ecosystem functions, goods and services. Ecol Econ 41:393–408

Duan H, Wang L, Zhang Y, Fu X, Tsang Y, Wu J, Le Y (2018) Variable decomposition of two plant litters and their effects on the carbon sequestration ability of wetland soil in the Yangtze River estuary. Geoderma 319:230–238. https://doi.org/10.1016/j.geoderma.2017.10.050

EPA (2000) Wastewater technology fact sheet—wetlands: subsurface flow (Gov No. EPA 832-F-00-023). Environmental Protection Agency Municipal Technology Branch

EPA (2015) A compilation of cost data associated with the impacts and control of nutrient pollution (No. EPA 820-F-15-096). Office of Water

FEMA (2013) Consideration of environmental benefits in the evaluation of acquisition projects under the hazard mitigation programs | FEMA.gov [WWW Document]. Consid. Environ. Benefits Eval. Acquis. Proj. Hazard Mitig. Programs FEMA. https://www.fema.gov/media-library/assets/documents/33314. Accessed 16 July 2020

FEMA (2016) Benefit-cost analysis tools for drought, ecosystem services, and post-wildfire mitigation for hazard mitigation assistance [WWW Document]. Hazard Mitig. Assist. Publ. https://www.fema.gov/hazard-mitigation-assistance-publications. Accessed 16 July 2020

Fischbach JR, Johnson DR, Ortiz DS, Bryant BJ, Hoover M, Ostwald J (2012) CLARA flood risk model supports Louisiana's coastal planning. RAND Corporation

Franklin A, Noon B, George T (2002) What is habitat fragmentation? Stud Avian Biol 25:20–29

Fromm O (2000) Ecological structure and functions of biodiversity as elements of its total economic value. Environ Resour Econ 16:303–328

García J, García-Galán MJ, Day JW, Boopathy R, White JR, Wallace S, Hunter RG (2020) A review of emerging organic contaminants (EOCs), antibiotic resistant bacteria (ARB), and antibiotic resistance genes (ARGs) in the environment: increasing removal with wetlands and reducing environmental impacts. Bioresour Technol 307. https://doi.org/10.1016/j.biortech.2020.123228

García-Llorente M, Martín-López B, Iniesta-Arandia I, López-Santiago CA, Aguilera PA, Montes C (2012) The role of multi-functionality in social preferences toward semi-arid rural landscapes: an ecosystem service approach. Environ Sci Policy 19–20:136–146. https://doi.org/10.1016/j.envsci.2012.01.006

Getzner M, Spash C, Stagl S (2006) Alternatives for environmental valuation. Routledge

Gramling R, Hagelman R (2005) A working coast: people in the Louisiana Wetlands. J Coast Res 112–133

Gren M, Folke C, Turner K, Batemen I (1994) Primary and secondary values of wetland ecosystems. Environ Resour Econ 4:55–74

Gross B, Montgomery-Brown J, Naumann A, Reinhard M (2004) Occurrence and fate of pharmaceuticals and alkylphenol ethoxylate metabolites in an effluent-dominated river and wetland. Environ Toxicol Chem 23:2074–2083. https://doi.org/10.1897/03-606

Haines-Young R, Potschin M (2010) The links between biodiversity, ecosystem services and human well-being. Ecosyst Ecol New Synth 1:110–139

Hinshaw SE, Tatariw C, Flournoy N, Kleinhuizen A, Taylor C, Sobecky PA, Mortazavi B (2017) Vegetation loss decreases salt marsh denitrification capacity: implications for marsh erosion. Environ Sci Technol 51:8245–8253. https://doi.org/10.1021/acs.est.7b00618

Hoegh-Guldberg O, Bruno JF (2010) The impact of climate change on the world's marine ecosystems. Science 328:1523–1528. https://doi.org/10.1126/science.1189930

Hsia-Kiung K, Reyna E, O'Connor T (2014) Carbon market California—a comprehensive analysis of the golden state's cap-and-trade program/year one 2012–2013. Environmental Defense Fund

Hunt LM, Camp E, van Poorten B, Arlinghaus R (2019) Catch and non-catch-related determinants of where anglers fish: a review of three decades of site choice research in recreational fisheries. Rev Fish Sci Aquac 27:261–286. https://doi.org/10.1080/23308249.2019.1583166

Hunter EA, Nibbelink NP, Cooper RJ (2017) Divergent forecasts for two salt marsh specialists in response to sea level rise. Anim Conserv 20:20–28. https://doi.org/10.1111/acv.12280

Hunter RG, Day JW, Lane RR, Shaffer GP, Day JN, Conner WH, Rybczyk JM, Mistich JA, Ko J-Y (2018) Using natural wetlands for municipal effluent assimilation: a half-century of experience for the Mississippi River Delta and surrounding environs. In: Nagabhatla N, Metcalfe CD (eds) Multifunctional wetlands: pollution abatement and other ecological services from natural and constructed wetlands, environmental contamination remediation and management. Springer International Publishing, Cham, pp 15–81. https://doi.org/10.1007/978-3-319-67416-2_2

Hurricane Laura Wreaks Havoc on Gulf of Mexico Gas Pipelines | Pipeline Technology Journal [WWW Document], n.d. https://www.pipeline-journal.net/news/hurricane-laura-wreaks-havoc-gulf-mexico-gas-pipelines. Accessed 12 June 2021

Iles T (2019) Waterfowl warriors. Louisiana Conservationist

Interis M, Petrolia D (2016) Location, location, habitat: how the value of ecosystem services varies across location and by habitat. Land Econ 92:292–307. https://doi.org/10.3368/le.92.2.292

Isle De Jean Charles Relocation [WWW Document], n.d. Lowl. Cent. https://www.lowlandercenter.org/isle-de-jean-charles-relocation. Accessed 12 June 2021

IWG (2015) Interagency working group on social cost of carbon, response to comments: social cost of carbon for regulatory impact analysis under executive order 12866

Jin X, Ma J, Cai T, Sun X (2016) Non-use value assessment for wetland ecosystem service of Hongxing National Nature Reserve in northeast China. J For Res 27:1435–1442. https://doi.org/10.1007/s11676-016-0264-8

Jordan E (2018) "Dead zone" worsens troubles for Louisiana shrimpers. The Gazette

Kenter J, Hyde T, Christie M, Fazey I (2011) The importance of deliberation in valuing ecosystem services in developing countries—evidence from the Solomon Islands. Glob Environ Change 21:505–521

Keul A (2011) Social spatialization in the Atchafalaya Basin. Florida State University

Khan S, Ahmad I, Shah MT, Rehman S, Khaliq A (2009) Use of constructed wetland for the removal of heavy metals from industrial wastewater. J Environ Manag 90:3451–3457. https://doi.org/10.1016/j.jenvman.2009.05.026

Ko J-Y, Day JW (2004) A review of ecological impacts of oil and gas development on coastal ecosystems in the Mississippi Delta. Ocean Coast Manag 47:597–623. https://doi.org/10.1016/j.ocecoaman.2004.12.004

Ko J-Y, Johnston SR (2007) The economic value of ecosystem services provided by the Galveston Bay/Estuary System. Texas Commission on Environmental Quality Galveston Bay Estuary Program

Lam NS-N, Cheng W, Zou L, Cai H (2018) Effects of landscape fragmentation on land loss. Remote Sens Environ 209:253–262. https://doi.org/10.1016/j.rse.2017.12.034

Land M, Granéli W, Grimvall A, Hoffmann CC, Mitsch WJ, Tonderski KS, Verhoeven JTA (2016) How effective are created or restored freshwater wetlands for nitrogen and phosphorus removal? A systematic review. Environ Evid 5:9. https://doi.org/10.1186/s13750-016-0060-0

Lane RR, Mack SK, Day JW, DeLaune RD, Madison MJ, Precht PR (2016) Fate of soil organic carbon during wetland loss. Wetlands 36:1167–1181. https://doi.org/10.1007/s13157-016-0834-8

Lane RR, Mack SK, Day JW, Kempka R, Brady LJ (2017) Carbon sequestration at a forested wetland receiving treated municipal effluent. Wetlands 37:861–873. https://doi.org/10.1007/s13157-017-0920-6

Lassuy DR (2001) Species profiles: life histories and environmental requirements (Gulf of Mexico) spotted seatrout (No. FWS/OBS 82/11.4). US Fish and Wildlife Service

LDWF (2019) Louisiana Department of Wildlife & Fisheries 2018–2019 Annual Report. Louisiana Department of Wildlife and Fisheries

Lee DE, Hosking SG, Preez MD (2015) Managing some motorised recreational boating challenges in South African estuaries: a case study at the Kromme River Estuary. South Afr J Econ 83:286–302. https://doi.org/10.1111/saje.12059

Lindstedt DM (2005) Renewable resources at stake: Barataria-Terrebonne estuarine system in Southeast Louisiana. J Coast Res 162–175

Louisiana Office of Cultural Development (n.d.) Louisiana Office of Cultural Development [WWW Document]. OCD-Homepage. https://DCRT-MAIN/cultural-development/index. Accessed 15 July 2020

Louisiana Sea Grant (n.d.) Wetlands [WWW Document]. http://www.laseagrant.org/education/topics/wetlands/. Accessed 2 Oct 2019

Lynne GD, Conroy P, Prochaska FJ (1981) Economic valuation of marsh areas for marine production processes. J Environ Econ Manag 8:175–186. https://doi.org/10.1016/0095-0696(81)90006-1

MA (2005) Millennium ecosystem assessment (Working group assessment report). Island Press, Washington D.C., USA

Mack SK, Yankel C, Lane RR, Day JW, Kempka D, Mack JS, Hardee E, LeBlanc C (2014) Carbon market opportunities for Louisiana's Coastal Wetlands. Tierra Resources, Entergy

Mantyka-pringle CS, Martin TG, Rhodes JR (2012) Interactions between climate and habitat loss effects on biodiversity: a systematic review and meta-analysis. Glob Change Biol 18:1239–1252. https://doi.org/10.1111/j.1365-2486.2011.02593.x

Marsooli R, Lin N, Emanuel K, Feng K (2019) Climate change exacerbates hurricane flood hazards along US Atlantic and Gulf Coasts in spatially varying patterns. Nat Commun 10. https://doi.org/10.1038/s41467-019-11755-z

Max-Neef M (1992) Human scale development: conception, application and further reflection. APEX PR

McGinnis JT, Ewing RA, Willingham CA, Rogers SE, Douglass DH, Morrison DL (1972) Environmental aspects of gas pipeline operations in the Louisiana Coastal Marshes (Final Report). Battelle Columus Laboratories, Columbus, OH

McLatchey GP, Reddy KR (1998) Regulation of organic matter decomposition and nutrient release in a wetland soil. J Environ Qual 27:1268–1274. https://doi.org/10.2134/jeq1998.00472425002700050036x

Milcu A, Hanspach J, Abson D, Fischer J (2013) Cultural ecosystem services: a literature review and prospects for future research. Ecol Soc 18. https://doi.org/10.5751/ES-05790-180344

Molinos-Senante M, Reif R, Garrido-Baserba M, Hernández-Sancho F, Omil F, Poch M, Sala-Garrido R (2013) Economic valuation of environmental benefits of removing pharmaceutical and personal care products from WWTP effluents by ozonation. Sci Total Environ 461–462:409–415. https://doi.org/10.1016/j.scitotenv.2013.05.009

Morrison ML, Marcot BG, Mannan RW (1999) Wildlife-habitat relationships: concepts and applications. NCASI Tech Bull 2

National Disaster Resilience—HUD Exchange [WWW Document], n.d. https://www.hudexchange.info/programs/cdbg-dr/resilient-recovery/. Accessed 12 June 2021

National Ecosystem Services Partnership (n.d.) Federal resource management and ecosystem services guidebook [WWW Document]. Fed. Resour. Manag. Ecosyst. Serv. Guideb. https://nespguidebook.com/. Accessed 14 July 2020

NOAA (2019) NOAA forecasts very large 'dead zone' for Gulf of Mexico | National Oceanic and Atmospheric Administration [WWW Document]. https://www.noaa.gov/media-release/noaa-forecasts-very-large-dead-zone-for-gulf-of-mexico. Accessed 24 Oct 2019

Nyman JA, Delaune RD, Patrick WH (1990) Wetland soil formation in the rapidly subsiding Mississippi River Deltaic Plain: mineral and organic matter relationships. Estuar Coast Shelf Sci 31:57–69. https://doi.org/10.1016/0272-7714(90)90028-P

O'Connell A, Hijuelos A, Sable S, Geaghan J (2017) Attachment C3–14: White Shrimp, Litopenaeus setiferus, habitat suitability index model. CPRA

Özengin N, Elmaci A (2016) Removal of pharmaceutical products in a constructed wetland. Iran J Biotechnol 14:221–229. https://doi.org/10.15171/ijb.1223

Pauly D, Yáñez-Arancibia A (2013) Fisheries in lagoon-estuarine ecosystems. In: Estuarine ecology. Wiley-Blackwell, pp 465–482

Petrolia DR, Interis MG, Hwang J (2014) America's wetland? A national survey of willingness to pay for restoration of Louisiana's Coastal Wetlands. Mar Resour Econ 29:17–37. https://doi.org/10.1086/676289

Rutherford JS, Day JW, D'Elia CF, Wiegman ARH, Wilson CS, Caffey RH, Shaffer GP, Lane RR, Batker D (2018) Evaluating trade-offs of a large, infrequent sediment diversion for restoration of a forested wetland in the Mississippi delta. Estuar Coast Shelf Sci 203:80–89. https://doi.org/10.1016/j.ecss.2018.01.016

Schaefer M, Goldman E, Bartuska AM, Sutton-Grier A, Lubchenco J (2015) Nature as capital: advancing and incorporating ecosystem services in United States federal policies and programs. Proc Natl Acad Sci 112:7383–7389. https://doi.org/10.1073/pnas.1420500112

Selman WJr, Linscombe J (2016) Long-term population and colony dynamics of Brown Pelicans (Pelecanus occidentalis) in rapidly changing Coastal Louisiana, USA. Waterbirds 39:45–57. https://doi.org/10.1675/063.039.0106

Sguotti C, Otto SA, Frelat R, Langbehn TJ, Ryberg MP, Lindegren M, Durant JM, Stenseth NC, Möllmann C (2019) Catastrophic dynamics limit Atlantic cod recovery. R Soc B. https://doi.org/10.1098/rspb.2018.2877

Shaffer GP, Day JW, Mack S, Kemp GP, van Heerden I, Poirrier MA, Westphal KA, FitzGerald D, Milanes A, Morris CA, Bea R, Penland PS (2009) The MRGO navigation project: a massive human-induced environmental, economic, and storm disaster. Coast Res 54:206–224. https://doi.org/10.2112/SI54-004.1

Smith MD, Oglend A, Kirkpatrick AJ, Asche F, Bennear LS, Craig JK, Nance JM (2017) Seafood prices reveal impacts of a major ecological disturbance. Proc Natl Acad Sci. https://doi.org/10.1073/pnas.1617948114

Snyder R, Williams D, Peterson G (2003) Culture loss and sense of place in resource valuation: economics, anthropology, and indigenous cultures. Indigenous peoples: resource management and global rights, pp 107–123

Stagg CL, Schoolmaster DR, Krauss KW, Cormier N, Conner WH (2017) Causal mechanisms of soil organic matter decomposition: deconstructing salinity and flooding impacts in coastal wetlands. Ecology 98:2003–2018. https://doi.org/10.1002/ecy.1890

Steven R, Smart JCR, Morrison C, Castley JG (2017) Using a choice experiment and birder preferences to guide bird-conservation funding. Conserv Biol 31:818–827. https://doi.org/10.1111/cobi.12849

Stevenson FJ, Cole MA (1999) Cycles of soil: carbon, nitrogen, phosphorus, sulfur, micronutrients, 2nd edn. River Edge, NJ, Wiley, London

Suárez S, Carballa M, Omil F, Lema JM (2008) How are pharmaceutical and personal care products (PPCPs) removed from urban wastewaters? Rev Environ Sci Biotechnol 7:125–138. https://doi.org/10.1007/s11157-008-9130-2

University of Hawaii Agriculture-Based Remediation Program (2001) Constructed wetlands [WWW Document]. Constr Wetl. http://www.hawaii.edu/abrp/Technologies/constru.html. Accessed 6 Aug 20

U.S. Environmental Protection Agency (EPA) (2018) Why are wetlands important? [WWW Document]. Why are Wetl. important. https://www.epa.gov/wetlands/why-are-wetlands-important

USACE (2016) Southwest Coastal Louisiana integrated final feasibility report and environmental impact statement (No. Annex A-1)

Valente JJ, King SL, Wilson RR (2011) Distribution and habitat associations of breeding secretive marsh birds in Louisiana's Mississippi Alluvial Valley. Wetlands 31:1–10. https://doi.org/10.1007/s13157-010-0138-3

Weitzman ML (1998) Why the far-distant future should be discounted at its lowest possible rate. J Environ Econ Manag 36:201–208. https://doi.org/10.1006/jeem.1998.1052

Whiting GJ, Chanton JP (2001) Greenhouse carbon balance of wetlands: methane emission versus carbon sequestration. Tellus B Chem Phys Meteorol 53:521–528. https://doi.org/10.3402/tellusb.v53i5.16628

Whitman CT, Mehan GT, Frace SE, Johnson RS (2002) Economic analysis of the final revisions to the national pollutant discharge elimination system regulation and the effluent guidelines for concentrated animal feeding operations (No. EPA-821-R-03-002). EPA

Widney S, Kanabrocki Klein A, Ehman J, Hackney C, Craft C (2017) The value of wetlands for water quality improvement: an example from the St. Johns River watershed, Florida. Wetl Ecol Manag 26:265–276. https://doi.org/10.1007/s11273-017-9569-4

Williams JA, Holt GJ, Robillard MMR, Holt SA, Hensgen G, Stunz GW (2016) Seagrass fragmentation impacts recruitment dynamics of estuarine-dependent fish. J Exp Mar Biol Ecol 479:97–105. https://doi.org/10.1016/j.jembe.2016.03.008

Wilson SG, Fischetti TR (2010) Coastline population trends in the United States: 1960 to 2008. US Census Bureau

Yeager LA, Keller DA, Burns TR, Pool AS, Fodrie FJ (2016) Threshold effects of habitat fragmentation on fish diversity at landscapes scales. Ecology 97:2157–2166. https://doi.org/10.1002/ecy.1449

Chapter 8
Restoring Coastal Ecosystems Impacted by Oil and Gas Activity

Charles Norman, John W. Day, and Rachael G. Hunter

8.1 Introduction

This chapter discusses restoration of wetlands degraded by oil and gas activities. Restoration involves a synergistic approach that tries to reverse the damage of oil and gas impacts and rebuild a functioning coastal wetland system. Current scientific and engineering knowledge about successful wetland restoration and creation has been documented in various publications (e.g., Kusler and Kentula 1990; Craft 2015). Knowledge about wetland restoration and creation varies with wetland function, type, and location.

A basic premise of wetland restoration is to restore the ecosystem to its original condition or to a more functional and sustainable state relative to some point in time. Restoring an ecosystem to its historical condition is often not possible, not always desirable, and may be difficult to sustain under future conditions (Clewell and Aronson 2013). Instead, the goal of restoration is to establish conditions and processes (e.g., physical, biogeochemical, and biological) that interact to develop and sustain the wetland against current and future global change and human stressors.

Restoration requires a damage assessment of the type of wetland (e.g., freshwater, brackish), the amount of wetland loss, causation of the damage, and analysis of methods to restore the wetland, including a cost/benefit analysis. A full functional

---

The original version of this chapter was revised: The author "Rachael G. Hunter's" affiliation has been updated. The correction to this chapter is available at https://doi.org/10.1007/978-3-030-94526-8_10

C. Norman
Charles Norman & Associates, Lake Charles, LA, USA

J. W. Day (✉)
Department of Oceanography and Coastal Sci, Louisiana State University, 2005 Olive St, Baton Rouge, LA 70806-, 6660, USA

R. G. Hunter
Comite Resources, Inc, Baton Rouge, LA, USA

assessment of the wetland should be conducted, describing or quantifying processes such as flood storage, waterfowl habitat, water quality improvement and groundwater recharge. Successful restoration projects have numerous stakeholders including landowners, local, state and federal government, non-governmental organizations and others.

A well-designed restoration plan should be based on information about the impacts of exploration and production during the life cycle of an oil and gas field (Day et al. 2020). A typical objective in wetland restoration in coastal Louisiana is to rebuild elevation loss due to both surface subsidence caused by altered hydrology and subsurface subsidence induced by oil and gas extraction. Subsurface induced subsidence cannot be reversed, so sediment must be added at the surface to offset both surface and subsurface subsidence. Options for restoration include (1) marsh creation and restoration using dredged sediments (DeLaune et al. 1990; Ford et al. 1999; Mendelssohn and Kuhn 2003; La Peyre et al. 2009); (2) use of all available sediment resources including diversions (e.g., Mississippi River, Gulf Intracoastal Waterway) and sediments resuspended during storms (Perez et al. 2000; Day et al. 2007); (3) hydrologic restoration such as backfilling canals and removing spoil banks and restoration of hydrologic networks (Turner and Streever 2002; DeLaune et al. 2003; Baustian and Turner 2006; Day et al. 2007; Weinstein and Day 2014); and (4) addressing contamination. Rebuilding natural levee ridges that have been severed by canals can also help restore natural hydrology to prevent saltwater intrusion and reduce increased hydrologic energy caused by more direct connections between interior parts of the delta and the Gulf of Mexico.

Wetland restoration involves either re-establishing or rebuilding a former wetland or rehabilitating or repairing a degraded wetland (USEPA 2007). For coastal marshes, this means choosing an appropriate location such as one that once supported or now supports a degraded tidal wetland and establishing processes that balance flooding and drainage, sedimentation and erosion so that the system can self-organize and become self-sustaining. Sometimes it is necessary to take an active role in ecosystem development such as promoting the appropriate vegetation community (e.g., planting or removing invasive species), adding sediment, and/or re-creating tidal networks to achieve intertidal elevations and tidal hydrology.

A restoration plan should include achievable and measurable goals and long-term ecosystem monitoring. Goal planning should involve key stakeholders and experts to set beneficial goals and to incorporate an ecosystem perspective. Individual project goals should be clearly stated, site specific, measurable and long-term—in many cases greater than 20 years, especially to incorporate potential climate change (RAE 1999). Site plans need to address off-site considerations, such as potential flooding and salt water intrusion, to be sure projects do not have negative impacts on nearby people and property. Ecological engineering practices should be applied to identify energy efficient restoration options to maximize the use of natural processes and to achieve self-sustaining habitats and landscapes.

Long-term monitoring is critical to determine the success of a restoration project. Scientifically based monitoring is essential to evaluate performance, conduct adaptive management and improve restoration science and techniques. Performance criteria for projects need to include both functional and structural elements and be linked to suitable, local habitats that are used as a reference for comparison of structure and function. Public access to restoration sites should be encouraged wherever appropriate, but designed to minimize impacts on the ecological functioning of the site. Below we discuss the steps involved in restoring coastal wetlands impacted by oil and gas activities.

8.2 Evaluation of Historical Conditions

The first step in restoration of a particular area is the evaluation of the historical condition of the land at a particular time and then identification of the use(s) that may have altered the condition. In other words, a geographical area or project area is identified where land change and land loss have occurred. Areas surrounding the project boundary must also be identified for potential contributions to the loss. Aerial photographs over time are extremely useful in identifying impacts and changes over time. The location of oil and gas wells, produced water storage basins, canal dredging, and spoil bank creation can be seen on these aerials. The routes of pipelines, flowlines, roads, and production facilities can also be identified. Site visits to the areas being evaluated for restoration are important to verify impacts identified on aerial images. After these steps are completed, field testing may be required to measure contaminant concentrations in soils and groundwater or to identify dominant vegetation species for example. Once the impacts of oil and gas activity and changes to an ecosystem have been identified, the restoration goals can be defined.

8.3 Definition of Restoration Goals

The second step in a restoration project is to determine the restoration goals, typically by describing the desired future conditions of the site and determining how best to restore the land to those conditions. Restoration goals should be measurable so that progress towards the goal can be determined. Common methods for coastal wetland restoration often include marsh creation using dredged material, hydrological restoration, or sediment diversions. Many different restoration projects have been completed within the Louisiana coastal zone already and these projects are useful in developing restoration goals and in evaluating restoration methods for a project. Louisiana's Coastal Protection and Restoration Authority (CPRA) has developed several monitoring programs, including the System-Wide Assessment and Monitoring Program (SWAMP), the Coastwide Reference Monitoring System (CRMS), and the Barrier Island Comprehensive Monitoring (BICM) program, "to ensure that a comprehensive network of coastal data collection activities is in place to support the planning, development, implementation, and adaptive management of the protection and restoration program and projects within coastal Louisiana" (Raynie et al. 2020).

In addition, CPRA has developed a data management system, the Coastal Information Management System (CIMS), that provides geospatial, tabular database and document access to CPRA's restoration projects, CRMS data, geophysical data, and coastal community resiliency information.

8.4 Restoration of Hydrology

The biggest drivers of marsh health in coastal Louisiana are salinity and water level. Semi-diurnal, monthly, and multidecadal variations in water levels drive exchange of fresh and saltwater, nutrients, and sediments between coastal wetlands and adjacent water bodies (Hiatt et al. 2019). The temporal and spatial pattern of soil salinity strongly limits the species of plants that can grow in coastal marshes and wetland health and productivity and largely reflects tidal influence, but it is also a product of evapotranspiration and net movement of water seaward. These factors affect marsh elevation relative to mean sea level, which is important to maintain a healthy and productive marsh (Mitsch and Gosselink 2015, e.g., Hackney et al. 1996). Sediments bring nutrients to the marsh and are captured by vegetation as flood waters slow, helping to maintaining optimal marsh elevation along coasts experiencing rising sea levels. Sediments lead to marsh accretion directly due to deposition, and indirectly by stimulating marsh productivity that leads to organic soil formation. Tidal restrictions interfere with marsh building processes, disadvantage native perennials, encourage invasive species, alter essential habitat and reduce biodiversity, interfere with fish passage and coastal food webs, and lead to loss of stored carbon resulting in subsidence. This has been especially extensive in the Mississippi delta (e.g., Couvillion et al. 2017; Day et al. 2019, 2020). Oil and gas impacts often limit tidal exchange. Tidal flow through creeks is is often cut off by spoil banks and tide gates or only partially accommodated by engineered structures such as bridges and culverts.

Hydrologic restoration, as a technique for improving marsh health, seeks to restore natural hydrologic patterns either by conveying fresh water to areas that have been isolated by man-made features, relieving unnatural impoundments, or by preventing the intrusion of salt water. Hydrological restoration often requires computer modeling to fully design a restoration plan that provides a well-functioning marsh.

8.4.1 Canal Backfilling and Spoil Bank Removal

Canals have been a significant cause of wetland loss in the MRD and removing spoil banks and backfilling canals could be an integral part of delta restoration. Turner and McClenachan (2018) advocate backfilling canals by dragging the remaining spoil material into canals as a cost-effective approach to slow and reverse wetland loss. Care must be taken in how this is done and which canals are chosen to be backfilled. The current volume of spoil banks is much less than when canals were first dredged.

Most canals were dredged prior to 1980 and the spoil material has dewatered and oxidized (Neill and Turner 1987). Baustian (2005) studied canal backfilling and reported that on average 58% of spoil bank area was restored to marsh and the restoration of marsh in the canals averaged 13%. Thus, backfilling would mostly result in shallower canals and linear strips of restored marsh adjacent to the canals. In many cases, such as the Leeville oil and gas field, much of the remaining subaerial land in many oil and gas fields is spoil bank with patches of remnant marsh (e.g., Leeville, Fig. 1A in Turner and McClenachan 2018). Many of the spoil banks in the Leeville field disappeared between 1989 and 2017 (Fig. 8.1), having subsided or eroded. Removal of all spoil banks in Fig. 1 to an elevation where marsh would grow would expose most of the narrow strips of marsh created by backfilling and remaining remnant marsh to wave attack, which is now mostly fetch-limited by spoil banks (e.g., Day et al. 2011). Thus, some spoil banks should be retained as buffers against wave attack. Backfilling should be coupled with marsh creation to restore marsh in the entire oil field.

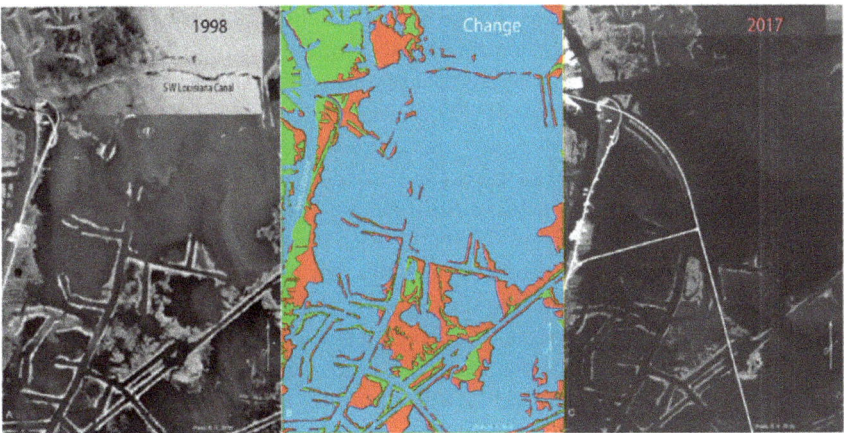

Fig. 8.1 Aerial imagery of a portion of the Leeville oil and gas field in southern Louisiana showing the disappearance of spoil banks and marsh between 1998 and 2017. Image A is a 1998 USGS LANDSAT satellite image downloaded from Google Earth Pro and georeferenced in ArcMap. Image B is an ArcGIS Shapefile of the Land–Water changes from 1998 to 2017. No new wetlands appeared during this interval. Image C is a 2017 USDA NAIP aerial image. Areas depicted in red in image B disappeared between the two dates. Red linear strips are spoil banks that disappeared. Green shows subaerial land, both spoil banks and marsh, that was still present in 2017. The Southwest Louisiana Canal at the top of image A was dredged in the late nineteenth century. In 1998, some spoil banks along the canal were still present, but by 2017 they had disappeared (Image C). Louisiana Highway 1 is shown at the upper left in image A. By 2017, a new elevated highway was constructed (white lines in Image C). Geospatial analysis by R. Hampton Peele (March 2019). The width of images is approximately 2.8 km. (For interpretation of the references to color in this figure legend, the reader is referred to the Web version of this article.)

8.4.2 Restoration of Tidal Exchange

Tidal flooding needs of tidal marshes were rarely considered until the second half of the twentieth century. This is especially the case for oil and gas development in the Mississippi River Delta where dramatic changes in surface and subsurface hydrology have occurred. Tidal exchange in many coastal marshes has been dramatically altered and re-establishing that natural exch is important. Hydrologic restoration can involve both increasing tidal exchange where it has been reduced or decreasing exchange where it has been amplified as for example by deep straight canals. One approach is to remove obstructions to allow tidal exchange or removing sections at the blocked tidal channels; replacing broken or undersized pipes or culverts; and replacing tide gates with large culverts. Some sites may require excavation of the main tidal channels to initiate channel flow and the development of secondary creeks, as accomplished at Delaware Bay coastal marshes.

Measurement of the tidal prism will help inform restoration of tidal connectivity (MacBroom and Schiff 2012). Examining 20 available projects in the Gulf of Maine, Konisky et al. (2006) found that only six tidal restoration projects had pre- and post-restoration hydrologic data, measuring an average increase in tidal range from 38 to 74% of the potential tide, suggesting many crossings may still be inadequate to fully support natural tidal dynamics. Thus, monitoring is necessary and should be included in restoration of hydrology in areas impacted by oil and gas activities.

Once tides are restored, fish access is immediate, but other responses require varying amounts of time to achieve full functionality of tidally restored marshes (Simenstad and Thom 1996; Burdick et al. 1997; Rogers et al. 1992). Natural revegetation typically occurs, but vegetation plantings may be necessary (Fig. 8.2; Wolters et al. 2005). Some marshes may have subsided such that they cannot support native vegetation and partial restoration to establish habitat and rebuild elevation (Smith et al. 2009) or sediment additions have been used (Carnu and Sadro 2002). Marsh creation is being widely used to restore wetlands where elevation is too low for natural regeneraton and this will be necessary for many areas in the Louisiana coastal zone. Where sediments are plentiful, as in the Achafalaya and Wax Lake deltas, natural sedimentation and channel development can 'self-design' marsh topography and hydrology, though natural colonization by vegetation and plant community development can take decades. Capture of high levels of resuspended sediments during frontal passages can help to sustain a restored marsh (Perez et al. 2000).

8.4.3 Development of Tidal Channel Networks in Restored Marshes

A tidal network needs to develop in restored marshes. All restoration plans need a component that focuses on natural hydrological functioning. A large-scale restoration project on Delaware Bay indicated that if a few primary channels were dredged, the

Fig. 8.2 Vegetative planting at a restored marsh in coastal Louisiana (CWPPRA)

system would naturally then develop a full tidal channel network in balance with the tidal prism (Teal and Weinstein 2002; Teal and Peterson 2005). The project involved restoration of tidal flooding by breeching dikes around formerly diked salt-hay farm and re-excavation of tidal creeks. Hydrologic design occurred in a "self-design" fashion after only initial cuts by construction of the largest (class 1) channels.

Results of the Delaware study were described in a number of publications (Teal and Weinstein 2002; Peterson et al. 2005; Teal and Peterson 2005). From a hydrodynamic perspective, in the Delaware marshes where tidal exchange was restored, there has been the development of an intricate tidal creek density. Figure 8.3 illustrates the development of a tidal stream network at two of the newly restored marsh sites. The "order" of the stream channels increased from 5 or less to well over 20 over 9 years (1996 through 2004) in one area and over 30 over 17 years (1997 through 2013) in a second site. The number of small tributaries increased from "dozens" to "hundreds" in the restored marshes that were reopened to tidal flushing. Hydrologic design occurred in a "self-design" fashion after only initial cuts by construction of the largest (class 1) channels. Re-establishment of *Spartina alterniflora* and other favorable vegetation has been rapid and extensive at the Delaware sites.

The Delaware study demonstrated that the speed with which salt marsh hydrological restoration takes place is dependent on three main factors:

1. The degree to which the tidal "circulatory system" works its way through the marsh
2. The size of the site being restored
3. The initial presence of *Spartina* and other desirable species to colonize the area.

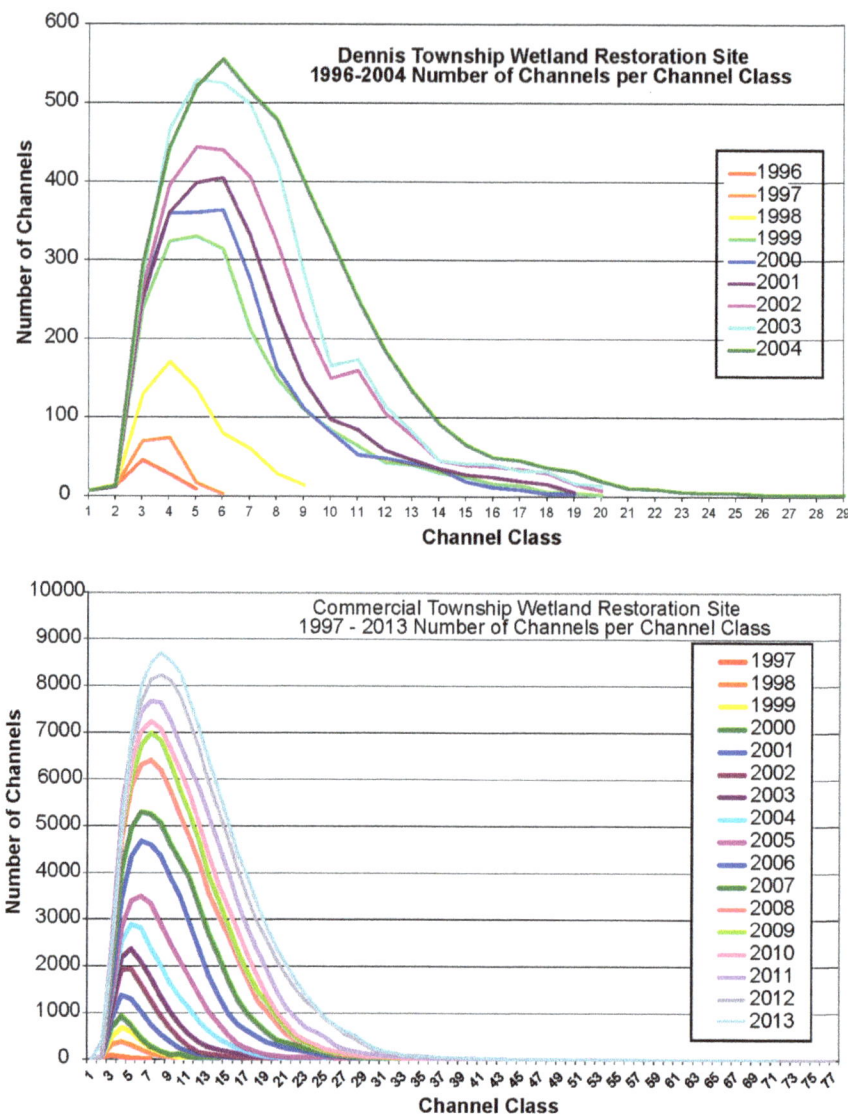

Fig. 8.3 Total number of stream channels by channel class at Dennis Township (1996–2004) and Commercial Township (1997–2013) restored salt marshes on Delaware Bay (Reprinted from Mitsch and Gosselink 2015, with permission from John Wiley & Sons, Inc. Hoboken N.J.)

No planting was necessary on these sites, as *Spartina* seeds arrive by tidal fluxes from nearby salt marshes, but the design of the sites to allow that tidal connectivity (and hence the importance of appropriate site elevations relative to tides) was critical. Self-design works when the proper conditions for propagule disbursement are provided. Extensive ponding in some areas of the marshes that have a high ratio of area to edge, initially impeded the reestablishment of *Spartina* in some locations (Teal and Weinstein 2002). Creating additional streams or waiting for the tidal forces to cause the same effect eventually allows these areas to develop tidal cycles and *Spartina* to establish itself.

Elevation can be built up by adding dredged material to shallow water bodies, some that were previously wetlands, to an elevation that will support marsh vegetation (Ford et al. 1999; Ray et al. 2007; Stagg and Mendelssohn 2010). Restoration of open water and degraded wetland areas to sustainable marsh are of tremendous importance in coastal Louisiana. The 2017 Coastal Master Plan (CMP) for restoration of the Mississippi River Delta dedicates nearly $18 billion to marsh creation.

8.5 Sediment Management

After natural hydrology has been re-established, full use of all available sediments should be incorporated into the restoration design. Important sources of sediments include the Mississippi/Atchafalaya Rivers as well as smaller rivers, sediment diversions, the Gulf Intracoastal Waterway, and sediments resuspended by storms (Perez et al. 2000). Water sources with appreciable levels of nutrients should be used where possible to enhance marsh production including upland runoff, agricultural drainage, and treated municipal effluent (Shaffer et al. 2018; Hunter et al. 2018).

Sediment addition increases marsh surface elevation, thus reducing flood stress to the plant community. DeLaune et al. (1990) showed that raising the surface of a deteriorating *Spartina alterniflora* salt marsh by 10 cm using dredge spoil resulted in a twofold increase in aboveground biomass production after the second growing season. Ford et al. (1999) found that increased elevation through the deposition of a 2 cm layer of dredged material in a deteriorated Louisiana marsh increased percent cover of *S. alterniflora* three-fold within one year. Sediment subsidy increases soil mineral matter, soil fertility, and marsh elevation, and thereby reduces nutrient deficiency, flooding, and interstitial sulfide stresses, generating a more favorable environment for plant growth and potentially, marsh sustainability (Mendelssohn and Kuhn 2003; La Peyre et al. 2009).

Stagg and Mendelssohn (2010) studied dead marshes near Leeville that were restored using dredged material and found that sediment-slurry addition increased the elevation of the marsh surface and alleviated stress associated with excessive inundation and high salinity. Primary production was highest at elevations ranging from 29 to 36 cm NAVD 88 (12–20 cm above ambient marsh), and decreased at elevations above 36 cm NAVD 88, where primary production was limited by insufficient flooding and low nutrient availability. Sediment subsidy increases soil mineral matter, soil fertility, and marsh elevation, and thereby reduces nutrient deficiency, flooding, and interstitial sulfide stresses, generating a more favorable environment for plant growth and potentially, marsh sustainability (Mendelssohn and Kuhn 2003; La Peyre et al. 2009).

Increased plant production as a result of increased sedimentation is due to several factors. The sediment will increase marsh surface elevation, thus reducing flood stress to the plant community (DeLaune et al. 1990). In dieback areas of salt marsh near Caminada Bay, Louisiana, *S. alterniflora* transplanted into elevated plots had more than twice the aboveground and belowground biomass than plants transplanted into non-elevated plots after three months of growth (Wilsey et al. 1992).

Mendelssohn and Kuhn (2003) studied a marsh near Venice, Louisiana where hydraulically dredged sediment (85% liquid and 15% solids) accidentally overflowed into an adjacent submerging salt marsh and resulted in added sediment from trace amounts to as much as 60 cm of sediment above the natural marsh surface. Sediment subsidy increased soil mineral matter, and, in turn, soil fertility and marsh elevation, and thereby reduced nutrient deficiency, flooding, and interstitial sulfide (Fig. 8.4). Thus, sediment subsidy generated a more favorable environment for plant growth and potentially, marsh sustainability.

Sediment addition confers a long-term positive impact. Kuhn (2003) and Slocum et al. (2005) reported that examined plant growth over a 7-year period (1993–1998). They found that found that the positive effects of increased elevation were longer lasting and that even after 7 years, sediment enriched areas had 55% cover compared to only 20% cover in areas that did not receive sediment (Slocum et al. 2005). They concluded that the sediment slurry addition countered the effects of subsidence and sea level rise, but not so much as to surpass the intertidal position to which *S. alterniflora* is best adapted.

La Peyre et al. (2009) studied deteriorating marsh and open-water pond habitats located in six brackish marshes (dominated by *Spartina patens* and *Schoenoplectus americanus*) in coastal Louisiana that were enhanced between 1999 and 2006 using a low-pressure hydraulic dredge to pipe a slurry (>80% water) of dredged material over the marsh surface.

Sediment additions, resulting in changes to substrate texture and composition, have also been shown to influence the chemical conditions of marsh soil. Slocum et al. (2005) documented reduced H_2S concentrations following sediment additions that resulted in improved aeration due to increased elevation and increased sand content in the soil. Koning (2004) found no significant change in oxidation–reduction potential for sandy sediment additions of up to 4 cm on wetland soils. Mendelssohn and Kuhn (2003) that sediment addition increased redox potential, bulk density and interstitial iron while sulfide decreased. Stagg and Mendelssohn (2010) reported that low and intermediate rates of sediment addition had the greatest benefit, decreasing flood duration and frequency and increasing drainage, redox potential, and above- and belowground biomass compared to untreated areas and areas receiving high rates of sediment additions.

8 Restoring Coastal Ecosystems Impacted by Oil and Gas Activity

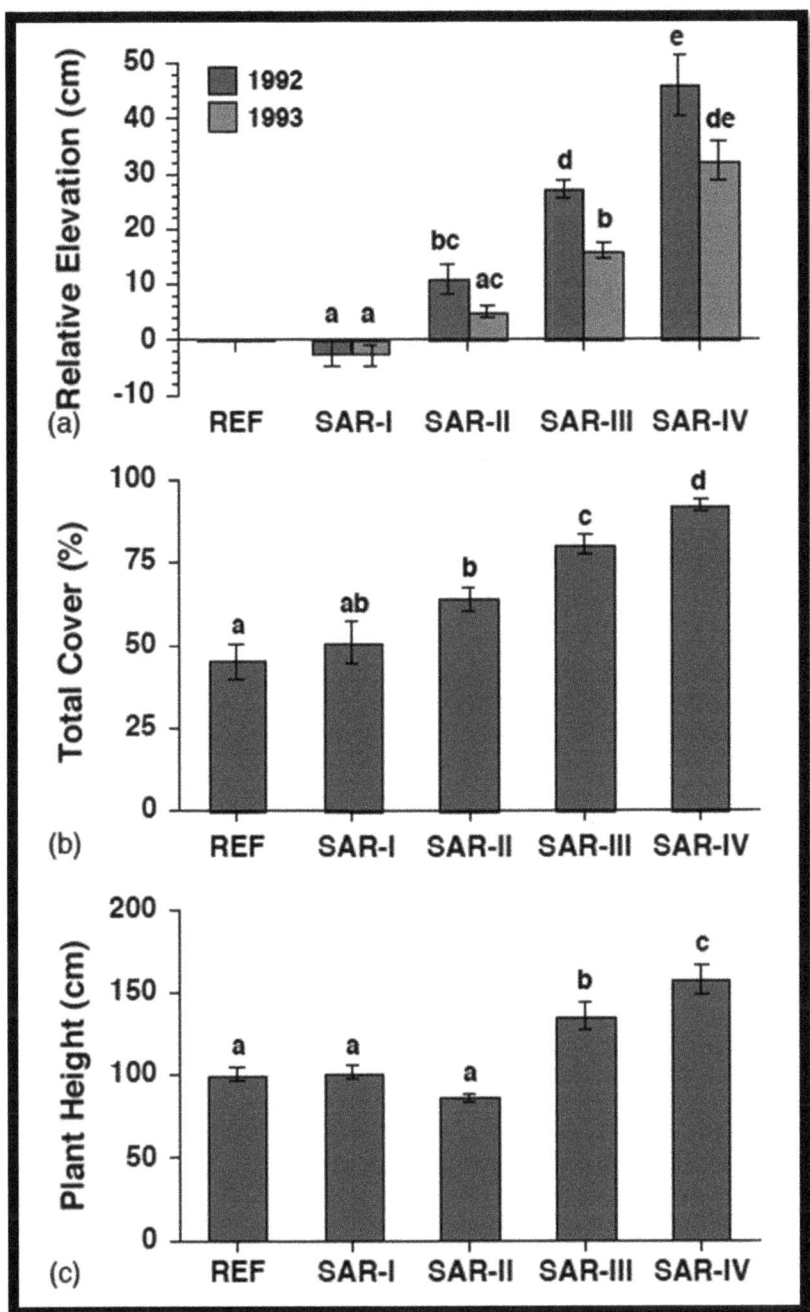

Fig. 8.4 The effect of sediment addition on **a** relative elevation in 1992 and 1993, **b** total vegetative cover, and **c** plant height (Mendelssohn and Kuhn 2003). SAR refers to sediment affected region

8.5.1 Riverine Input

River input introduces sediments, nutrients and freshwater that support the function and maintenance of coastal wetlands. Re-introduction of riverine input is critical for the restoration of deltas and specifically for the MRD (Syvitski et al. 2009; Day et al. 2011, 2016a, b, 2019). The near elimination of riverine input in many deltas is a result of levees constructed for flood protection. The result has been high rates if wetland loss, both in terms of area and ecological functions and services (Tockner et al. 2000; Giosan et al. 2014; Syvitski et al. 2009; Day et al. 2019). The MRD is one of the best described examples of the impact of restriction of riverine input by levees (Britsch and Dunbar 2006; Couvillion et al. 2017). Under natural conditions, sediment input to the larger delta occurs by direct riverine input in the deltaic plain and by down drift of riverine sediments that were then incorporated into the coastal basins of the Chenier Plain. To the extent possible, riverine sediment sources should be utilized in the restoration of coastal marshes damaged by oil and gas activities.

8.5.2 River Diversions

In areas where coastal marshes have been isolated from riverine input by levees and dikes, diversions of river water back into coastal marshes is being proposed and implemented in the MRD. River diversions are an important component of the large-scale restoration of the MRD (CPRA 2017). Diversions are also being planned for the Sacramento-San Juaquin delta, the Ebro delta and other areas. Pont et al. (2017) reported on two large floods in the Rhone delta where levees failed leading to widespread flooding of the delta with deposition of up to 10 cm. The authors proposed that these floods served as an example of how diversions could be used to sustain the Rhone delta. Where possible, sediments from river diversions should be considered in the restoration of marshes destroyed by oil and gas activities.

8.5.3 Marsh Creation Through Sediment Addition

Large, contiguous marsh creation projects are more cost-effective than many small, isolated marsh creation projects. Elevation can be built up by adding dredged material to shallow water bodies, some that were previously wetlands, to an elevation that will support marsh vegetation (Ford et al. 1999; Ray et al. 2007; Stagg and Mendelssohn 2010; CPRA 2017). Restoration of open water and degraded wetland areas to sustainable marsh is of tremendous importance in coastal Louisiana.

Marsh creation is a widely used technique and is an important part of the Coastal Master Plan (CMP 2017; Fig. 8.5). It is also known as thick sediment addition. In contrast to thin layer deposition, the water is too deep and significant filling

Fig. 8.5 Marsh creation area in coastal Louisiana (Photo taken September 12, 2017 by Patrick M. Quigley III)

must take place in order for elevation of the sediment surface to support marsh vegetation (Fig. 8.6). With this approach, up to a meter of dredged sediment is deposited in shallow water bodies to create intertidal habitat at appropriate elevations

Fig. 8.6 Dredged material discharged from a pipe raising elevation during a marsh creation project (Photo from CPRA)

for marsh establishment (Streever 2000; Edwards and Proffitt 2003; Staver 2015). These projects often use dredge spoil and may be large and require planting (Streever 2000). The loss of the original marsh may be due to erosive forces and as such, these areas are often armored where erosion remains an issue (Poplar Island, MD: Staver 2015; and Jamaica Bay, NY: Frame et al. 2006). Once a marsh is established, the development of a tidal network is essential so that proper drainage can take place and so that sediments can be introduced on the flood tide.

Marsh creation (the largest investment in a single type of restoration project) and many of those projects include both a short-term and a long-term phase (CPRA 2017). The short-term phase focuses on immediate actions needed to protect vulnerable marshes from the proximal causes of loss (saltwater intrusion, erosion, and other consequences of significant hydrologic modifications) using a combination of restoration techniques (especially hydrologic restoration and marsh creation using dredged sediments). This is especially important for marshes degraded by oil and gas activities. Successful implementation of short-term strategies reduces rates of wetlands loss and provides the foundation for longer-term strategies. The long-term phase focuses on wetlands gains through Mississippi River sediment diversions and capture of sediments advected over the marsh surface, with the intent of encouraging development of a sustainable wetland ecosystem (https://www.lacoast.gov/new/about/Basin_data/te/Default.aspx).

Large scale sediment diversions-using new channels and/or structures to divert sediment and fresh water from the Mississippi and Atchafalaya Rivers into adjacent basins-are a cornerstone of the CMP. Marsh creation in the influence area is not a substitute for the long-term benefits of sediment diversions, and is not consistent unless specifically identified in the CMP.

Sediment can be pumped long distances, as is the case for many 2017 CMP projects (Cahoon and Cowan 1988; Slocum et al. 2005; Croft et al. 2006). Containment dikes are utilized to confine unconsolidated material until it settles and dewaters (Turner and Streever 2002). Then these dikes can be degraded or gapped to the same elevation as the marsh platform in order to allow water exchange and marsh drainage, to improve productivity and aeration, and increase capture of suspended sediments that are advected over the marsh surface.

After containment dikes are gapped, restored marsh will have connectivity with the surrounding ecosystem and vegetation will naturally recruit to the area or can be planted. Although the created marsh will be initially higher than surrounding marshes, it will settle and have a higher bulk density. It is likely that these created marshes will have a higher productivity and production of organic matter than surrounding marshes. The plant species richness should the same as surrounding marshes, which is not very high in brackish and salt marshes of coastal Louisiana compared to fresh marshes. Plant diversity and soil organic matter content are higher in brackish marsh than in salt marsh. Brackish marsh is typically dominated by *Spartina patens* (marshhay cordgrass). Other significant associated species include *Distichlis spicata* (salt grass), *Schoenoplectus olneyi* (three-cornered grass), *S. robustus* (salt marsh bulrush), *Eleocharis parvula* (dwarf spikesedge), *Ruppia maritima* (widgeon grass), *Paspalum vaginatum* (seashore paspalum), *Juncus roemerianus* (black rush), *Bacopa monnieri* (coastal water hyssop), *S. alterniflora* (smooth cordgrass), and *S. cynosuroides* (big cordgrass) (Lester et al. 2005). Salt marsh has the least plant diversity of any marsh type. The community is often totally dominated by smooth cordgrass. Significant associate species includes marshhay cordgrass, salt grass, black rush, and *Batis maritima* (salt wort).

8.5.4 Terraces

Marsh terracing is a relatively new technique and has become a common feature of coastal restoration efforts in the northern Gulf of Mexico. Marsh terraces are segmented ridges of bare soil and emergent marsh constructed from excavated subtidal substrates in shallow, open water areas (Fig. 8.7). Marsh terraces function similarly to other restoration techniques such as breakwater and sediment retention structures (e.g., Christmas tree fences; (Boumans et al. 1997), by reducing fetch and wave energy, thereby increasing the potential for sediment deposition in the leeward side of the structure (Steyer 1993, Turner and Streever 2002; Brasher 2015). However, if there is not a source of sediments to marsh terraces, they will likely stay the same or decrease in size, reducing the benefit of the restoration project.

Terraces are most often constructed in large water bodies that were once emergent marsh but have converted to open water as a result of a variety of factors. Terraces are

Fig. 8.7 Marsh terraces in coastal Louisiana (Photo by CPRA)

usually designed and constructed with a crown height equal to surrounding marsh elevation to enable periodic tidal inundation of the terraces (Turner and Streever 2002; Brasher 2015). Vegetation may be planted on the perimeter of the terraces following construction to accelerate colonization and to provide immediate protection from erosion (Brasher 2015).

8.5.5 Thin Layer Sediment Addition

"Thin-layer placement," also known as "thin-layer sediment addition" and "marsh nourishment," is a process where sediment removed from navigation channels during dredging is transported to a marsh by pipeline or barge and applied to the surface of the marsh by spraying a slurry of water, sand, and silt (VIMS 2014). Dredging can also be done specifically to obtain sediment for thin-layer placement. The term thin-layer placement itself has been used to describe thicknesses ranging from less than 1.0–>30 cm. Since the ecological impact of sediment thickness differs among habitats, Wilber (1992a) concluded that the best definition of thin-layer placement would be placement of a thickness of dredged material that does not transform the receiving habitat's ecological functions (Ray 2007). La Peyre et al. (2006) suggested that the proper thickness can easily be determined by calculating how much sediment needs to be added to return the deteriorated marsh back to the elevation of nearby healthy marsh.

Thin-layer deposition of dredged material has been proposed as a means of restoring marsh elevations because, in addition to minimizing spoil bank impact (Cahoon and Cowan 1988), thin-layer deposition has the potential to restore soil elevations in subsiding coastal marshes (Wilber 1992b, c). When done correctly, dredged sediment addition to marshes can be beneficial both as a dredge material disposal site and as a mechanism for increasing marsh resilience (VIMS 2014). Thin-layer placement of dredge material is used to increase soil surface elevation to reduce waterlogging and porewater H_2S and to increase soil redox potential and vegetation stem density, productivity and nutrient uptake (DeLaune et al. 1990; Mendelssohn and Kuhn 2003; Croft et al. 2006). The benefits of thin layer placement vary depending on the initial elevation of the marsh and the amount of sediment added (Craft 2016).

For thin-layer addition, existing live plants can grow through sediment additions of less than 10 cm, but sediment additions much greater than 10–15 cm require re-seeding from adjacent marshes (Wilbur 1993) or by planting. In all projects where it has been examined, sediment compaction has been reported, with thicker deposits leading to greater compaction of added sediment as well as underlying marsh sediments (Wilbur 1993, Edwards and Mills 2005; Slocum et al. 2005) to the point of no added elevation in one study within a brackish marsh (Graham and Mendelssohn 2013). Even in sites with good to excellent plant regeneration, it is not clear how much this new sediment adds to elevation and leads to long-term sustainability (Ford et al. 1999; La Peyre et al. 2009; Graham and Mendelssohn 2013) and monitoring

Fig. 8.8 High-pressure spray disposal of dredged material (Photo by Dredge America)

is recommended (Edwards and Proffitt 2003). Marshes in the network of canals and spoil banks in many oil and gas fields are excellent candidates for thin layer deposition.

The primary method of thin-layer placement is to deposit thin layers of sediment by spraying sediment slurry under high pressure over the marsh surface. The technique is essentially a modification of existing hydraulic dredging methods in which sediments are hydraulically dredged, liquefied, and then pumped through a high-pressure spray nozzle (Fig. 8.8). Spray disposal is capable of handling a variety of soil types ranging from sands to heavy clays and organic sediments. Its operation can be modified to target specific sites and avoid sensitive areas (Ray 2007). The placement of materials can be controlled to a great extent by proper operation of the spray nozzle, however, the proximity of habitats sensitive to turbidity and sedimentation such as oyster beds or seagrass meadows should be a critical concern. For example, seagrasses and other submerged aquatic vegetation are generally thought to be inhibited by suspended sediment concentrations above 15 mg/l (Chesapeake Bay Program 2000) and accumulation of as little as 1 mm of sediment can prevent oyster larvae from settling (Ray 2007).

High-pressure spray disposal does have some limitations (Ray 2007). Shafer (2002) noted that site access is a primary consideration and that the receiving area must be close to a body of water deep enough to provide access for the dredging equipment. Although high pressure spraying has a greater physical range of disposal than traditional methods, placement is still limited to an area less than 130 m from the spray equipment (Cahoon and Cowan 1987, 1988).

The effect on vegetation must be considered in thin-layer placement. Ford et al. (1999) examined the impact of spray dredging on vegetated marsh in a 0.5-ha *Spartina alterniflora*-dominated salt marsh in coastal Louisiana. Measurements made immediately following spraying in July 1996 revealed that stems of *S. alterniflora* were knocked down by the force of the spray and covered with 23 *mm* of dredged material. However, vegetation soon recovered and by July 1997 the percent cover of *S. alterniflora* had increased three-fold over pre-project conditions. Thus, the layer of dredged material was thin enough to allow for survival of the *S. alterniflora* plants, with no subsequent colonization by plant species typical of higher marsh zones (Ford et al. 1999). Cahoon and Cowan (1987, 1988) evaluated the response of Louisiana coastal wetlands to thin-layer placement by high-pressure spraying at Lake Coquille and Terrebonne Parish wetlands (Dog Lake). Sediments were deposited 10–15 cm deep and up to 70 m away from the water's edge at Dog Lake and 18–38 cm deep and up to 79 m from the water's edge at Lake Coquille. Fourteen months after placement, vegetation was still smothered at both sites; however, recolonization by typical salt marsh plant species was underway. The authors estimated it would take three years for full re-vegetation but the lack of pre-placement elevation data limited interpretation of their data.

A second method of thin-layer placement uses a low-pressure hydraulic dredge to pipe a slurry of dredged material over the marsh surface. The sediment slurry consists of a high water to solids ratio (>80% water) piped over the marsh so that sediments sheet flow and settle across the marsh and pond surfaces. In low-pressure hydraulic dredging, the heavier, coarse (sand) fraction piles up at the outfall while the fine-grained material floats away, resulting in an uneven distribution of particle size fractions and a sediment deposit of uneven thickness (Cahoon and Cowan 1988). The advantage of high-pressure hydraulic dredging over low-pressure hydraulic dredging is that the dredged soils are thoroughly mixed and sprayed in a thin layer onto the marsh surface with a fire-hose nozzle apparatus. Consequently, the coarse and fine fractions of the soil remain thoroughly mixed and do not segregate after deposition, ensuring a uniform distribution of all size fractions throughout the sprayed area (Ford et al. 1999). Rather than targeting specific sediment addition levels, piping slurry during thin-layer placement involves setting target elevations and allowing the sediment to settle across the area in variable thicknesses to reach the mean target elevation for the site (La Peyre et al. 2009). Croft et al. (2006) recommended that slurry be piped at high tide when sufficient water is on the marsh surface to reduce the effect on vegetation and to promote uniform distribution.

Thin-layer placement of dredged sediment to restore coastal marshes can vary widely in costs, ranging from $8,103/acre in a marsh in the Gulf of Mexico to $500,000/acre in Big Egg Marsh in New York City (USA-ERDC 2016). Primary costs may include transport of dredged material to the marsh location, removal of contaminants in the sediment, preparation of the site to reduce wave erosion, studies of environmental impacts, and planting marsh vegetation seedlings on the newly deposited sediment (Ray 2007). To be cost-effective, the dredging site should be relatively close to the placement area; however, if transport costs are not a limiting factor, it should be possible to barge or pump sediments over far greater distances (Ray 2007).

The amount of dredge material added to a marsh affects the rate of ecosystem development, dependent upon the thickness of the placed sediment, vegetation type, and amount of ecosystem degradation (Craft 2016) and resulting elevation compared to nearby reference marshes (La Peyre et al. 2009). In many studies, thicknesses have ranged from a few millimeters to over 30 cm (Cahoon and Cowan 1988; Wilber 1992a, b, c; Ford et al. 1999; Slocum et al. 2005; Tong et al. 2013). Wilber (1993) suggested that thickness of the sediment deposited controls the mechanism of re-colonization. Generally, deposits <15 cm thick are re-colonized by re-sprouting of in situ vegetation, while deposits >15 cm can smother and kill the vegetation, necessitating invasion by new plant material. Croft et al. (2006) concluded that up to 10 cm of sediment could be added without causing adverse effects to non-deteriorated marsh. Reimold et al. (1978) found that *S. alterniflora* stems covered with a layer of sediment up to 23 cm deep could grow through the sediment, but plants covered with greater than 60 cm of sediment did not recover. The duration of the initial recovery period varies according to the thickness of sediment placement and the extent of soil modificati) (Wilber 1993). If a thinner layer of sediment is deposited, vegetation can recover more quickly via the production of new shoots from surviving roots and rhizomes; however, a thicker layer of sediment must be re-colonized by seeds from adjacent marshes and, thus, would require a much longer recovery period (Langlois 2011). Other studies have shown that marsh nourishment requires an initial recovery period; however, these studies also found that the deposition of a thin layer of sediment in deteriorated marshes resulted in an increase in plant biomass, percent cover, and primary productivity (DeLaune et al. 1990; Wilsey et al. 1992; Wilber 1992a, b, c; Ford et al. 1999; Kuhn and Mendelssohn 1999; Leonard et al. 2002; Mendelssohn and Kuhn 2003; Slocum et al. 2005; Schrift et al. 2008; Stagg and Mendelssohn 2010; Tong et al. 2013).

8.6 Shoreline Protection

Shoreline protection involves the installation of rock revetments and breakwaters to reduce wave energies along shorelines in open bays, lakes, sounds, channels, barrier islands and bayous (Fig. 8.9). In Louisiana, bags of oyster shells may also be used as a revetment to protect shorelines (Fig. 8.10).

Creation and restoration of dune, beach, and back barrier marsh to restore or augment Louisiana's barrier islands and headlands is a critical part of the CPRA Coastal Master Plan. Dredging and placement of sediment, to achieve these goals for the barrier islands identified in the CMP will be considered consistent. In cases where engineering and technical analysis show that the inclusion of structural features is beneficial to long-term project performance (e.g., terminal groins, breakwaters, etc.), the feature may be considered.

Fig. 8.9 Breakwaters (top) and rock revetment (bottom) used to reduce shoreline erosion (Photos by Phillip L. Trosclair III)

Early knowledge from marsh creation through use of dredge spoil laid the foundations for a new effort to restore or create new marshes between partially hardened (using a variety of sill materials) lower shorelines and uplands susceptible to erosion termed 'living shorelines' (NOAA 2015)—a shoreline managed to control erosion that has one or more living components (e.g., seagrass, oysters, marsh, dune grass). Armoring shorelines with rip-rap and seawalls that extend vertically from or just above the high tide line prevents marshes from being able to migrate landward as sea levels rise and eliminates the high diversity plant community at the upper marsh edge (Bozek and Burdick 2005). Instead, the living shoreline concept as applied to salt

Fig. 8.10 Oysters shells used to create a revetment to protect shorelines from erosion (North Carolina Coast Federation)

marshes moves the erosion resistant material seaward to the base of the marsh and creates a gradual vegetated slope to the upland edge, using the vegetation to protect higher elevations from erosion (Morgan et al. 2009). As landowners become more concerned with erosion and conversion of their upland property to intertidal, the living shoreline approach could become a significant mechanism to promote marsh habitat that can effectively protect shorelines (Gittmann et al. 2014) and provide other benefits (e.g., habitat: Gittman et al. 2016; blue carbon: Davis et al. 2015).

Artificial or bioengineered oyster reef projects, in which reefs are created using shell or engineered products to provide substrate for oyster recruitment, have become an increasingly popular restoration technique over the last decade. The primary goal of these projects is coastal restoration, not management or enhancement of the oyster fishery. With increasing rates of sea-level rise and significant losses of salt marshes reported across the US (Day et al. 2007; Couvillion et al. 2017), there are medium and large-scale efforts to build and restore marshes using both thin and thick sediment addition in several regions: California, (e.g., Elkhorn Slough); Chesapeake Bay (e.g., Poplar Island); Delaware Bay (Prime Hook National Wildlife Refuge); Rhode Island (three national wildlife refuges). In Louisiana the Coastal Wetland Planning, Protection and Restoration Act has funded several such projects (LCWCRTF 2008) with similar projects as part of a $50 billion plan (CPRA 2012).

8.7 Geology

Restoration involves geology. It's the foundation and the framework of any restoration project. And the geology of the Mississippi River Delta is well known—there are literally thousands of well, hundreds of studies and papers, and the role of geology is recognized in the continuing work to evaluate present day surface and near surface processes. In addition, restoration projects are preceded by soil borings and other site specific measurements.

Each restoration project should evaluate and plan for the factors that affect the displacement rate at the site. Sea level rise is one, but that is complicated by the particular location and contributing factors that make the impact the relative sea level rise. It makes little sense to restore an area, only to lose it again a few years later. For example, subsidence increases relative to the location of the Delta load and decreases across the Chenier Plain. Compaction will be greater for sites with a clay (muck) substrate, and less for those over sands or next to sands related to distributary channels.

Active faults affect the local geology all across the Delta. Where these faults happen, there is a clear choice about where to place restoration resources. In each case, the upthrown side is stable and the downthrown side represents a fatal impact on a restoration. An example restoration choice would be to site a retention levee along the upthrown side of a fault and pump spoil behind it rather than across the fault or on its downthrown side. Failure to consider faulting in developing projects involved with restoration has been the subject of recent discussion. The Lake Hermitage marsh creation project involves sediment placement the downthrown side of the Magnolia fault as described by Bullock (2018), apparently without consideration of the active fault displacement or role of the peat accumulation associated with the fault environment. The diversion project planned for the Mid-Barataria region contemplates an outlet structure that will apparently cross the Ironton fault as mapped by Wallace (1966) and Bridgeman (2018). In these cases and others, recommendations by the geologic community (e.g., Mclindon 2020) should be considered at the outset.

8.8 Decontamination

Decontamination is an integral part of restoration. In many cases, toxins can be buried in place but in some instances the contaminants have to be removed (Ko and Day 2004). Contamination to soils, groundwater, and, in some cases, underground sources of drinking water has occurred from oil and gas activities and restoration often requires extensive testing. Dealing with contamination can involve removal of the surface waste, excavation of 'hot spots' of concentrated pollution (i.e., chemicals of concern, COC), COC treatment and disposal, and pumping and treatment of contaminated groundwater. The goal of the remediation effort is to return the soil and groundwater to original conditions according to an oil and gas mineral lease or agreed-upon standards such as background soil standards contained in the Louisiana Department of Natural Resources (LDNR)/OC regulations known as 29B (the section of the rules) and the Louisiana Department of Environmental Quality (LDEQ) Risk Evaluation/Corrective Action Program (RECAP) for both soil and groundwater. A mix of the standards is sometimes used. The term 'agreed upon' is used because cleanup is usually done in the context of a legal action with the involvement of the landowner-plaintiff, the oil companies(s), LDNR, LDEQ, and the Court. The LDNR 29B (Title 43, Part XIX, statewide order 29B) standards were/are heavily influenced by the oil and gas exploration and production industry (E&P), while the LDEQ RECAP standards are based on risk science and provide numerical screening standards for many chemicals in soil and groundwater (LAC 33, 2003; https://www.deq.louisiana.gov/assets/docs/Land/RECAP/RECAPfinal.pdf).

Dealing with contamination involves a number of approaches. In the case of contamination and debris, the source of contamination must be determined (e.g., produced water, oil and pollutants and infrastructure (e.g., equipment, pipeline, and facilities). Causation analysis relating contamination to sources is necessary to establish damages related to oil and gas activities as compared to natural or other causes.

Surface restoration includes removing abandoned or remaining equipment, facilities, wells, flowlines, pits, roads, and sunken barges. Subsurface restoration includes the proper plugging and abandonment of all wells to ensure they do not leak or seep to the Underground Source of Drinking Water (USDW), soils or groundwater. In other words, prior to developing a plan to restore land loss, all contamination, sources of contamination and abandoned oil and gas equipment must be removed.

Cleanup of oil and gas operations and the land restoration requires permits by numerous government agencies including land owner(s). The permitting process is often quite lengthy and time consuming. The government agencies that issue permits in Louisiana include LDNR, LDEQ, US Army Corps of Engineers, Louisiana Department of Wildlife and Fisheries, and local governments.

Once plans have been developed to identify and remove contamination and debris, the restoration plan for wetland loss can be developed and implemented. Land loss must be identified over time to determine causation and what loss is associated with oil and gas activities. Land loss from other important variables that contribute to land loss must also be determined, including sea level rise, hurricanes and major weather events, saltwater intrusion, altered hydrology, enhanced erosion, and non-oil and gas projects such as dredging for drainage.

If the polluted sediments are removed and replaced with clean fill to appropriate elevations or if pollutants can be capped with clean fill to substantially eliminate leaching of contaminants, restoration is possible. Since marshes will accrete through deposition of fine-grained sediments and in situ organic soil formation, any remaining contaminant may be buried and release becomes less likely over time. Documentation of such projects is not easily found in the literature (e.g. lead: Housatonic River CT: https://darrp.noaa.gov/hazardous-waste/lordship-point; mercury: South Bay Salt Ponds, San Francisco Bay CA: Valoppi 2018; heavy metals: Jamaica Island, Piscataqua River NH: https://www3.epa.gov/region1/superfund/sites/portsmouth/277529.pdf.pdf).

8.9 Coordination with Other Projects

In planning for restoration of degraded wetlands in oil and gas fields, there should be coordination with other projects, especially those included in the CPRA Coastal Master Plan. For example, beneficial use of dredged sediments, where available, can be used to raise elevation. There are numerous wetland restoration projects taking place in the Mississippi Delta. These include marsh creation, river diversions, hydrologic restoration, barrier island restoration, rebuilding of distributary ridges, and shoreline protection (CPRA 2017; Wiegman et al. 2018). Restoration of the impacts of oil and gas activities involves to a lessor or greater degree all of these activities. For example, where sediment diversions are planned in the vicinity of an oil and gas field that is being restored, maximum use of diverted sediments should be a goal of the project. As noted above, beneficial use of sediments dredged for other activities such as navigation canal maintenance should be considered for marsh creation in restoring wetlands degraded by oil and gas activities. Hydrologic restoration and natural levee restoration in the vicinity of a restoration project in an oil and gas field should be done in a way that compliments the larger efforts. In this way, the restoration of and oil and gas field can be done to increase benefits over a larger area. Addressing land loss and environmental cleanup must be done at the same time to fully restore the land.

8.10 Evaluation of Success

Wetland restoration and creation projects must be evaluated for success. This is based on stated objectives, monitoring, and performance to demonstrate success. Following are guidelines in analyzing successful restoration.

- Defining successful parametrics (removal of contamination, elevation of land, hydrology requirements, vegetation types, land recovery, soil and water chemistry, complying with regulations, etc.) as it relates to wetland restoration.
- Methods for monitoring the restoration, i.e. site inspections, surveys, field testing with instruments, laboratory testing, aerial surveillance, etc.
- Analyzing data and determining reasons for an unsuccessful project.

Evaluating restoration success is important not only to see if project goals are accomplished, but to help in the design, implementation, and monitoring of future projects. Defining success of a wetland restoration project can be made difficult due to a lack of clearly stated objectives, lack of long-term monitoring of previous restoration efforts, and the subjective point of view of the definer (Stelk et. al 2017).

One way of measuring success is comparing the land gained by completion of a restoration project in relation to the "cost of doing nothing" as determine by Louisiana Coastal Master Plan 2050 modeling (Zinn 2004). This method provides a comparison of a particular area to itself should no action be taken in restoring the wetland. Another way of measuring success is the comparison of the flora and fauna ecosystem of a completed restored project and that of a nearby natural wetland (USEPA 2000). Long term monitoring of a restoration site is also critical in determining the success of a project. Typically, 3–5 years is the timeframe post project completion. However, this is often not enough time to truly determine whether a wetland restoration project has been successful. (Stelk et al. 2017).

Numerous variables can influence the success or failure of future wetland restoration projects. During the design phase objectives must be clearly stated and the design plan must be presented in a way that it can be easily understood by contractors and their personnel. Due to the often complicated nature of the design plan; project designers or someone familiar with the design must be present during the implementation phase to ensure that the project is built as designed. A well-designed restoration plan is of little value if the implementation does not align with the restoration plan.

The monitoring phase can often be overlooked or undervalued. During budgeting of a project sufficient resources should be dedicated to monitoring both in the form of monitoring infrastructure but as well as long-term personnel. Monitoring not only helps determine if a restoration project has reached its objectives, but it also provides data that can be valuable in the design and implementation of future projects. In Louisiana, CRMS is comprised of 391 monitoring stations across coastal Louisiana which gather pedologic, spatial, hydrologic, and vegetative data.

8.11 Post-project Monitoring

Methods for monitoring wetland restoration include site inspections, surveys, field testing with instruments, laboratory testing, aerial surveillance, etc. The results are compiled in different formats, and these include:

- Acres created, protected, restored

- Acres enhanced
- Total acres benefited
- Cost per benefited acres $/acre
- Average Annual Habitat Units (AAHU)
- Cost effectiveness $/AAHU
- Total costs
- Percent completion versus time.

Also included is aerial photography analysis, vegetation growth, elevation of land and hydrology. Each of these can be monitored yearly or at earlier intervals.

References

Baustian JJ (2005) Restoration success of backfilling canals in Coastal Louisiana marshes. M.S. Thesis, LSU, Baton Rouge, LA

Baustian JJ, Turner RE (2006) Restoration success of backfilling canals in Coastal Louisiana marshes. Restor Ecol 14:636–644

Brasher (2015) Review of the benefits of marsh terraces in the Northern Gulf of Mexico. Ducks Unlimited, Lafayette, LA. 22 pp.

Bridgeman JG (2018) Understanding Mississippi Delta subsidence through stratigraphic and geotechnical analysis of a continuous Holocene core at a subsidence superstation. M.S. Thesis, School of Science and Engineering, Tulane University, New Orleans, LA, 96 pp.

Britsch LD, Dunbar JB (2006) Land loss in Coastal Louisiana: 1930s to 2001. Technical Report TR-05-13. Engineer Research and Development Center, Vicksburg, MS

Boumans RM, Day JW, Kemp GP, Kilgen K (1997) The effect of intertidal sediment fences on wetland surface elevation, wave energy and vegetation establishment in two Louisiana coastal marshes. Ecol Eng 9:37–50

Bozek CM, Burdick DM (2005) Impacts of seawalls on saltmarsh plant communities in the Great Bay Estuary, New Hampshire, USA. Wetlands Ecol Manage 13:553–568

Bullock JS, Kulp MA, McLindon CD (2018) Evaluation of the Magnolia growth fault, Plaquemines Parish, southeastern Louisiana, poster session, GSA Annual Meeting, Indianapolis, Indiana.

Burdick DM, Dionne M, Boumans RM, Short FT (1997) Ecological responses to tidal restorations of northern New England salt marshes. Wetlands Ecol Manage 4:129–144. https://doi.org/10.1007/BF01876233

Cahoon DR, Cowan JH (1987) Spray disposal of dredged material in coastal Louisiana: habitat impacts and regulatory policy implications. Louisiana Sea Grant College Program, Center for Wetland Resources, LSU, Baton Rouge, LA. 27 pp.

Cahoon DR, Cowan JH (1988) Environmental impacts and regulatory policy implications spray disposal of dredged material in Louisiana wetlands. Coast Manage 16:341–362

Cornu CE, Sadro S (2002) Physical and functional responses to experimental marsh surface elevation manipulation in Coos Bay's south slough. Restor Ecol 10:474–486

Clewell AF, Aronson J (2013) Ecological restoration: Principles, values and structure of an emerging profession. Island Press, Washington, D.C., p 303

Couvillion BR, Beck H, Schoolmaster D, Fischer M (2017) Land area change in coastal Louisiana (1932 to 2016). Pamphlet to accompany U.S. Geological Survey Scientific Investigations Map 3381

Coastal Protection and Restoration Authority (CPRA) (2012) Louisiana's comprehensive master plan for a sustainable coast, 188 pp. CPRA, Baton Rouge, Louisiana

CPRA (2017) Louisiana's comprehensive master plan for a sustainable coast. In: 2017 Coastal master plan. Louisiana Coastal Protection and Restoration Authority, Baton Rouge

Craft C (2015) Creating and restoring wetlands: from theory to practice. Elsevier Press, Amsterdam, Netherlands

Craft CB (2016) Ch. 8: Tidal marshes. In Creating and Restoring Wetlands from Theory to Practice 195–232. Elsevier, Amsterdam

Croft AL, Leonard LA, Alphin T, Cahoon LB, Posey M (2006) The effects of thin layer sand renourishment on tidal marsh processes: Masonboro Island, North Carolina. Estuaries Coasts 29:737–750

Davis JL, Currin CA, O'Brien C, Raffenburg C, Davis A (2015) Living shorelines: coastal resilience with a blue carbon benefit. PLoS ONE 10:e0142595

Day JW, Boesch DF, Clairain EJ, Kemp GP, Laska SB, Mitsch WJ, Orth K, Mashriqui H, Reed DJ, Shabman L, Simenstad CA, Streever BJ, Twilley RR, Watson CC, Wells JT, Whigham DF (2007) Restoration of the Mississippi delta: lessons from hurricanes Katrina and Rita. Science 315:1679–1684

Day JW, Kemp GP, Reed DJ, Cahoon DR, Boumans RM, Suhayda JM, Gambrell R (2011) Vegetation death and rapid loss of surface elevation in two contrasting Mississippi delta salt marshes: the role of sedimentation, autocompaction and sea-level rise. Ecol Eng 37:229–240

Day JW, Cable JE, Lane RR, Kemp GP (2016a) Sediment deposition at the Caernarvon crevasse during the great Mississippi Flood of 1927: implications for coastal restoration. Water 3(38):1–12

Day J, Lane R, D'Elia C, Wiegman A, Rutherford J, Shaffer G, Brantley C, Kemp G (2016b) Large infrequently operated river diversions for Mississippi delta restoration. Estuar Coast Shelf Sci. https://doi.org/10.1016/j.ecss.2016.05.001

Day JW, Shaffer G, Cahoon D, DeLaune R (2019) Canals, backfilling and wetland loss in the Mississippi Delta. Estuar Coast Shelf Sci. https://doi.org/10.1016/j.ecss.2019.106325

Day J, Clark H, Chang C, Hunter R, Norman C (2020) Life cycle of oil and gas fields in the Mississippi River Delta: a review. Water 12:1492. https://doi.org/10.3390/w12051492

DeLaune RD, Pezeshki SR, Pardue JH, Whitcomb JH, Patrick WH Jr (1990) Some influences of sediment addition to a deteriorating salt marsh in the Mississippi River deltaic plain: a pilot study. J Coastal Res 6:181–188

DeLaune RD, Jugsujinda A (2003) Denitrification potential in Louisiana wetland receiving diverted Mississippi river water. Chemistry and Ecology 19:411–418

Edwards KR, Proffitt CE (2003) Comparison of wetland structural characteristics between created and natural salt marshes in southwest Louisiana, USA. Wetlands 23(2):344–356

Edwards KR, Mills KP (2005) Aboveground and belowground productivity of Spartina alterniflora (Smooth cordgrass) in natural and created Louisiana salt marshes. Estuaries 28:252–265

Ford MA, Cahoon DR, Lynch JC (1999) Restoring marsh elevation in a rapidly subsiding salt marsh by thin-layer deposition of dredged material. Ecol Eng 12:189–205

Frame GW, Mellander MK, Adamo DA (2006) Big Egg Marsh experimental restoration in Jamaica Bay, New York. In: Harmon D (ed) People, places, and parks: proceedings of the 2005 george wright society conference on parks, protected areas, and cultural sites. The George Wright Society, Hancock, Michigan, pp 123–130

Giosan L, Syvitski J, Constantinescu SD, Day J (2014) Protect the world's deltas. Nature 516:31–33

Gittman RK, Popowich AM, Bruno JF, Peterson CH (2014) Marshes with and without sills protect estuarine shorelines from erosion better than bulkheads during a category 1 hurricane. Ocean Coast Manag 102:94–102

Gittman RK, Peterson CH, Currin CA, Fodrie FJ, Piehler MF, Bruno JF (2016) Living shorelines can enhance the nursery role of threatened estuarine habitats. Ecol Appl 26:249–263

Graham SA, Mendelssohn IA (2013) Functional assessment of differential sediment slurry applications in a deteriorating brackish marsh. Ecol Eng 51:264–274

Hackney CT, Brady S, Stemmy L, Boris M, Dennis, CHancock, T (1996) Does intertidal vegetation indicate specific soil and hydrologic conditions. Wetlands 16:89–94

Hiatt MGA, Snedden JW, Day RV, Rohli JA, Nyman LAS, Lane RR (2019) Drivers and impacts of water level fluctuations in the Mississippi River delta: implications for delta restoration. Estuar Coast Shelf Sci 224:117–137

Hunter RG, Day JW, Lane RR, Shaffer GP, Day JN, Conner WH, Rybczyk JM (2018) Using natural wetlands for municipal effluent assimilation: A half-century of experience for the Mississippi River Delta and surrounding environs. In: Metcalf C (ed) Pollution abatement using natural and constructed Wetlands. Springer Publishing Co

Ko JY, Day JW (2004) A review of ecological impacts of oil and gas development on coastal ecosystems in the Mississippi Delta. Ocean Coast Manag 47:597–623

Koning CO (2004) Impacts of small amounts of sandy sediment on wetland soils and vegetation: results from field and greenhouse studies. Wetlands 24:295–308

Konisky RA, Burdick DM, Dionne M, Neckles HA (2006) A regional assessment of saltmarsh restoration and monitoring in the Gulf of Maine. Restor Ecol 14:516–525. https://doi.org/10.1111/j.1526-100X.2006.00163.x/epdf

Kuhn NL, Mendelssohn IA (1999) Halophyte sustainability and sea level rise: Mechanisms of impact and possible solutions. In: H. Lieth et al. (eds), Halophyte uses in different climates. Backhuys Publishers, Leiden, The Netherlands. p 13

Kusler JA, Kentula ME (1990) Wetlands creation and restoration: the status of the science. Island Press, Washington, D.C.

Langlois SM (2011) Ecological Review—Grand Liard Marsh and Ridge Restoration. Coastal Protection and Restoration Authority of Louisiana. Baton Rouge, Louisiana. p 27

La Peyre MK, Gossman B, Piazza BP (2009) Short- and long-term response of deteriorating brackish marshes and open-water ponds to sediment enhancement by thin-layer dredge disposal. Estuaries Coasts 32:390–402

LaPeyre M, Piazza B, Gossman B (2006) Short and long-term effects of thin layer deposition of dredged material on marsh health (1434-05HQRU1561, RWO No. 77). NMFS – USGS Interagency Agreement No. HC-119. Year 1 Report. p 32. + appendices

LCWCRTF (2008) 18th Priority Project List Report (Appendices). Louisiana Department of Natural Resources. Baton Rouge, LA. p 305

Leonard LA, Posey M, Cahoon L, Alphin T, Laws R, Croft A, Panasik G (2002) Sediment recycling: marsh renourishment through dredged material disposal. The NOAA/UNH Cooperative Institute for Coastal and Estuarine Environmental Technology (CICEET). p 49

Lester GD, Sorensen SG, Faulkner PL, Reid CS, Maxit IE (2005) Louisiana comprehensive wildlife conservation strategy. Louisiana Department of Wildlife and Fisheries, Baton Rouge, Louisiana, 455 pp

MacBroom JG, Schiff R (2012) Predicting the Hydrologic Response of Salt Marshes to Tidal Restoration. In: Roman CT, Burdick DM (eds), Tidal Marsh Restoration: A Synthesis of Science and Management. Island Press, Washington, DC. https://doi.org/10.5822/978-1-61091-229-7_2

Mendelssohn IA, Kuhn NL (1999) The effects of sediment addition on salt marsh vegetation and soil physico-chemistry. In: Rozas LP, Nyman JA, Proffitt CE, Rabalais NN, Reed DJ, Turner RE (eds) Recent research in Coastal Louisiana: natural system function and response to human influence. Louisiana Sea Grant College Program, pp 55–61

McLindon C (2020) Louisiana's oil and gas industry—the missing link in coastal sustainability [WWW Document]. McLindon Geosciences, LLC. https://www.mcgeo.me/blog/louisianas-oil-and-gas-industry-the-missing-link-in-coastal-sustainability. Accessed 1 March 2021

Mendelssohn IA, Kuhn NL (2003) Sediment subsidy: effects on soil plant responses in a rapidly submerging coastal salt marsh. Ecol Eng 21:115–128

Mitsch WJ, Gosselink JG (2015) Wetlands. John Wiley & Sons, Hoboken, N.J

Morgan PA, Burdick DM, Short FT (2009) The functions and values of fringing salt marshes in northern New England, USA. Estuaries Coasts 32:483–495

Neill C, Turner RE (1987) Backfilling canals to mitigate wetland dredging in Louisiana coastal marshes. Environ Manage 11:823–836

Pont D, Day JW, Ibanez C (2017) The impact of two large floods (1993–1994) on sediment deposition in the Rhone delta: Implications for sustainable management. Sci Total Environ 609:251–262

Perez B, Day J, Rouse L, Shaw R, Wang M (2000) Influence of Atchafalaya River discharge and winter frontal passage on suspended sediment concentration and flux in Fourleague Bay, Louisiana. Estuar Coast Shelf Sci 50:271–290

Peterson SB, JM Teal, and WJ Mitsch (eds), (2005) Deleware Bay salt marsh restoration. Ecological Engineering, Special Issue 25:199–314

RAE (1999) Principles of Estuarine Habitat Restoration. Arlington Virginia. https://www.edc.uri.edu/restoration/html/resource/rae-erf.pdf

Ray GL (2007) Thin layer placement of dredged material on coastal wetlands: a review of the technical and scientific literature. USACE ERDC/EL TN-07-1

Raynie RC, Syed SM, Villarrubia C, Haywood E (2020) Coastal monitoring and data management for restoration in Louisiana. Shore & Beach 88:92–101

Reimold RJ, Hardisky MA, Adams PC (1978) The effects of smothering a Spartina alterniflora salt marsh with dredged material. report prepared for the USACE Office, Chief of Engineers, Washington D.C. p 114

Rogers DR, Rogers BD, Herke WH (1992) Effects of a marsh management plan on fishery communities in coastal Louisiana. Wetlands 12:53–62

Schrift AM, Mendelssohn IA, Materne MD (2008) Salt marsh restoration with sediment-slurry amendments following a drought-induced large-scale disturbance. Wetlands 28:1071–1085

Shaffer GP, Day JW, Lane RR (2018) Optimal use of fresh water to restore baldcypress-water tupelo swamps and fresh marshes and protect against saltwater intrusion: a case study of the Lake Pontchartrain Basin. In: Day JW, Erdman JA (eds) Restoration of the Mississippi Delta: pathways to a sustainable future, pp 61–76. Springer

Simenstad CA, Thom RM (1996) Functional equivalency trajectories of the restored Gog-Le-Hi-Te estuarine wetland. Ecol Appl 6:38–56

Slocum MG, Mendelssohn IA, Kuhn NL (2005) Effects of sediment slurry enrichment on salt marsh rehabilitation—plant and soil responses over seven years. Estuaries 28:519–528

Smith SM (2009) Multi-decadal changes in salt marshes of Cape Cod, MA: photographic analyses of vegetation loss, species shifts, and geomorphic change. Northeastern Naturalist 16:183–208

Stagg CL, Mendelssohn IA (2010) Restoring ecological function to a submerged salt marsh. Restor Ecol 18:10–17

Staver LW (2015) Ecosystem dynamics in tidal marshes constructed with fine grained, nutrient rich dredged material. Dissertation, University of Maryland, College Park, MD

Steyer GD (1993) Sabine terracing project, final report. Unpublished report, Louisiana Department of Natural Resources, Baton Rouge, Louisiana, USA

Streever WJ (2000) Spartina alterniflora marshes on dredged material: A critical review of the on-going debate over success. Wetlands Ecol Mngt 8:295–316.

Spartina alterniflora marshes on dredge material: a critical review of the ongoing debate over success. Wetlands Ecol Manage 8:295–316

Stelk MJ, Christie J, Weber R, Lewis RR, Zedler J, Micacchion M, Merritt J (2017) Wetland restoration: contemporary issues and lessons learned. Association of State Wetland Managers, Windham, Maine

Syvitski J, Kettner A, Overeem I, Hutton E, Hannon M, Brakenridge R, Day J, Vorosmarty C, Saito Y, Giosan L, Nichols R (2009) Sinking deltas due to human activities. Nat Geosci. https://doi.org/10.1038/NGE0629

Teal JM, Weinstein MP (2002) Ecological engineering, design, and construction considerations for marsh restorations in Delaware Bay, USA. Ecol Eng 18:607–618

Teal JM, Peterson SB (2005) Introduction to the Delaware Bay salt marsh restoration. Ecol Eng 25:199–203

Tockner K, Baumgartner C, Schiemer F, Ward JV (2000) Biodiversity of a Danubian floodplain: structural, functional and compositional aspects. In: Gopal B et al (ed) Biodiversity in Wetlands: assessment, function and conservation, vol 1, pp 141–159. Backhuys Publications

Tong C, Baustian JJ, Graham SA, Mendelssohn IA (2013) Salt marsh restoration with sediment-slurry application: effects on benthic macroinvertebrates and associated soil-plant variables. Ecol Eng 51:151–160

Turner RE, McClanachan G (2018) Reversing wetland death from 35,000 cuts: opportunities to restore Louisiana's dredged canals. PLoS ONE 13. https://doi.org/10.1371/journal.pone.0207717

Turner RE, Streever B (2002) Approaches to coastal wetland restoration: Northern Gulf of Mexico. Kugler Publications, p 147

USACE-ERDC (2016) Jamaica Bay—Big Egg Marsh thin-layer placement factsheet. USACE Engineer Research and Development Center, Dredging Operations Technical Support Program

USEPA (2000) Principles for the ecological restoration of aquatic resources. EPA841-F-00-003. Office of Water (4501F), United States Environmental Protection Agency, Washington, DC, p 4

USEPA, USACE (2007) Identifying, planning, and financing beneficial use project using dredge material. Beneficial Use Planning Manual. US Environmental Protection Agency and US Army Corp of Engineers. p 114

USGS (2017) Louisiana rate of land loss continues to slow. https://www.usgs.gov/news/usgs-louisiana-s-rate-coastal-wetland-loss-continues-slow. Accessed online June 2021

VIMS (2014) Thin-layer sediment addition of dredge material for enhancing marsh resilience. Virginia Institute of Marine Science, College of William and Mary. https://doi.org/10.21220/V5X30S

Wallace WE (1966) Fault and salt map of South Louisiana. GCAGS TRANS 16:2

Weinstein M, Day J (2014) Restoration ecology in a sustainable world. Ecol Eng. https://doi.org/10.1016/j.ecoleng.2014.02.001

Wiegman A, Rutherford J, Day J (2018) The costs and sustainability of ongoing efforts to restore and protect Louisiana's coast. In: Day J, Erdman J (eds) Mississippi delta restoration pathways to a sustainable future. Springer, Cham, Switzerland, pp 93–111

Wilber P (1992a) Thin-layer disposal: concepts and terminology. Environmental effects of dredging information exchange bulletin D-92-1. Vicksburg, MS, U.S. Army Engineer Waterways Experiment Station

Wilber P (1993) Managing dredged material via thin-layer disposal in coastal marshes. Environmental Effects of Dredging Technical Bulletin, EEDP-01-32. Waterway Experiment Station, U. S. Army Corps of Engineers. p 14

Wilber P (1992b) Thin-layer dredged material disposal in coastal marshes-case studies. In: Ports '92, proceedings of the conference. July 20–22 1992, Seattle, WA. Washington, DC: Permanent International Association of Navigation Congresses

Wilber P (1992c) Case studies of the thin-layer disposal of dredged material: Gull Rock, North Carolina. Environmental effects of dredging technical bulletin D-92-3. Vicksburg, MS: U.S. Army Engineer Waterways Experiment Station

Wilsey BJ, McKee KL, Mendelssohn IA (1992) Effects of increased elevation and macro- and micronutrient additions on Spartina alterniflora transplant success in salt-marsh dieback areas in Louisiana. Environ Manage 16:505–551

Wolters M, Bakker JP, Bertness MD, Jefferies RL, Möller I (2005) Salt-marsh erosion and restoration in south-east England: squeezing the evidence requires realignment. J Appl Ecol 42. https://doi.org/10.1111/j.1365-2664.2005.01080.x

Zinn J (2004) Coastal Louisiana: attempting to restore an ecosystem. CRS Report for Congress, pp 18–20

Chapter 9
Summary and Conclusions

John W. Day, Rachael G. Hunter, and H. C. Clark

Oil and gas activity has been pervasive in the Mississippi River Delta and both production and environmental impacts follow a predictable life cycle. There are hundreds of O&G fields as well as a dense network of canals associated with drilling access, navigation, and pipelines. The production history for individual fields can last 40–60 years with production rising rapidly to a peak (around 1970 in the MRD) and then declining. Since most drilling in the MRD started in the 1940s and 1950s, most wells are no longer producing or are in the final stages of production and this cycle conclusion holds true for aggregate MRD production. Most fields had very low levels of production by the 2000s. Produced water generally lagged O&G production and was generally higher during declining O&G production.

Oil and gas activities have contributed in three major ways to environmental impacts on coastal ecosystems and specifically to wetland loss. These include alteration of surface hydrology due to canal dredging and spoil placement, induced subsidence and fault re-activation due to fluids withdrawal, and toxic stress due to pollution by spilled oil and produced water. Wetland loss due to O&G gas activity is initially due mainly to direct impacts of canal dredging and spoil placement, but grows over time due to cumulative and interactive effects. Wetland loss also increases over time to encompass much of the field and adjacent areas. This is due to loss of natural channels in areas of high canal density. Networks of interconnected canals form new patterns of water flow and often lead to salt water intrusion. Interestingly, spoil banks are not necessarily a permanent landscape feature and have a life cycle of their own.

J. W. Day (✉)
Department of Oceanography & Coastal Sciences, Louisiana State University, 2005 Olive St, Baton Rouge, LA 70806-, 6660, USA

R. G. Hunter
Comite Resources, Inc., Baton Rouge, LA, USA

H. C. Clark
Department of Earth, Environmental and Planetary Sciences, Rice University, Houston, TX, USA
e-mail: hcclark@rice.edu

They subside and compact over time and a quarter to a third of spoil banks likely have no subaerial expression.

Fluid withdrawal from O&G formations leads to induced subsidence and fault activation. Induced subsidence occurs in two phases. Withdrawal of O&G, and produced water, induce reservoir compaction resulting in a reduction of reservoir thickness. A slow drainage of pore pressure in the bounding shale mainly due to water pumping induces time-delayed shale compaction and subsidence can continue for decades after most O&G has been produced. This results in subsidence over much of the oil fields that can be greater than surface subsidence due to altered hydrology.

Produced water from O&G fields is water brought to the surface during O&G extraction and generally includes a mixture of liquid or gaseous hydrocarbons, high salinity produced water, dissolved or suspended solids, produced solids such as sand or silt, and injected fluids and additives associated with exploration and production activities. Produced water has been shown to be toxic to many estuarine organisms including marsh grass and consumers. Spilled oil has been shown to have lethal and sub-lethal effects of a wide range of estuarine organisms. The three main types of impact of O&G activities act in cumulative, synergistic, and indirect ways that lead to greater overall impact. Restoration of wetlands lost due to O&G activities will involve a synergistic approach that deals with the damage of O&G impacts and rebuilds a functioning coastal wetland system.

Restoration should be the final stage in the life cycle of O&G fields. A central objective of restoration is restoring lost elevation. Options for restoration include marsh creation and restoration using dredged sediments, full use of all available sediment resources, and hydrologic restoration. Decontamination of restored sites is an integral part of restoration.

Correction to: Energy Production in the Mississippi River Delta

J. W. Day, Rachael G. Hunter, and H. C. Clark

Correction to:
J. W. Day et al. (eds.), *Energy Production in the Mississippi River Delta*, **Lecture Notes in Energy 43,**
https://doi.org/10.1007/978-3-030-94526-8

In the original version of the book, the following belated corrections are to be incorporated: In chapters 1, 2, 3, 4, 5, 6, 7 and 8, the author's affiliation have been amended and Index has been included. The erratum book has been updated with the changes.

The updated versions of these chapters can be found at
https://doi.org/10.1007/978-3-030-94526-8_1
https://doi.org/10.1007/978-3-030-94526-8_2
https://doi.org/10.1007/978-3-030-94526-8_3
https://doi.org/10.1007/978-3-030-94526-8_4
https://doi.org/10.1007/978-3-030-94526-8_5
https://doi.org/10.1007/978-3-030-94526-8_6
https://doi.org/10.1007/978-3-030-94526-8_7
https://doi.org/10.1007/978-3-030-94526-8_8

© The Author(s), under exclusive license to Springer Nature Switzerland AG 2022
J. W. Day et al. (eds.), *Energy Production in the Mississippi River Delta*,
Lecture Notes in Energy 43, https://doi.org/10.1007/978-3-030-94526-8_10

Appendix

See Tables A1, A2 and A3.

Table A1 Canal widening rates for navigation and oil and gas canals in coastal Louisiana

Location	Parish	Basin	Station #	Original canal width (ft)	Measured canal width (ft)	Increase in canal width (ft/yr)	Increase of canal width (%/yr)	Years for data	Source
Bayou St. Denis		Barataria					8.2	1926–1976	Davis (1973)
Humble Canal	Cameron	Vermillion					8.3	1953–1958	Nichols (1958)
Humble Canal	Cameron	Vermillion					7.5	1953–1958	Nichols (1958)
Humble Canal	Cameron	Vermillion					6.9	1953–1958	Nichols (1958)
Humble Canal	Cameron	Vermillion					6.5	1953–1958	Nichols (1958)
Superior Canal	Cameron	Vermillion					14.8		Nichols (1958)
Superior Canal	Cameron	Vermillion					12.4		Nichols (1958)
Superior Canal	Cameron	Vermillion					13.9		Nichols (1958)
Superior Canal	Cameron	Vermillion					12.4		Nichols (1958)
Humble Canal	Cameron	Vermillion	1	65	137		5.3	1940–1961	Nichols (1961)
Humble Canal	Cameron	Vermillion	2	65	156		6.7	1940–1961	Nichols (1961)
Humble Canal	Cameron	Vermillion	3	65	162		7.1	1940–1961	Nichols (1961)
Humble Canal	Cameron	Vermillion	4	65	130		4.8	1940–1961	Nichols (1961)
Humble Canal	Cameron	Vermillion	5	65	134		5.1	1940–1961	Nichols (1961)
Humble Canal	Cameron	Vermillion	6	65	136		5.2	1940–1961	Nichols (1961)
Humble Canal	Cameron	Vermillion	7	65	109		3.2	1940–1961	Nichols (1961)
Superior Canal—Deep Lake Field	Cameron	Vermillion	1	65	171		18.1	1952–1961	Nichols (1961)
Superior Canal—Deep Lake Field	Cameron	Vermillion	2	65	153		15.0	1952–1961	Nichols (1961)
Superior Canal—Constance Bayou	Cameron	Vermillion	1	65	159		20.7	1954–1961	Nichols (1961)

(continued)

Appendix 227

Table A1 (continued)

Location	Parish	Basin	Station #	Original canal width (ft)	Measured canal width (ft)	Increase in canal width (ft/yr)	Increase of canal width (%/yr)	Years for data	Source
Superior Canal—Constance Bayou	Cameron	Vermillion	2	65	147		18.0	1954–1961	Nichols (1961)
Golden Meadow	LaFourche	Barataria					2.0	1940–1953	Craig et al. (1979)
Golden Meadow	LaFourche	Barataria					3.7	1940–1953	Craig et al. (1979)
Golden Meadow	LaFourche	Barataria					4.0	1953–1969	Craig et al. (1979)
Golden Meadow	LaFourche	Barataria					2.0	1953–1969	Craig et al. (1979)
Golden Meadow	LaFourche	Barataria					3.0	1953–1969	Craig et al. (1979)
Golden Meadow	LaFourche	Barataria					4.6	1953–1969	Craig et al. (1979)
Golden Meadow	LaFourche	Barataria					3.0	1953–1969	Craig et al. (1979)
Golden Meadow	LaFourche	Barataria					2.0	1953–1969	Craig et al. (1979)
Falgout Canal	LaFourche	Barataria					4.6	1953–1969	Craig et al. (1979)
SW LA Canal	LaFourche	Barataria	1	30	130	1.4	4.8	1880–1953	Doiron and Whitehurst (1974)
SW LA Canal	LaFourche	Barataria	2	30	120	1.2	4.3	1880–1953	Doiron and Whitehurst (1974)
SW LA Canal	LaFourche	Barataria	3	30	130	1.4	4.8	1880–1953	Doiron and Whitehurst (1974)
SW LA Canal	LaFourche	Barataria	4	30	130	1.4	4.8	1880–1953	Doiron and Whitehurst (1974)
SW LA Canal	LaFourche	Barataria	5	30	130	1.4	4.8	1880–1953	Doiron and Whitehurst (1974)
SW LA Canal	LaFourche	Barataria	6	30	110	1.1	3.8	1880–1953	Doiron and Whitehurst (1974)
SW LA Canal	LaFourche	Barataria	7	30	140	1.5	5.2	1880–1953	Doiron and Whitehurst (1974)
SW LA Canal	LaFourche	Barataria	8	30	145	1.6	5.5	1880–1953	Doiron and Whitehurst (1974)
SW LA Canal	LaFourche	Barataria	9	30	120	1.2	4.3	1880–1953	Doiron and Whitehurst (1974)

(continued)

Table A1 (continued)

Location	Parish	Basin	Station #	Original canal width (ft)	Measured canal width (ft)	Increase in canal width (ft/yr)	Increase of canal width (%/yr)	Years for data	Source
SW LA Canal	LaFourche	Barataria	10	30	140	1.5	5.2	1880–1953	Doiron and Whitehurst (1974)
SW LA Canal	LaFourche	Barataria	11	30	120	1.2	4.3	1880–1953	Doiron and Whitehurst (1974)
SW LA Canal	LaFourche	Barataria	12	30	110	1.1	3.8	1880–1953	Doiron and Whitehurst (1974)
SW LA Canal	LaFourche	Barataria	13	30	120	1.2	4.3	1880–1953	Doiron and Whitehurst (1974)
SW LA Canal	LaFourche	Barataria	14	30	130	1.4	4.8	1880–1953	Doiron and Whitehurst (1974)
SW LA Canal	LaFourche	Barataria	15	30	125	1.3	4.5	1880–1953	Doiron and Whitehurst (1974)
SW LA Canal	LaFourche	Barataria	16	30	140	1.5	5.2	1880–1953	Doiron and Whitehurst (1974)
SW LA Canal	LaFourche	Barataria	17	30	150	1.6	5.7	1880–1953	Doiron and Whitehurst (1974)
SW LA Canal	LaFourche	Barataria	18	30	150	1.6	5.7	1880–1953	Doiron and Whitehurst (1974)
SW LA Canal	LaFourche	Barataria	1	30	200	4.4	3.4	1953–1969	Doiron and Whitehurst (1974)
SW LA Canal	LaFourche	Barataria	2	30	185	4.1	3.4	1953–1969	Doiron and Whitehurst (1974)
SW LA Canal	LaFourche	Barataria	3	30	200	4.3	3.4	1953–1969	Doiron and Whitehurst (1974)
SW LA Canal	LaFourche	Barataria	4	30	270	2.4	6.7	1953–1969	Doiron and Whitehurst (1974)
SW LA Canal	LaFourche	Barataria	5	30	300	4.3	8.2	1953–1969	Doiron and Whitehurst (1974)
SW LA Canal	LaFourche	Barataria	6	30	200	4.7	5.1	1953–1969	Doiron and Whitehurst (1974)
SW LA Canal	LaFourche	Barataria	7	30	155	0.9	0.7	1953–1969	Doiron and Whitehurst (1974)
SW LA Canal	LaFourche	Barataria	8	30	220	4.6	3.2	1953–1969	Doiron and Whitehurst (1974)
SW LA Canal	LaFourche	Barataria	9	30	150	1.8	1.6	1953–1969	Doiron and Whitehurst (1974)
SW LA Canal	LaFourche	Barataria	10	30	240	6.2	4.5	1953–1969	Doiron and Whitehurst (1974)

(continued)

Appendix

Table A1 (continued)

Location	Parish	Basin	Station #	Original canal width (ft)	Measured canal width (ft)	Increase in canal width (ft/yr)	Increase of canal width (%/yr)	Years for data	Source
SW LA Canal	LaFourche	Barataria	11	30	210	5.6	4.7	1953–1969	Doiron and Whitehurst (1974)
SW LA Canal	LaFourche	Barataria	12	30	215	6.5	6.0	1953–1969	Doiron and Whitehurst (1974)
SW LA Canal	LaFourche	Barataria	13	30	250	8.1	6.8	1953–1969	Doiron and Whitehurst (1974)
SW LA Canal	LaFourche	Barataria	14	30	250	7.5	5.8	1953–1969	Doiron and Whitehurst (1974)
SW LA Canal	LaFourche	Barataria	15	30	230	6.5	5.3	1953–1969	Doiron and Whitehurst (1974)
SW LA Canal	LaFourche	Barataria	16	30	250	6.8	4.9	1953–1969	Doiron and Whitehurst (1974)
SW LA Canal	LaFourche	Barataria	17	30	270	7.1	5.0	1953–1969	Doiron and Whitehurst (1974)
SW LA Canal	LaFourche	Barataria	18	30	270	7.5	5.0	1953–1969	Doiron and Whitehurst (1974)
SW LA Canal	LaFourche	Barataria	1	30	240	10.0	5.0	1969–1973	Doiron and Whitehurst (1974)
SW LA Canal	LaFourche	Barataria	2	30	240	13.6	7.4	1969–1973	Doiron and Whitehurst (1974)
SW LA Canal	LaFourche	Barataria	3	30	240	10.0	5.0	1969–1973	Doiron and Whitehurst (1974)
SW LA Canal	LaFourche	Barataria	4	30	320	12.5	4.6	1969–1973	Doiron and Whitehurst (1974)
SW LA Canal	LaFourche	Barataria	5	30	350	12.5	4.2	1969–1973	Doiron and Whitehurst (1974)
SW LA Canal	LaFourche	Barataria	6	30	280	20.0	10.0	1969–1973	Doiron and Whitehurst (1974)
SW LA Canal	LaFourche	Barataria	7	30	190	8.8	5.6	1969–1973	Doiron and Whitehurst (1974)
SW LA Canal	LaFourche	Barataria	8	30	320	25.0	11.4	1969–1973	Doiron and Whitehurst (1974)
SW LA Canal	LaFourche	Barataria	9	30	200	10.0	8.3	1969–1973	Doiron and Whitehurst (1974)
SW LA Canal	LaFourche	Barataria	10	30	310	18.5	7.3	1969–1973	Doiron and Whitehurst (1974)
SW LA Canal	LaFourche	Barataria	11	30	270	15.0	7.1	1969–1973	Doiron and Whitehurst (1974)

(continued)

230

Table A1 (continued)

Location	Parish	Basin	Station #	Original canal width (ft)	Measured canal width (ft)	Increase in canal width (ft/yr)	Increase of canal width (%/yr)	Years for data	Source
SW LA Canal	LaFourche	Barataria	12	30	300	21.2	9.9	1969–1973	Doiron and Whitehurst (1974)
SW LA Canal	LaFourche	Barataria	13	30	320	18.5	7.0	1969–1973	Doiron and Whitehurst (1974)
SW LA Canal	LaFourche	Barataria	14	30	330	20.0	8.0	1969–1973	Doiron and Whitehurst (1974)
SW LA Canal	LaFourche	Barataria	15	30	320	22.5	9.8	1969–1973	Doiron and Whitehurst (1974)
SW LA Canal	LaFourche	Barataria	16	30	300	12.0	5.0	1969–1973	Doiron and Whitehurst (1974)
SW LA Canal	LaFourche	Barataria	17	30	360	20.0	8.3	1969–1973	Doiron and Whitehurst (1974)
SW LA Canal	LaFourche	Barataria	18	30	360	22.5	8.3	1969–1973	Doiron and Whitehurst (1974)
SW LA Canal	LaFourche	Barataria	1	30	310	17.5	7.3	1974–1978	Johnson and Gosselink (1982)
SW LA Canal	LaFourche	Barataria	2	30	290	12.5	5.2	1974–1978	Johnson and Gosselink (1982)
SW LA Canal	LaFourche	Barataria	3	30	360	30.0	12.5	1974–1978	Johnson and Gosselink (1982)
SW LA Canal	LaFourche	Barataria	4	30	410	22.5	7.0	1974–1978	Johnson and Gosselink (1982)
SW LA Canal	LaFourche	Barataria	5	30	460	27.5	7.8	1974–1978	Johnson and Gosselink (1982)
SW LA Canal	LaFourche	Barataria	6	30				1974–1978	Johnson and Gosselink (1982)
SW LA Canal	LaFourche	Barataria	7	30	300	27.4	14.4	1974–1978	Johnson and Gosselink (1982)
SW LA Canal	LaFourche	Barataria	8	30	410	22.5	7.0	1974–1978	Johnson and Gosselink (1982)
SW LA Canal	LaFourche	Barataria	9	30				1974–1978	Johnson and Gosselink (1982)
SW LA Canal	LaFourche	Barataria	10	30	330	5.0	1.6	1974–1978	Johnson and Gosselink (1982)
SW LA Canal	LaFourche	Barataria	11	30	360	22.5	8.3	1974–1978	Johnson and Gosselink (1982)
SW LA Canal	LaFourche	Barataria	12	30	300	0.0	0.0	1974–1978	Johnson and Gosselink (1982)

(continued)

Table A1 (continued)

Location	Parish	Basin	Station #	Original canal width (ft)	Measured canal width (ft)	Increase in canal width (ft/yr)	Increase of canal width (%/yr)	Years for data	Source
SW LA Canal	LaFourche	Barataria	13	30	380	15.0	4.7	1974–1978	Johnson and Gosselink (1982)
SW LA Canal	LaFourche	Barataria	14	30	370	10.0	3.0	1974–1978	Johnson and Gosselink (1982)
SW LA Canal	LaFourche	Barataria	15	30	330	2.5	0.8	1974–1978	Johnson and Gosselink (1982)
SW LA Canal	LaFourche	Barataria	16	30	300	0.0	0.0	1974–1978	Johnson and Gosselink (1982)
SW LA Canal	LaFourche	Barataria	17	30	360	0.0	0.0	1974–1978	Johnson and Gosselink (1982)
SW LA Canal	LaFourche	Barataria	18	30	360	0.0	0.0	1974–1978	Johnson and Gosselink (1982)
Freshwater Bayou Canal	Vermillion			173	583	12.5	3.7	1967–1992	Brown and Root (1992)
Houma Navigation Canal	Terrebonne			301	579	12.6	4.5	1965–1987	T. Baker Smith and Sons (2002)
Houma Navigation Canal	Terrebonne			579	666	7.9	7.7	1987–1998	T. Baker Smith and Sons (2002)
						Mean	6.1		

Table A2 Sedimentation rates in impounded (I), semi-impounded (SI), and non-impounded (NI) wetlands in coastal Louisiana

Location	Parish	Basin	Impounded, semi, non-impounded	Sedimentation (g/m^2/day)	Years for data	Source
Rockefeller Wildlife Refuge		Mermentau	I	0.6		Boumans and Day (1994)
Rockefeller Wildlife Refuge		Mermentau	NI	1		Boumans and Day (1994)
Fina-Laterre Marsh Mngt Area		Terrebonne	I	1.7		Boumans and Day (1994)
Fina-Laterre Marsh Mngt Area		Terrebonne	NI	3.8		Boumans and Day (1994)
Leeville	Lafourche	Terrebonne	SI	0.41		Kuhn et al. (1999)
Leeville	Lafourche	Terrebonne	NI	1.19		Kuhn et al. (1999)
Little Lake		Barataria	I	0.24	1993–1994	Reed et al. (1997)
Little Lake		Barataria	NI	0.87	1993–1994	Reed et al. (1997)
Three Bayou		Barataria	I	0.45	1993–1994	Reed et al. (1997)
Three Bayou		Barataria	NI	2.13	1993–1994	Reed et al. (1997)
Jug Lake		Terrebonne	I	0.50	1993–1994	Reed et al. (1997)
Jug Lake		Terrebonne	NI	2.00	1993–1994	Reed et al. (1997)
Otter Bayou		Terrebonne	I	0.86	1993–1994	Reed et al. (1997)
Otter Bayou		Terrebonne	NI	3.07	1993–1994	Reed et al. (1997)
LL/3B mean		Barataria	I	0.89	1992–1993	Reed et al. (1997)
LL/3B mean		Barataria	NI	1.89	1992–1993	Reed et al. (1997)
JL/OB mean		Terrebonne	I	0.79	1992–1993	Reed et al. (1997)
JL/OB mean		Terrebonne	NI	4.60	1992–1993	Reed et al. (1997)
LL/3B mean		Barataria	I	0.78	1994–1995	Reed et al. (1997)
LL/3B mean		Barataria	NI	2.13	1994–1995	Reed et al. (1997)
JL/OB mean		Terrebonne	I	1.00	1994–1995	Reed et al. (1997)

(continued)

Table A2 (continued)

Location	Parish	Basin	Impounded, semi, non-impounded	Sedimentation (g/m^2/day)	Years for data	Source
JL/OB mean		Terrebonne	NI	2.37	1994–1995	Reed et al. (1997)
		Mean	NI	2.28		
		Mean	I/SI	0.75		

Table A3 Accretion rates in impounded (I), semi-impounded (SI), and non-impounded (NI) wetlands in coastal Louisiana

Location	Parish	Basin	Station #	Impounded, semi, non-impounded	Total accretion (mm)	Accretion (mm/yr)	Years for data	Source
Jean Lafitte National Park		Barataria		I		1.88	1986–1987	Taylor et al. (1989)
Jean Lafitte National Park		Barataria		NI		7.23	1986–1987	Taylor et al. (1989)
Jean Lafitte National Park		Barataria		Inland		2.83	1986–1987	Taylor et al. (1989)
Marsh Island Unit 1			A	I	90	2.37	1956–1994	Bryant and Chabreck (1998)
Marsh Island Unit 1			A	NI	360	9.47	1956–1994	Bryant and Chabreck (1998)
Marsh Island Unit 1			B	I	100	2.63	1956–1994	Bryant and Chabreck (1998)
Marsh Island Unit 1			B	NI	380	10.00	1956–1994	Bryant and Chabreck (1998)
Sabine NWR Unit 3			A	I	270	6.28	1951–1994	Bryant and Chabreck (1998)
Sabine NWR Unit 3			A	NI	590	13.72	1951–1994	Bryant and Chabreck (1998)
Sabine NWR Unit 3			B	I	290	6.74	1951–1994	Bryant and Chabreck (1998)
Sabine NWR Unit 3			B	NI	600	13.95	1951–1994	Bryant and Chabreck (1998)
Rockefeller Unit 14		Mermentau	A	I	100	2.33	1951–1994	Bryant and Chabreck (1998)
Rockefeller Unit 14		Mermentau	A	NI	280	6.51	1951–1994	Bryant and Chabreck (1998)
Rockefeller Unit 14		Mermentau	B	I	100	2.33	1951–1994	Bryant and Chabreck (1998)
Rockefeller Unit 14		Mermentau	B	NI	300	6.98	1951–1994	Bryant and Chabreck (1998)
Rockefeller Unit 15		Mermentau	A	I	130	3.02	1951–1994	Bryant and Chabreck (1998)
Rockefeller Unit 15		Mermentau	A	NI	290	6.74	1951–1994	Bryant and Chabreck (1998)
Rockefeller Unit 15		Mermentau	B	I	50	1.16	1951–1994	Bryant and Chabreck (1998)
Rockefeller Unit 15		Mermentau	B	NI	300	6.98	1951–1994	Bryant and Chabreck (1998)
Salt marsh	Lafourche	Terrebonne		I		6.6	1986–1987	Cahoon and Turner (1989)

(continued)

Table A3 (continued)

Location	Parish	Basin	Station #	Impounded, semi, non-impounded	Total accretion (mm)	Accretion (mm/yr)	Years for data	Source
Salt marsh	Lafourche	Terrebonne		SI		6.0	1986–1987	Cahoon and Turner (1989)
Salt marsh	Lafourche	Terrebonne		NI		9.9	1986–1987	Cahoon and Turner (1989)
Salt marsh	Cameron	Calcasieu/Sabine		I/SI		3.5	1986–1987	Cahoon and Turner (1989)
Salt marsh	Cameron	Calcasieu/Sabine		I/SI		4.3	1986–1987	Cahoon and Turner (1989)
Salt marsh	Cameron	Calcasieu/Sabine		NI		11.3	1986–1987	Cahoon and Turner (1989)
Fina La Terre		Terrebonne		I		1.0	1989–1990	Cahoon (1994)
Fina La Terre		Terrebonne		NI		5.0	1989–1990	Cahoon (1994)
Rockefeller, Unit 4		Mermentau		I		0.8	1989–1990	Cahoon (1994)
Rockefeller, Unit 4		Mermentau		NI		11.1	1989–1990	Cahoon (1994)
				Mean	NI	9.15		
				Mean	I/SI	3.40		

Index

A
Abandoned infrastructure, 93
Accretion, 100, 103, 105, 107, 109, 112, 114, 116
Alteration of hydrology, 24
Altered hydrology, 194, 216
American Petroleum Institute, 85, 87
Amphipods, 145
Anoxia, 101, 104, 116
Anoxic soils, 20
Anticline, 51–53, 56, 57, 64. *See also* rollover anticline
Arenchyma tissue, 22
Arsenic, 84
Asthenosphere, 41, 43, 54
Atchafalaya, 94, 116, 123, 201, 206
Average Annual Habitat Units (AAHU), 218

B
Bacopa monnieri, 207
Barataria basin, 94, 95, 110
Barium, 84
Barrier Island Comprehensive Monitoring (BICM), 195
Bastian Bay, 63
Batis maritime, 207
Batis sp., 96
Bayou Chitigue, 13, 17, 116
Bayou Lafourche, 59, 97, 100
Benthic organisms, 141, 143
Benzene, 87
Bioavailability, 137, 145
Bivalves, 145
Boat wakes, 110, 112, 113

Boron, 138, 139
BP oil spill, 181
Brackish Marsh, 20, 21, 22, 24
Breakwater, 207, 211, 212
Breton sound, 15, 95, 97, 104
Brine, 133, 134, 138–144
Bruun Rule, 111
Bulk density, 101, 102, 111, 112
Bully Camp, 123, 126

C
Cable tool drilling, 41, 50, 57, 68–77. *See also* drilling
Campaign contributions, 86, 88, 89
Canal, 1, 3, 4, 15, 20–22, 60, 62, 71–77, 93, 94, 96–101, 104, 105, 107–110, 114, 116, 118, 121, 123–126, 194–198, 209, 216. *See also* canal dredging
Canal backfilling, 196, 197
Canal density, 94, 99
Canal dredging, 93, 94, 96, 110
Canal impacts, 94
Canal widening, 98, 109
Cancer Alley, 89
Carbon sequestration, 155, 162, 166, 175, 176
Channel theft, 99
Chemical pollution, 4
Chenier Plain, 13–15, 22, 40, 43, 46, 48, 54, 60, 62, 63, 65, 70. *See also* Chenier
Chrysenes, 135, 138, 139
CITES, 156
Climate, 156, 159, 162, 166, 172, 174–176, 178
Climate stability, 162, 174–176

Index

Coastal Information Management System (CIMS), 196
Coastal land loss, 94, 99
Coastal Master Plan, 201, 204, 206, 211, 216
Coastal Protection and Restoration Authority (CPRA), 195, 196, 204–207, 211, 214, 216
Coastal wetlands, 1, 2, 4, 133, 140
Coastal Wetland Sustainability, 18
Coastal Zone Management Act 1972, 83
Coastwide Reference Monitoring System (CRMS), 22, 195, 196, 217
Collapse, 60, 62, 63, 65
Commercial fisheries, 169, 170
Compaction, 42, 54–60, 63–68, 77, 98, 100, 106, 116, 117, 119, 121, 123, 208, 214
Consolidation, 100, 106, 116
Containment dike, 206
Contamination, 194, 215, 217
Core, 46, 48, 51, 57, 59, 65. *See also* coring
Cost/benefit analysis, 193
Cretaceous, 42, 43, 55
Crevasse formation, 8, 10
Crude oil, 133–138, 140, 141, 143–146
Crust, 40, 42, 43, 54. *See also* mantle
Cultural services, 159, 162
Cumulative impacts, 166, 167
Cytochrome-450, 145

D

Damage assessment, 193
Decomposition, 109, 116, 121
Decontamination, 4, 215, 224
Deepwater horizon, 178, 181
Delacroix, 63
Delta, 1–4
Delta lobe, 2, 8–11, 39–52, 54–67, 70, 77. *See also* delta; Mississippi River Delta (MRD)
Depocenter, 39–44, 46, 48–52, 54–60, 63, 70, 76, 77. *See also* sediment; sedimentation; deposition
Deposition, 42–44, 48, 50, 54, 55, 57, 59
Detachment, 44, 45
Differential stress, 63
Direct impacts, 3, 84, 88
Distichlis spicata, 206
Distributaries, 8–11
Down to the coast normal fault, 64
Drainage, 99, 100, 103, 105, 118

Dredged sediments, 194, 202, 205, 206, 208, 210, 216
Drilling, 41, 50, 57, 68–77. *See also* cable tool drilling; rotary drilling; drilling barge
Drilling barge, 72, 74
"D" shape water bodies, 63

E

Ecological function, 137
Ecosystem assets, functions, and processes, 158
Ecosystem Goods and Services (EGS), 4, 155, 157–161, 163, 164–169, 184
Ecosystem monitoring, 194
Ecosystems, 155–169, 173, 177, 181, 184
Ecosystem services, 160–169, 177, 181
Ecosystem service valuation, 164, 165, 167
Ekofisk, 63
Eleocharis parvula, 206
Elevation, 194, 196–198, 200–206, 208, 210, 211, 213, 216–218
Empire, 63
Emulsion, 135–137, 145
Environmental management, 83
Environmental setting, 3
Eocene, 43, 45, 55–57
Erosion, 96, 98, 106, 109–114, 116, 194, 206, 208, 210, 212, 213, 216
Exposure, 133–135, 140, 141, 144–146, 148
Externalities, 84, 88
Exxon Valdez, 178, 181

F

Fault, 40, 43–46, 51–61, 63–68, 70, 73, 76, 77, 214. *See also* radial fault down to the coast fault normal fault growth fault; fault activation; antithetic fault
Fault activation, 46, 63, 64, 121, 224
FEMA, 164, 172
Fina-La Terre, 104
Fish, 140, 141, 145, 146
Floating marsh, 101–103
Flooding, 101, 104, 109, 121
Flowline, 76
Fresh marsh, 22, 24
Functional assessment, 194

G

Gentilly oil field, 104

Index

Geologic, 39–44, 46, 48–51, 54, 55, 57–60, 63, 64, 77. *See also* geologic cross-section; geologic section
Geologic faulting, 3
Geologic section, 40, 43, 46, 49, 51, 54, 55, 57, 60, 77
Geologist, 50, 51, 74
Geophysical, 50, 51, 53, 69, 70, 73, 74
Giliasso, 72, 74. *See also* drilling barge
Golden Meadow, 63, 66, 68
Goose Creek, 63
Grand Chenier, 15
Gravity, 43, 45, 70. *See also* gravity torsion balance
Greenhouse gas emissions, 24
Groin, 211
Growth fault, 40, 44, 45, 51, 55, 60, 63
Gulf killifish, 145, 146
Gulf of Mexico, 4, 39–41, 43, 44, 46, 48–52, 54, 56, 133, 137, 139, 140, 143, 145. *See also* Gulf opening; Gulfward

H
Habitat, 157, 159–163, 168, 169, 171, 175, 182
Herbivory, 22
Hidden Subsidies, 88–90
Holocene, 43, 46, 48, 49, 55, 58–60
Hurricane, 3, 11, 13–19, 25, 102, 123
Hurricane Andrew, 16, 17
Hurricane Katrina, 13, 15, 16
Hydrocarbons, 84
Hydrogen sulfide, 108
Hydrologic restoration, 194, 196, 198, 206, 216
Hydrology alteration, 7, 20, 24, 60, 93, 94, 121, 123, 124, 194, 216, 223, 224
Hypersaline brines, 84, 86
Hypoxic, 139, 142, 143

I
Impoundment, 1, 96, 100, 101, 103–107, 109
Indirect impacts, 3, 86, 88
Induced subsidence, 1, 4, 93, 94, 117, 118, 121, 123, 124
Infrastructure, 215, 217
Interdistributary basins, 8
Intermediate Marsh, 20, 22, 24
International Whaling Commission, 156

Inundation, 11, 15, 18, 20, 22, 100, 101, 103–105, 110, 114, 116, 118, 121
Inverse problem, 51
Ironton, 63
Isostatic, 54–57, 59, 60. *See also* load isostatic adjustment; isostatic adjustment; glacial isostatic adjustment
Iva frutescens, 96

J
Jennings, 51, 68, 69
Juncus roemerianus, 207
Jurassic, 42, 43, 45, 55

L
Lake Washington, 52, 53
Land contamination, 166
Land loss, 46, 54, 60–64, 66, 67, 77, 155, 157, 160, 163, 166–169, 171, 172, 175, 178, 182–184. *See also* marsh loss
Lapeyrouse, 66, 68, 76
Law of the Sea, 156
Leeville, 98, 99, 118, 121, 122, 197, 201
Legal framework, 83–90
Levees, 1
Life cycle, 50, 61, 62, 67, 68, 76
Linear pond, 97
Lirette, 52, 53
Little Chenier, 63
Living shoreline, 212, 213
Lobbyists, 83, 86, 88, 89
Localized subsidence, 97
London Dumping Convention, 156
Louann salt, 41, 44, 45, 50, 55
Louisiana Coastal Resources Management Act 1978, 83
Louisiana coastal zone management program, 83
Louisiana Department of Environmental Quality, 84
Louisiana Department of Natural Resources, 84, 87
Louisiana Office of Conservation, 84, 87
Louisiana Statewide Order 29-B, 84

M
Macondo, 133–135, 137, 138, 141, 143–145
Macrophytes, 140, 141, 145

Magnetic, 43
Magnolia, 63
Mangroves, 137, 138
Marsh creation, 194, 195, 197, 198, 201, 204–206, 212, 214, 216
Marsh edge erosion, 109, 110, 112, 114
Marsh management, 104
Marsh nourishment, 208, 211
Marsh terracing, 207
Martin API report 1932, 85
Meander, 46, 47
Mercury, 84
Millennium Ecosystem Assessment (MEA), 159–161
Miocene, 43, 45, 52, 53, 55–57, 60, 63
Mississippi River, 7–11, 13, 15, 17, 19, 21, 23
Mississippi River Delta, 1, 39, 47, 52, 64, 65, 77
Monitoring, 194, 195, 198, 208, 216, 217
Multiphase flow, 139, 141

N
Naphthalenes, 135, 136, 138, 139, 145
Natural capital, 156–158, 167, 178
Natural hydrological flows, 1
Naturally Occurring Radioactive Material (NORM), 86, 134, 139
Navigation, 1
Nonaqueous Phase Liquids (NAPL), 141
Nonhazardous Oilfield Waste (NOW), 86, 87
Non-impounded, 107
Northernmost banana republic, 89

O
Offshore Pipeline Committee, 155
Oil and gas activity, 9, 13
Oil and gas field, 2, 3, 43, 49, 52, 53, 60, 61, 63, 65–71, 76
Oil and gas industry, 83, 76–88
Oil and gas production, 1, 4, 93, 117, 118
Oilfield radiation, 86
Old Oyster Bayou, 116
Oligocene, 43, 45, 56
Organic matter accretion, 107
Organic soil formation, 8, 18, 20
Overbank flooding, 8
Overpressure, 50, 57, 64
Oxidation-reduction potential, 202
Oxygen, 137, 139–144

P
Paleocene, 45, 57, 60
Paleo Hurricanes, 17
Paleotempestology, 17
Peat, 46, 48, 49, 59, 60, 77. *See also* basal peat; peat collapse
Peat Collapse, 22–24
Performance criteria, 194
Periwinkle snail, 146
Permeability, 57
Phase 1 metabolism, 145
Phase 2 metabolism, 145
Phenanthrenes, 135, 138, 139, 144, 145
Phytotoxins, 139–141
Pipeline canal, 97, 99, 109
Pit leakage, 85
Pleistocene, 43, 46, 48, 49, 55, 57, 58, 60
Pliocene, 57, 58
Plugging, 43, 51, 52, 57, 76. *See also* oil and gas well; drilling, completion, plugging
Political Power, 89
Pollution, 88, 90
Pollution control, 88
Pollution effects on health, 88
Pollution subsidy, 88
Polycyclic Aromatic Hydrocarbons (PAHs), 133, 135, 136–140, 144, 145
Polynuclear aromatic compounds, 87
Poroelastic stress, 42, 54–60, 62–68, 77. *See also* compaction, collapse
Porosity, 54, 56, 57
Poverty, 88, 90
Pressure, 44, 49–51, 54, 56, 57, 59, 60, 63–65, 76
Produced water, 1, 3, 84–86, 93, 94, 121, 133–135, 138–141, 143, 145
Production, 50, 51, 60–65, 68, 69, 71–73, 76, 77. *See also* peak oil and gas production
Profit margin, 88
Provisioning goods and services, 159

R
Radioactive materials, 86
Radioisotopes, 140
Radionucleids, 86
Radium, 86, 87
Radium-226, 134, 139
Radium-228, 139
Recreation, 155, 158–160, 162, 163, 170, 171, 174, 177
Red River, 13

Reference wetland, 215
Regulating services, 159, 162
Regulatory framework, 83–90
Relative Sea Level Rise (RSLR), 11, 19, 22, 106
Reservoir, 42, 49–53, 56, 57, 60, 62–68, 70, 76. *See also* trap
Reservoir collapse, 42, 54–60, 62–68, 77. *See also* compaction; collapse
Resource Conservation and Recovery Act (RCRA), 86–88
Restoration, 4, 193–199, 201, 204, 206, 207, 211, 214–217
Restoration goals, 195
Restoration plan, 194, 196, 198, 215, 217
Risk Evaluation/Corrective Action Program (RECAP), 215
River diversion, 204, 216
Rockefeller Refuge, 104, 105, 114
Rock revetment, 211, 212
Rotary drilling, 71
Ruppia maritime, 207

S

Saline Marsh, 18, 20, 21, 24
Salinity, 133, 139–144
Salt, 40, 41, 43–46, 50–55, 57, 59, 60, 63, 68, 70, 73, 76, 77. *See also* dome; Louann; diapirism; weld; detachment
Salt dome, 43, 44–46, 51–53, 55, 60, 63, 68, 70, 73. *See also* salt diapirism; salt weld; gravity; density difference
Salt stress, 20, 101, 104, 116
Saltwater intrusion, 15, 18, 22, 24, 99, 108–110, 121, 194, 206, 216
Salt weld, 45, 55
Schoenoplectus olneyi, 206
Sea level rise, 2, 39, 46, 48, 49, 60, 106, 110, 111, 123, 202, 214, 216. *See also* sea level curve
Seaside sparrow, 148
Sediment, 39–44, 46, 48–52, 54–60, 63, 70, 76, 77
Sediment addition, 198, 201–204, 208, 210, 214
Sedimentation, 105, 107, 108, 194, 194, 202, 209
Sediment deposition, 8, 13, 15, 16, 107, 108
Sediment diversion, 195, 201, 206, 216
Sediment oxygen demand, 139
Sediment slurry, 201, 202, 209, 210

Seismic reflection, 43, 70
Seismic, 43, 70. *See also* reflection; refraction
Seismic refraction, 70
Shale, 49, 56, 57, 64–67. *See also* shale dewatering; bounding shale
Shoreline erosion, 212
Shoreline protection, 211, 216
Social capital, 89, 90
Soil oxidation, 100
Soil strength, 19, 21, 23, 116
Spartina, 140, 145
Spartina alterniflora, 22, 104, 116, 199, 201, 210
Spartina patens, 22, 104, 202, 206
Spoil, 93, 94, 96–105, 107–109, 114, 121, 123–125
Spoil bank, 15, 20–22, 93, 94, 96–105, 107–109, 114, 121, 123–125
St. Bernard, 94, 99
Storm protection value of wetlands, 171
Stratigraphy, 41, 43, 49–51, 57, 59, 65. *See also* biostratigraphic stratigraphic
Stratigraphic, 41–43, 49–51, 65. *See also* stratigraphy geostratigraphic; bounding shale
Structure, 46, 52, 53, 66, 70, 73, 76, 77. *See also* structural block structural contour
Subsidence, 39, 40, 54–60, 62–68, 77, 78, 93, 94, 97, 98, 101, 104–106, 110, 114, 116–124, 194, 196, 202, 214. *See also* surface subsidence
Subsidies, 84, 88–90
Subsurface injection, 85
Sulfate, 139, 141
Sulfide, 201, 202
Sulfide toxicity, 20, 101, 104, 116
Supporting services, 159, 162, 168
Surface hydrology, 93, 94, 121, 124
Survey, 51, 63, 65, 66, 70, 73, 74, 77
Sustainability, 201, 202, 208
Systemic impacts from oil and gas, 182
System-Wide Assessment and Monitoring Program (SWAMP), 195

T

Terraces, 207, 208
Terrebonne, 63, 94, 99, 107, 118, 123
Terrebonne Basin, 16, 17
Thin layer sediment addition, 208
Tidal channel, 96, 99, 101, 121

Tidal exchange, 109, 196, 198, 199
Tidal network, 194, 198, 206
Tide range, 104
Toluene, 87
Toxic impacts, 93, 121
Toxic pollution, 4
Toxins, 93, 108
Trap, 49–53, 57, 60. *See also* trapping
Triassic, 43

U
Unemployment, 88, 90
Unlined pits, 84–86

V
Vegetation productivity, 100, 103, 107, 116, 124
Vegetation waterlogging stress, 114
Venice, 202
Venice salt dome, 123
Voting rates, 89, 90

W
Water-accommodated fraction, 145
Water logging, 20
Water quality, 159, 161, 162, 166, 172–174
Wax Lake, 198
Wealth leakage, 89
Well, 41, 43, 46, 50–53, 57–59, 61, 62, 65–74, 76. *See also* wellhead; plugged well; oil and gas well; derrick
Wetland deterioration, 18
Wetland loss, 1–4, 7, 8, 11, 15, 16, 18, 25, 61, 65, 93–97, 99, 109, 116–118, 120, 121, 123, 125
Wetland mortality, 22
Wetlands, 1–4
Wilmington, 63

Z
Zooplankton, 145

GPSR Compliance

The European Union's (EU) General Product Safety Regulation (GPSR) is a set of rules that requires consumer products to be safe and our obligations to ensure this.

If you have any concerns about our products, you can contact us on

ProductSafety@springernature.com

In case Publisher is established outside the EU, the EU authorized representative is:

Springer Nature Customer Service Center GmbH
Europaplatz 3
69115 Heidelberg, Germany

www.ingramcontent.com/pod-product-compliance
Ingram Content Group UK Ltd.
Pitfield, Milton Keynes, MK11 3LW, UK
UKHW021257180426
11947UKWH00011B/818